Exemplar Creation

Credible Origins for the Next Millennium

J. Gene White

About the Front Cover

The front cover is a composite photograph of two unusual galaxies. The largest object is the Sombrero Galaxy (also known as M104), consisting of a disk of stars with a bright central core surrounded by a thin, uniform, concentric ring of dust. The galaxy is an estimated 50,000 light years in diameter and 29 million light years away. Classified as a spiral galaxy, the disk shape and spiral arms cannot be seen in this photograph, being dominated by the unusually large and bright central bulge consisting of tens of billions of stars. However, the perfectly aligned, precision ring of rust colored dust surrounding the entire galaxy is clearly visible.

Hoag's Object (located in the lower right hand corner) is classified as a ring galaxy, estimated to be 600 million light years away. As viewed from Earth, a nearly perfect ring of hot blue and white stars surrounds a spherical core of cooler yellow stars. The ring has an estimated outer diameter of 121,000 light years (slightly larger than the Milky Way Galaxy) and contains tens of billions of stars.

These two galaxies dramatically illustrate design and construction on a galactic scale and are prime examples of a universe due to intelligent cause. Evolutionary ideas have been unable to remotely explain these remarkable structures. Natural effects due to black holes, self-gravity, explosions, galactic collision, centrifugal force, momentum and random movement cannot account for their unique configuration. The heavens truly declare the handiwork of God.

ISBN 978-0-615-26728-9

Sunnybrooke Publications
St. Louis, MO 63301

Table of Contents

Acknowledgements

I am deeply indebted to my wife, Karen, for her support during this entire project. She was a sounding board for many of my thoughts. Her candid comments and influence brought clarity and civility into the discussion of several thorny issues within this book.

I wish to thank Justin Watkins for editing and proofreading the initial manuscript. I am forever grateful to Rhonda Barfield for the second editing and proofreading in addition to professional coaching on a variety of literary, publishing and marketing issues. Appreciation is extended to the following individuals for reading and critically reviewing selected chapters and in some cases the entire manuscript. (Their assistance does not necessarily indicate endorsement of the material presented herein.)

Ron and Serece Baker, career public high school science teachers, Kansas City, MO.

Robert D. Bergen, Professor of Old Testament and Biblical Languages, Hannibal LaGrange College, Hannibal, MO.

Dr. Pat Campbell, Senior Pastor, Ridgecrest Baptist Church, St. Charles, MO.

Gordon H. Johnston, Assistant Professor of Old Testament Studies, Dallas Theological Seminary.

Ray T. Matheny, Professor of Anthropology, Brigham Young University.

Ben Merold, Senior Minister, Harvester Christian Church, St. Charles, MO.

Prattle P. T. Pun, Professor of Biology, Wheaton College, Wheaton IL.

Rev. Jeff Spaulding, Senior Pastor, Faith United Methodist Church, St. Charles, MO.

Gary D. Stormo, Professor of Genetics, Washington University, St. Louis, MO.

Dr. Roger C. Wiens, Physics, Space & Atmospheric Sciences Group at the Los Alamos National Laboratory.

Wilbur G. Williams, Professor of Biblical Literature and Archaeology, Indiana Wesleyan University.

Abraham Winitzer, Visiting Assistant Professor, Department of Theology, Notre Dame University.

Prelude

Within scientific, philosophical and religious circles, origins is a controversial subject. On the surface, science appears to be in irreconcilable conflict with the Genesis creation story. Some believe science totally discredits the Bible account about the origin of humankind, life, Earth and the heavens. To others, God's supernatural abilities easily overcome any possible objections. Within the biblical community, there is no consensus regarding the Genesis creation story. Belief ranges from traditional young-Earth creation theology to conviction that the entire Genesis story is mythology. The reigning paradigm is confusion, conflict and chaos.

The rise of intelligent design adds an additional dimension to the above controversy. Intelligent design focuses on identifying design in living things, nature and the universe at large evidenced by information, complex functional systems, mathematically improbable structures and fine-tuned arrangements. Natural cause is severely limited and cannot account for these phenomena. Among those who accept intelligent design, views on the Bible vary. Intelligent design does not require identifying a designer and can accept scientific evidence supporting a long age for the Earth and universe. While members of the intelligent design community readily agree about design, their views of the Bible and biblical creation theology vary significantly.

The path to writing this book began more than thirty-five years ago when I developed an interest in the evolution-creation controversy soon after graduating from college. During most of this time, I aligned myself with proponents of traditional young-Earth creation theology and provisionally accepted information from their books, magazines, video tapes and DVDs. I was aware of the scientific problems with this theology, but held to my private beliefs, hoping that one day some brilliant theologian might develop a viable solution capable of withstanding the light of scientific inquiry, yet respecting the Bible as the inspired word of God.

A few years ago a teenager in my Sunday school class said, "The Genesis account doesn't make sense." This was reinforced by a presentation to an adult class exploring contemporary religious issues. These events and others motivated me to seriously examine the Genesis creation story to determine if a credible solution was remotely possible.

When carefully read, traditional English translations of Genesis Chapter 1 have several major issues—specifically the sequence of creation events and inadequate time. Overwhelming scientific evidence shows the Earth and universe are much older than a 4004 BC creation date. If God created the universe and inspired the written account, they should agree. Since they do not, then possible sources of error include: (1) God incorrectly inspired the writer of Genesis. (2) Scientific evidence has been falsified. (3) We have misinterpreted the Genesis story. The latter seemed to be the most probable source of inaccuracy.

As a starting point, I decided to translate the Hebrew text of the Genesis creation story into English. From my professional training and experience, going back to the original source of information is important in conflicts of this nature—secondary knowledge is always suspect. An intensive study of Hebrew grammar progressed to the purchase of software containing *The Lexham Hebrew-English Interlinear Bible* by Logos Research Systems of Bellingham, Washington. The digital format of the Lexham version gives a clear word-by-word translation of the original Hebrew Old Testament text. The powerful Boolean search capability permits studying Hebrew words of interest and allows checking for the consistent translation of specific words—an absolute necessity based on my investigative approach. The Lexham Bible allows serious students to examine in detail the transformation of the Old Testament into English through the eyes of skilled Hebrew scholars. It allows their work to be checked for consistency and accuracy using twenty-first century computer technology.

I initially translated the Genesis story on a verse-by-verse basis from the Hebrew Bible without any interpretative framework in mind. There was no attempt to connect verses together or develop a rational flow of events. Early in the process, I realized something was amiss with traditional translations. After completing the first two chapters and smoothing out the wording, a strong explanatory framework emerged. The Genesis creation story could easily be harmonized with empirical science in regard to time. I found that approximately 98% of the words in traditional translations of Genesis Chapters 1 and 2 are satisfactory, but the remaining 2% seriously distort the original story. At that point, my personal Bible study expanded rapidly and writing this book became inevitable.

Hebrew scholars and theologians are naturally skeptical that a person with strictly scientific training could discover a solution to the Genesis creation problem—and while improbable, not impossible. It is interesting to note that the majority of books written within the past thirty years in an attempt to resolve the theological and scientific issues of Genesis have been authored by scientists, engineers, physicians and lawyers, all speaking partially outside their professional specialties.

From personal experience, I know that advancements within specialized areas of knowledge normally come from formally trained professionals, but that is not an absolute requirement. Several individuals have made significant contributions to the advancement of science, technology and theology outside their areas of formal training. Phillip Johnson, a law professor, essentially started the modern intelligent design movement by writing *Darwin on Trial*. Thomas Young (1773-1829), a physician and physicist best known for Young's modulus, made an important contribution to deciphering the Rosetta Stone. Henry M. Morris, a PhD in hydraulic engineering, completely revitalized the young-Earth creation movement in America. Two bicycle mechanics, Orville and Wilbur Wright, bested the most brilliant scientific minds of the time and built the first successful airplane. Bill Gates, a college drop out, started Microsoft Corporation with no additional formal scientific or business training and became one of the wealthiest people in America.

In regard to the Genesis account, it is easier for individuals with a science background to cross over into theology than for the reverse to occur. Information found in the Bible about the creation story is quite limited, while related science is broad and covers many specialized disciplines including biology, genetics, chemistry, physics, geology, archaeology, paleontology, anthropology and astronomy. Hebrew scholars and theologians whose knowledge is primarily confined to scripture can become quickly entangled and overwhelmed by the scientific complexities involved.

This book represents my personal study of the Bible and scientific evidence related to origins. The heart of my effort is reflected in a new interpretation and translation of the Genesis creation story which respects the integrity of the Bible as God's inspired word, yet harmonizes with empirical science.

While writing this book, I realized that the real power of Genesis lies in its ability to explain our beginnings from the perspective of God as our creator and to provide a framework for the evidence of science. A book strictly focused on a new translation of Genesis, confined to analyzing Hebrew verbs, nouns and translation issues would probably have little impact. For this reason, I enlarged the original scope of the book to integrate scripture and science. In regard to origins, my aim is to show that the historical record of the Bible and science complement each other where they intersect.

In order to build a solid foundation for discussion, I include an early chapter addressing science, empirical science and scientific thought. A subsequent chapter discusses creation theology, its history and importance. One chapter is devoted to developing design theory as a legitimate scientific idea. Several chapters give a critical analysis of chemical, biological, cosmological and human evolution. As a wrap-up, three leading ideas of origins are comparatively analyzed.

Throughout this book I use the term "Bible students and Christians" in an attempt to be as inclusive as possible. A Bible student can be anyone—Jew, Catholic, evangelical Protestant, non-evangelical Protestant, Mormon, Muslim, Buddhist, Hindu, agnostic or atheist—with a desire to know more about the Bible. By definition, anyone who wants to know more about the Genesis account by reading this book is a Bible student.

I have written this book primarily to members of the biblical community rather than the secular. It is written to those who say they believe in God, profess acceptance of the Bible or claim they are Christian. The material herein is for people seeking a credible answer to the origins problem who are open to a biblical solution.

My motivating principle for writing this book is an unrestricted search for truth. If the Bible is inspired and God created the heavens and Earth, then special revelation and general revelation should agree and we need not fear the truth. From my examination of the facts, the original biblical creation story readily harmonizes with the solid evidence of empirical science where they intersect. The Bible gives an abbreviated, non-technical, historical account of beginnings that is absolutely true.

J. Gene White
February 12, 2009

Visit the author at http://www.jgenewhite.com

Delight yourself in the Lord, and he will give you the desires of your heart. (Psalm 37:4) ESV

Chapter 1

Exemplar Creation

According to traditional interpretations of the Genesis account, God created the Earth around 4004 BC by speaking it into existence. Initially the Earth was without form and void (it was a mixed up ball of water and dirt). In the first three days, God created light, formed the sky, separated land from water and created plants to cover the Earth. On the fourth day the Sun, Moon and stars were spoken into existence. On the fifth and sixth days God created marine life, flying creatures and land dwelling animals. Lastly, man was created in the image of God out of dust from the ground.

While deeply engrained in biblical teaching, in the twenty-first century we are faced with addressing the truthfulness of the traditional biblical creation narrative. Is the Genesis account really true? If not true, then what does the Bible say about origins? In light of the controversy over this issue, we must be absolutely sure we have correctly interpreted the text that God inspired Moses to write.

The *exemplar* translation is a careful rendering of the original Old Testament Hebrew text of Genesis Chapters 1 and 2. The word *exemplar* means "original" in Latin, and refers to the original interpretation of the Genesis creation story recorded within inspired scripture. *Exemplar* creation focuses on harmonizing *special revelation* and *general revelation.* Special revelation is God's inspired word, the Bible, while general revelation is God's creation, the physical world.

The *exemplar* translation of the Genesis creation story is slightly different from traditional translations such as the King James Version. Small corrections permit a significantly different explanation of the biblical creation account, yet respect the integrity of God inspired scripture.

The primary differences between the *exemplar* translation and traditional translations of the first two chapters of Genesis are the treatment of several key words and verses. Verses starting with the expression, "And God said" in Genesis 1:3, 6, 9, 11, 14, 20 and 24 are normally explained as commands. However, based on the original Hebrew text, these seven verses are best viewed as descriptive statements. Other important changes include Genesis 1:2 where the illegitimate term "without form" is replaced by "vacant" as a correct translation. In Genesis 1:9, the illegitimate phrase "let the waters be gathered together" is replaced by "the waters wait." These changes and others lead to a substantially different creation story.

Exemplar creation addresses the critical issue of harmonizing inspired scripture and physical evidence. Two important features of *exemplar* creation are (1) acceptance of the Genesis story as a God inspired historical account that is truthful and trustworthy and (2) the harmonization of special revelation and general revelation. God's word and His work should not conflict.

The first three verses of Genesis are fundamental to understanding the creation account. In *exemplar* creation, Genesis 1:1 describes when God created the heavens and Earth in the beginning. Genesis 1:2 describes the Earth at an undefined point in time sometime after its creation. Between Genesis 1:2 and 1:3, the fossil record is laid down, extensive geologic change occurs and a significant portion of Earth's history takes place. At Genesis 1:3, the transition from undefined time to 24-hour days takes place.

The six-day Genesis chronicle is recorded in Genesis 1:3-31. During the first five days and part of the sixth (Gen. 1:3-25) God does not create, but reviews, explains and names what He created prior to Genesis 1:3. The Earth and heavens He describes have existed for some time. Between Genesis 1:26-31, God begins creating again and forms Adam out of dust from the ground. Genesis 2:4-25 contain a detailed summary of the sixth day, when God creates Adam and Eve, plants the Garden of Eden and forms a small number of animals to live within the garden.

In the remainder of this chapter, the *exemplar* translation of Genesis Chapters 1 and 2 is given followed by commentary. Words printed in italics are not in the Hebrew text and are added for grammatical and explanatory purposes, similar to the King James Version. All added words with a significant effect on verse meaning are in italics. Accuracy from the inspired text is favored over style and some verses may be slightly repetitive or awkward due to an emphasis on literal translation. (Before continuing, reading the *exemplar* translation without commentary in Appendix A is recommended.)

Genesis Chapter 1

Genesis 1:1 In the beginning, God created the heavens and the Earth.

Genesis 1:1 is an absolute statement acknowledging God as creator, when His actions started it all. As a powerful intelligence, He created the heavens and Earth. The following verses in the Old and New Testament are linked with this verse and recognize God as creator.

- Psalm 33:6 By the word of the LORD the heavens were made, And all the host of them by the breath of His mouth. (NKJV)

• Psalm 102:23 Of old You laid the foundation of the earth, And the heavens are the work of Your hands. (NKJV)
• Nehemiah 9:6 You alone are the LORD; You have made heaven, the heaven of heavens, with all their host, the earth and everything on it, the seas and all that is in them. (NKJV)
• Isaiah 42:5 Thus says God the Lord, Who created the heavens and stretched them out, Who spread forth the earth and that which comes from it. (NKJV)
• Isaiah 45:18 For thus says the Lord, Who created the heavens, Who is God, Who formed the earth and made it, Who has established it, Who did not create it in vain, Who formed it to be inhabited. (NKJV)
• Jeremiah 10:12 He has made the earth by His power, He has established the world by His wisdom, And has stretched out the heavens at His discretion. (NKJV)
• John 1:1-3 In the beginning was the Word, and the Word was with God, and the Word was God. He was in the beginning with God. All things were made through Him, and without Him nothing was made that was made. (NKJV)
• Acts 4:24 So when they heard that, they raised their voice to God with one accord and said: "Lord, You are God, who made heaven and earth and the sea, and all that is in them." (NKJV)
• Colossians 1:16 For by Him all things were created that are in heaven and that are on earth, visible and invisible, whether thrones or dominions or principalities or powers. All things were created through Him and for Him. (NKJV)
• Ephesians 3:9 And to make all see what is the fellowship of the mystery, which from the beginning of the ages has been hidden in God who created all things through Jesus Christ. (NKJV)
• Hebrews 1:10 And: "You, LORD, in the beginning laid the foundation of the earth, And the heavens are the work of Your hands." (NKJV).
• Revelation 4:11 Thou are worthy, O Lord, to receive glory and honor and power, for You created all things, and by Your will they exist and were created. (NKJV)
• Revelation 10:6 And swore by Him who lives forever and ever, who created heaven and the things that are in it, the earth and the things that are in it, and the sea and the things that are in it. (NKJV)
• Revelation 14:7 Saying with a loud voice, "Fear God and give glory to Him, for the hour of His judgment has come; and worship Him who made heaven and earth, the sea and springs of water." (NKJV)

Genesis 1:1 clearly states that God created the heavens and Earth. He existed from before the beginning and made them both; however, scripture does not give the details of how this was done. The exact time, sequence and method God used to create the heavens and Earth are not revealed, and we can only speculate why this information was withheld. Perhaps it was due to the limited ability of the Hebrew language for describing technical processes and procedures. Perhaps God knew man would be incapable of understanding the complex physics of creation. Regardless, these details are unknown and man is left to his own resources to fill in the missing information.

Evidence discovered by scientists indicates the Earth and universe are very old. The Earth is theorized to be 4.6 billion years old based on radiometric dating and isotope decay. The heavens are believed to be even older based on spectral redshift from distant galaxies.

Scientific evidence points to our planet being completely molten at some time in the distant past. A molten condition permitted the formation of Earth into an almost perfect sphere with a smooth surface due to self-gravity (almost as round and smooth as a billiard ball). A molten state over time also allowed differentiation of the elements according to density into the crust, mantle, outer core and inner core.[1] The density of Earth's crust is 3,300 kg/m^3 (200 lb/ft^3) which increases approximately four times at the core.[2] Density increases progressively from the outside skin to the center.

Earth's crust and interior are significantly different. The top eight elements in Earth's relatively light crust are oxygen (in the form of oxides and other compounds), silicon, aluminum, iron, calcium, sodium, magnesium and potassium.[3] Materials that make up the mantle and core are not firmly known, although they are believed to be primarily iron or nickel. Seismic tomography has confirmed the existence of three distinct regions within the interior of the Earth.[4, 5] Based on seismic evidence, the outer core is believed to be molten while the inner core and mantle are theorized to be solid.

Exemplar creation proposes that special and general revelation combined best explain how the Earth came about. Genesis 1:1 boldly proclaims that our planet was created by a powerful intelligence rather than natural cause. God was directly responsible for initially forming the Earth as a molten ball. Once all the raw materials were collected together and melted, self-gravity caused Earth's spherical shape and internal density changes.

Scientists have discovered that Earth is one planet of eight (Pluto was demoted) orbiting a star that is one among an estimated 400 billion visible stars in the Milky Way Galaxy.[6, 7] Our galaxy in turn is one of approximately 125 billion galaxies in the visible universe.[8] Earth is neither the center of our solar system nor the center of our galaxy, but is in the Orion Arm of the Milky Way an estimated 27,700 light years from its center.[9] Earth is an infinitesimal dot in the cosmos, but to our knowledge is unique. The many conditions responsible for making our planet a habitat for life are extraordinary.

By clearly identifying God as the one who created the heavens and Earth, Genesis 1:1 rejects the pagan beliefs of early cultures. This verse stands in stark contrast to the polytheistic beliefs of the Egyptians, Sumerians, Babylonians and Assyrians prevalent during the time of Abraham and Moses. These cultures worshiped many pagan gods and goddesses. For example, the Amduat, an ancient Egyptian religious book, lists 908 Egyptian gods and goddesses. All Egyptian books of the underworld combined list a total of around 2,500 deities.[10]

In the twenty-first century, Genesis 1:1 refutes philosophical naturalism and ancillary ideas of the Big Bang, evolutionary star formation and the solar nebular hypothesis. Natural cause is rejected as an all-encompassing explanation of origins.

Genesis 1:2A And the Earth was vacant and empty *of life.*

Genesis 1:2 describes the Earth sometime after its initial creation. Although Genesis 1:1 and 1:2 occur in a chronological sequence, the time between them is undefined and could be one day or billions of years. At this point, the Earth is vacant and empty. God has not yet created plants and animals.

Scientific evidence indicates that as the Earth slowly cooled over tens of thousands of years from a red hot molten ball, it was completely sterile. Not a single microscopic organism existed on the entire globe. It contained no bacteria, mold spores, blue-green algae, protozoa, single-celled life or multicellular life. Our planet's surface consisted exclusively of solidified igneous rock, minerals, elements, simple molecules and water.

Exemplar creation proposes that special and general revelation combined best describe the absence of life on early Earth sometime after its formation. The words "vacant" and "empty" in Genesis 1:2 emphasize the complete absence of life. With a starting temperature well above 800° F, our planet was completely sterile as it cooled from its molten state. Earth was totally devoid of all biological life.

Genesis 1:2B And darkness *was* over the surface of the deep *waters.*

Evidence discovered by scientists indicates the Earth gained an enormous supply of surface water early in its history. The most likely water source appears to be from materials out of which the Earth was made, now consisting of the mantle and core. Some mantle rock may still contain a small percentage of water.[11] The most plausible theory is that prior to solidification of the crust, prolonged outgassing of water vapor from molten material within the Earth provided water for its oceans.[12] The effects of heat, pressure, density and gravity forced water out of molten rock to the surface. All water on the Earth was initially in vapor form (hot steam). As our planet slowly cooled, water condensed out of the atmosphere into liquid form and covered its surface with one large ocean.

Planetary surface water is a rarity in our solar system, and Earth is unique because it has abundant water compared to other planets and moons. Earth's oceans, seas, lakes and rivers presently contains an estimated 1,386,000,000 cubic kilometers (332,500,000 cubic miles) of water.[13] As a smooth sphere, enough water exists to cover the Earth to a depth of approximately 2.72 kilometers (1.69 miles). The only other significant surface water in the solar system is on Jupiter's moon, Europa, covered with ice (its depth is unknown). Surface water is hypothesized to have existed at one time on Mars due to visible erosion features.

Exemplar creation proposes that special and general revelation combined best describe the surface water on early Earth. Genesis 1:2 testifies to the existence of deep water over the planet. This water was probably due to the release of moisture from molten material within the Earth. We can hypothesize that originally one large ocean covered the entire planet with no significant upraised land (volcanic cones and upraised meteor impact craters may have existed). Early Earth was covered with deep water.

The Genesis 1:2 description of darkness over the surface of the water appears to be supported by other scripture. In Job 38:4-11, God metaphorically discusses the Earth's creation. In Job 38:8 the sea bursts from the womb and comes out (water escaped from the Earth's interior). In Job 38:9 the sea is described as covered by a garment of clouds and swaddled with thick darkness. In Job 38:10, the appointed place is established by breaking up (God broke up Earth's crust, raised the continents and confined the waters). In Job 38:11, the sea and its waves are held within set boundaries established by the land.

- Job 38:4 Where were *you* when *I* established *the* Earth? Tell *me*, if *you* know understanding.
- 38:8 And *who* covered *the* sea with doors, when it burst from *the* womb *and* came out?
- 38:9 When I made clouds *for its* garment and thick darkness *for its* swaddling band.
- 38:10 And broke up for my decreed *place*, and set bars and doors.
- 38:11 And *I* said, *"The sea may* come until here and not more, and against here *shall* your proud waves set."

Scientific evidence indicates that conditions on early Earth were radically different from today due to meteor impacts and outgassing. After formation, our planet and the Moon were subject to extensive meteor bombardment during roughly the first billion years.[14] Some meteors created impact craters more than 50 km (31 miles) in diameter. Significant outgassing occurred in the form of steam and smoke as Earth cooled from its molten state. Our planet's atmosphere was heavily polluted by both meteor impacts and planetary outgassing.

Exemplar creation proposes that special and general revelation combined best describe the surface conditions of early Earth. Genesis 1:2 says that darkness was over the face of the deep waters. Meteor impacts threw immense quantities of dust and debris into the atmosphere. Planetary outgassing added steam and smoke. Earth's hot conditions (imagine an ocean temperature of 200° F or more) kept a substantial amount of water vapor in the atmosphere. All of these effects together created dark, continuous, heavy cloud cover over the entire globe. Conditions were probably similar to those found on Venus today, consisting of complete cloud cover, high temperatures and crushing atmospheric pressure.

Genesis 1:2C And the Spirit of God hovered over the surface of the waters.

Scientists have discovered that Earth is fine-tuned for life. Our planet has a consistent, yet slightly elliptical orbit around the Sun. Our Sun is not too large, not too small and quite stable for a G2 yellow dwarf star. Earth's orbital distance from the Sun appears just right. It is neither too close to be boiling hot nor too distant to be freezing cold. Earth's strong magnetic field protects it from dangerous solar winds. An ozone layer high in the atmosphere protects life from intense ultraviolet radiation produced by the Sun. Our planet has no significant harmful gases or minerals distributed in its atmosphere, soil and oceans. Earth's axial tilt creates a greater habitable zone for life and results in four seasons with more variable planetary weather. The Moon is just the right mass and distance to provide strong tidal action as an aid in mixing the ocean waters. Earth rotates at an optimum speed to moderate day and night temperatures.

Exploration has revealed that metals and minerals, vital to human technology, are conveniently concentrated at specific locations on Earth. Critical metals include copper, iron, nickel, titanium, aluminum, chromium, cadmium, lead, zinc, gold, silver and uranium. Important mineral deposits include boron, fluorspar, graphite, bromine, sodium sulfate, soda ash, salt and barite.[15] If these metals and minerals were spread uniformly throughout the Earth's crust, their mining and extraction would be prohibitive. They would probably be little more than scientific curiosities.

Clean air, soil and water should not be taken for granted. The Earth's crust contains several harmful minerals and metals such as antimony, arsenic, barium, cadmium, chromium, copper, lead, mercury, selenium, thallium and zinc. Fortunately, these are concentrated at specific locations rather than being evenly distributed throughout the Earth's soil and water. As an added benefit, concentrated deposits permit the mining of these metals for technological uses.

Exemplar creation proposes that special and general revelation combined best explain the conditions on Earth favorable for life and human technology. Genesis 1:2 describes the Spirit of God continuously hovering over our planet. This verse subtly alludes to the active part God had in shaping Earth. God did not abandon our planet and leave it to evolve in some mindless fashion, but developed it according to His plan. The fine-tuned features that make Earth habitable for mankind and favorable to human technology are the result of God's creative hand. He is responsible for the clean air, soil and water that make abundant life possible. Metal and mineral concentrations may be viewed as quasi-natural deposits orchestrated by God to aid man in the development of technology.

Between Genesis 1:2 and 1:3, *exemplar* creation proposes that God directed action and limited natural change shaped the Earth. As time passed, our planet continued to cool and its crust grew progressively thicker. Intense meteor bombardment began tapering off. We can hypothesize that due to the enormous amount of water vapor in the atmosphere, Earth was in the grip of a massive self-sustaining greenhouse effect, requiring God's intervention to stop (water vapor is a stronger greenhouse gas than carbon dioxide). Conditions slowly improved, and eventually water covering the Earth cooled to a point where life was possible.

Between Genesis 1:2 and 1:3, when conditions were right, God began creating life. He first created small, one-celled organisms such as algae and placed them in Earth's single vast ocean. As the planet continued to change and conditions became more favorable, God created additional marine life. The first major creation event is recorded in the fossil record as the Cambrian explosion. During this period, God created complex animals with eyes, skeletons, hard outer shells, limbs, internal organs, spinal columns and body symmetry. Earth's waters teemed with all kinds of complex marine life.

Between Genesis 1:2 and 1:3, God continued to work as He raised portions of the Earth's crust out of the water to create the continents. The original arrangement of the continents was different from today—they have drifted over time. Continents, sharply defined by the edge of the continental shelf, may be considered unnatural features and the result of God's creative power.

Earth's raised continents appear to be unique when compared with Mercury, Mars, Venus and the 18 spherical moons in our solar system. Venus and Mars have high and low areas, but appear to lack clearly defined continents bordered by a sharp edge as found on Earth. Further exploration may clarify this issue; however, at this time we may view Earth's original continents as being directly formed by God rather than due to natural cause. Earth's continents as observed today may be considered a combination of God directed change and natural change over time.

The raising of the continents appears to be supported by other scripture. Psalm 104 is commonly known as the "creation psalm." *The Treasury of David* by Charles Spurgeon views Psalm 104:5-9 as a metaphorical description about the creation of Earth. Psalm 104:6 describes the mountains covered with water like a garment (meaning the entire planet was covered). God then rebukes the water in verse seven, raises the mountains in verse eight and promises that water will never again cover the Earth in verse nine. The raising of the mountains and lowering of the plains in Psalm 104:8 can be explained in reference to the time between Genesis 1:2 and 1:3 when God raised the continents out of the water to create dry land. The *exemplar* translation given below may be compared with traditional translations.

- Psalm 104:5 *He* established *the* Earth over *its* foundation, not *to* be moved forever.
- 104:6 Like the garment *of the* deep, He covered over *the* mountains *with* standing water.

• 104:7 From your rebuke *the water* fled, from *the* voice *of* your thunder *the water* ran away.
• 104:8 *The* mountains went up *and the* plains went down—to this place established for them.
• 104:9 *Water shall* not pass *this* set boundary, *water shall* not again cover the Earth.

Between Genesis 1:2 and 1:3, God created a number of plants and animals to fill the land, air and water. Earth's fossil record clearly indicates the sudden appearance of many plant and animal species. As a part of His plan, God developed a complex system of dependent, symbiotic and synergistic relationships between plants and plants, plants and animals, and animals and animals. Dinosaurs and the great prehistoric animals were eventually created, followed by many of the species familiar today. Some life became extinct due to catastrophic events, environmental changes, disease, predators and other phenomena. When necessary, God restored Earth's livable conditions and created new species to populate the planet.

Between Genesis 1:2 and 1:3, God worked to create the oil and gas deposits beneath Earth's surface. Their existence, concentration and location are often taken for granted, yet if hydrocarbon deposits were distributed throughout the air, soil and water, Earth might not be habitable. For example, Titan, a moon of Saturn, has an atmosphere polluted by methane. Yet on Earth, virtually all oil, gas and methane are safely tucked away hundreds and thousands of feet underground in concentrations suitable for drilling and extraction. Earth's petroleum reserves may be considered quasi-natural deposits orchestrated by God for the future needs of humanity to aid in the expansion of technology and civilization.

Between Genesis 1:2 and 1:3, coal deposits were formed. Some of these deposits were created over long periods of time in peat bogs, marshes and swamps. As plant material slowly accumulated, it turned into coal. Other coal was formed by local and regional water catastrophes. In some instances, flood waters stripped vegetation from the land, concentrated it in low-lying areas and then covered it with sediments where it eventually turned into coal. Huge coal reserves were developed for the future energy needs of humankind and may be viewed as a combination of natural processes and God directed action over time.

Between Genesis 1:2 and 1:3, limited undirected change occurred due to the laws of chemistry and physics within systems established by God. The planet cooled due to the laws of thermodynamics. Earth's continents randomly drifted. Meteor impacts, volcanoes, earthquakes, tsunamis, floods, erosion, sedimentation, uplifting, settling, glacial action, hurricanes, tornados, fires and environmental change transformed Earth in a natural fashion. The fossil record was laid down during many local and regional catastrophic events. Valleys, canyons and river channels were formed by the action of water.

Exemplar creation views Genesis 1:3 as a critical dividing line in the Genesis creation story. The events prior to this verse have a different time frame from those that follow. Prior to Genesis 1:3, the verses are chronological but the length of time between them is undefined by scripture. The time could be one day or billions of years. Between Genesis 1:3 and 1:31, the six-day Genesis chronicle takes place and time is defined in normal, 24-hour, sequential days. Table 1.1 illustrates the timeline of Genesis Chapter 1.

Table 1.1 – The Timeline of Genesis Chapter 1

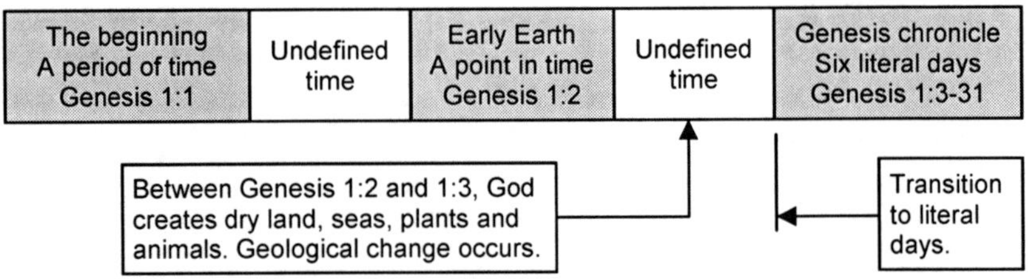

Exemplar creation considers the sequence that God used to create things prior to Genesis 1:3 as different from the sequence used to describe the completed work in Genesis 1:3-25. For example, a house is built starting with the foundation, floor, walls and roof. A general sequence is then followed for installing the heating and air conditioning ducts, drain lines, water supply pipes, electrical wiring, windows, exterior doors, insulation, sheetrock, cabinets, interior trim and carpeting. In describing the same house when completed, we might mention the date it was finished, its square footage and that it has two stories, a detached garage, a brick veneer exterior, hardwood floors and three baths. Items typically mentioned in describing a completed house are almost totally disconnected from the details of its construction. Similarly, in Genesis 1:3-25 God is describing His completed creation.

Exemplar creation regards Genesis 1:3-25 as an abbreviated description of the heavens and Earth, arranged in a six-day format. It is a story describing a completed Earth following a long series of God directed changes (creative events) and limited natural change—all prior to Genesis 1:3. It is a story of God reviewing, explaining and naming His creative work immediately before the creation of man on the sixth day. It is a story of God approving what He created as good and very good. It is a story identifying God as creator in contrast to the polytheistic beliefs of the day.

Genesis 1:3 And God said, "There is light, and light exists."
1:4 And God saw that the light *was* good. And God distinguished between the light and between the darkness.
1:5 And God called the light "day" and called the darkness "night." And evening was, and morning was—day one.

Genesis 1:3 begins day one of the six-day Genesis chronicle. As proposed by *exemplar* creation, on this day God describes the existence of light coming from the Sun, created during the beginning in Genesis 1:1. Often taken for granted, sunlight represents a unique phenomenon. It is completely unnatural and without it our solar system would be cold, dark and lifeless.

In Genesis 1:4-5, God distinguishes between light and dark by naming light "day" and darkness "night." One cycle of light and darkness establish the basis for one day. Through astronomy, scientists have come to understand how this is achieved. Earth rotates on its axis to produce day and night as one hemisphere is illuminated with light from the Sun while the other is covered in darkness.

This is the extent of God's reviewing and describing on day one. God introduces the terms "day" and "night" to describe light and darkness. Evening and morning represent one full day. This is the first of several naming events where God begins building a descriptive language. He establishes nomenclature to describe these phenomena for later use by Adam and his descendants.

Genesis 1:6 And God said, "There is an expanse in the middle of the water, and it exists to separate between water *above* from water *below*."
1:7 And God *is the one who* made the expanse and separated between the waters which *were* under the expanse and between the waters which *were* over the expanse. And *it* was so.
1:8 And God called the expanse "sky." And evening was, and morning was—second day.

Genesis 1:6 begins the second day of the six-day Genesis chronicle. As proposed by *exemplar* creation, on this day God describes the existence of Earth's expanse, created in past undefined time prior to Genesis 1:3. Sometimes expanse is translated from Hebrew into English as "dome." As viewed from the surface of the Earth, there is an expanse of clear space stretching from horizon to horizon. Water vapor and clouds may be explained as the water above the expanse and the oceans and seas are the water below. God names the expanse "sky."

Genesis 1:6-7 narrates events structured in a style identical to the fourth, fifth and sixth days. In verse six, a direct quotation by God describes the expanse as existing to separate water from water, and in verse seven He is identified as its maker. These verses represent a sequence of events starting with (1) creation of the expanse prior to Genesis 1:3, (2) the existence of an expanse in Genesis 1:6 and (3) identifying God as maker of the expanse in Genesis 1:7.

Notice that in Genesis 1:6, the words "create, make, form, plant, grow and build" are not used. This verse is a description of what has already been created. It should be obvious that if God describes the expanse as present in Genesis 1:6, He cannot create it in Genesis 1:7 since it already exists.

Genesis 1:7 identifies God as maker of the expanse in contrast to the many pagan gods. After being slaves in Egypt for several hundred years, the Hebrew people were thoroughly familiar with the many gods of the Egyptians. When Moses was inspired by God to write the Pentateuch, he rejected the polytheistic beliefs of the day. This rejection applied to the Egyptian gods, in addition to the Babylonian, Assyrian, Phoenician and Canaanite gods they would become familiar with in the future. These cultures created gods of the Sun, Moon, sky, air, weather, water, ground, mountains and rivers for people to worship.

In the present, Genesis 1:7 is a refutation of philosophical naturalism and ancillary ideas about how the Earth's sky was formed due to natural cause. This philosophy is rejected and God is identified as the one who selected nitrogen and oxygen as the major atmospheric gases. He kept harmful hydrocarbons out of the air. God fine-tuned water and its evaporation, condensation and precipitation to permit a transparent atmosphere. The sky we observe today is a direct result of God's creative hand.

Earth and its sky contrast sharply with other planets in our solar system. Earth is the only one with a significant atmosphere that permits viewing the stars from its surface. Other planets such as Venus, Uranus, Saturn and Jupiter have an atmosphere consisting of essentially a non-transparent soup.

> **Genesis 1:9 And God said, "The waters under the sky wait in one place and the dry ground is seen." And *it* was so.**
> **1:10 And God called the dry ground "land" and called the collection of waters "seas." And God saw that it *was* good.**
> **1:11 And God said, "The land is sprouting sprouts: plants seeding seeds and the fruit trees making fruit—to which kind is in the seed—over the land." And *it* was so.**
> **1:12 "And the land brings forth sprouts: plants seeding seeds *according* to *their* kind, and the trees making fruit—in which seed is to kind." And God saw that *it was* good.**
> **1:13 And evening was, and morning was—third day.**

Genesis 1:9 begins the third day of the six-day Genesis chronicle. As proposed by *exemplar* creation, on this day God describes Earth's waters as waiting tranquilly in one place. Sometime prior to Genesis 1:3, God formed depressions in the Earth's crust to create ocean basins while raising its continents. The collected waters in this verse are the oceans, seas and lakes of the world. In Genesis 1:10, God names the dry ground "land" and the collected waters "seas" as He continues to establish nomenclature.

At first glance, the author of Genesis has repeated himself in Genesis 1:11-12, albeit with some slight variation. On closer examination, the Hebrew words in these verses distinguish between *sprouting* and *growing* (bringing forth) and contrast two totally different plant phenomena. This perceived repetitive statement is actually a sophisticated observation of two distinct phases of plant maturation.

"Sprouting" is the key word in Genesis 1:11. God observes the land sprouting sprouts. When a seed is put into the soil, it must first sprout. Sprouting involves a series of complex mechanisms within the embryo of a seed generally triggered when the correct temperature, moisture and sunlight are sensed. Some seeds can sense very specific requirements before sprouting, such as the presence of friendly fungi or the absence of embryo inhibitors. When the right conditions are detected, hormones produced within the seed turn on DNA that creates specific proteins and enzymes. One of the critical enzymes produced, amylase, converts starch, stored in the endosperm of the seed, into sugar. Sugar is transported from the endosperm back to the embryo to fuel more growth. The endosperm contains all the stored energy for the seed during sprouting. When a seed sprouts, chlorophyll-rich green growth emerges and travels upwards toward the sunlight.

"Bring forth" is the key term in Genesis 1:12. God observes sprouts growing from the land—a different process than sprouting. After sprouting, a plant switches over from using internally stored nutrients to gathering them from its surroundings. A sprout sends out roots to take in moisture and minerals from the adjacent soil. The sprout uses energy from sunlight and photosynthesis to create the materials needed for plant growth. The sprout may die if the soil is extremely poor, moisture is lacking or the sunlight is inadequate.

God observes a land covered with vegetation, created in past undefined time between Genesis 1:2 and 1:3. His carefully designed process of sprouting, growth and plant reproduction is working. Seeds spread over the land carrying the genetic code of the parent plant. As seeds are buried and sprout, they faithfully reproduce to continue the cycle of life. God observes the plant world functioning as designed, and declares it good.

Genesis 1:14 And God said, "There are luminaries in the expanse of the heavens to distinguish between the day and between the night, and *they* are a sign *for* appointed times and days and years."
1:15 "And luminaries exist in the expanse of the heavens to give light over the Earth." And *it* was so.
1:16 And God *is the one who* made two great luminaries: the greater luminary to rule the day and the smaller luminary to rule the night and the stars.
1:17 And God gave them in the expanse of the heavens to give light over the Earth.
1:18 And to rule in the day and in the night and to distinguish between the light and between the darkness. And God saw that *it* was good.
1:19 And evening was, and morning was—fourth day.

Genesis 1:14 begins the fourth day of the six-day Genesis chronicle. As proposed by *exemplar* creation, on this day God describes two luminaries in the heavens (the Sun and Moon) that He created during the beginning in Genesis 1:1. The greater and the smaller luminary are the main focus of Genesis 1:14-19. No statement is made about the creation of stars in these verses. Stars are only mentioned in Genesis 1:16 as being dominated by the smaller luminary during the night.

Genesis 1:16 is connected with two other verses in the Bible that discuss how luminaries rule in the heavens.

- Psalm 136:9 The moon and stars to rule over the night, for his steadfast love endures forever. (ESV)
- Jeremiah 31:35 Thus says the LORD, who gives the sun for light by day and the fixed order of the moon and the stars for light by night. (ESV)

The order of events in Genesis 1:14-16 is important. In verses fourteen and fifteen, a direct quotation by God describes luminaries in the heavens that distinguish between day and night and give light on the Earth. In verse sixteen, He is identified as their maker. These verses represent a sequence of events starting with (1) the creation of these luminaries prior to Genesis 1:3, (2) the existence of luminaries in Genesis 1:14-15 and (3) identifying God as maker of these two great luminaries in Genesis 1:16.

Notice that in Genesis 1:14-15, the words "create, make, form, plant, grow and build" are not used. These verses describe what has already been created. It should be obvious that if God describes the luminaries as present in Genesis 1:14-15, He cannot create them in Genesis 1:16 since they already exist.

As mentioned in Genesis 1:14, heavenly luminaries determine times, days and years. Earth's rotation and orbit around the Sun establish our days and years. The orbit of the Moon around the Earth is an additional method of marking time. Earth's orbital position about the Sun marks the Winter Solstice, Spring Equinox, Summer Solstice and Fall Equinox—the beginning days of the four seasons. The relationship of the Sun, Earth and Moon determine the occurrence of eclipses.

Genesis 1:16 and 17 identify God, in contrast to the many pagan gods, as the *maker* and *giver* of the two great luminaries. The inspired words written by Moses are a direct refutation of the polytheistic beliefs of the Egyptians. Moses spent the first forty years of his life living in Egypt among royalty and was no doubt thoroughly educated in regard to their religious practices. Genesis 1:16 rejects any belief the Israelites may have held regarding Re, the Egyptian sun god, and the other Egyptian gods of Amon-Re, Re-Harakhty, Sobek-Re and Khunm-Re.[16] It also rejects belief in the Egyptian moon god, Thoth, as creator of the Moon.[17]

In the twenty-first century, Genesis 1:16 is a direct refutation of philosophical naturalism and the solar nebular hypothesis as an explanation for the origin of the Sun, Moon, Earth and our solar system. By identifying God as creator, natural cause (evolution) is rejected as an all-encompassing explanation for these objects. God is responsible for the origin, functional complexity and fine-tuning of the Sun, including its size, stability, temperature and distance from Earth. The Sun is a sophisticated fusion energy device built to run for millions of years and appears to operate unattended and without need of repair or replenishment. God is responsible for the creation and fine-tuning of the Moon in regard to its size, mass, distance from Earth and slightly elliptical orbit.

In Genesis 1:18, God observes the operation of the Sun and Moon and pronounces them good. When measured by the standards of twenty-first century science, the Sun and Moon are truly good for our planet.

> **Genesis 1:20 And God said, "The waters swarm *with* swarms of living creatures, and fliers flying over the land and over the surface of the expanse of the sky."**
> **1:21 And God *is the one who* created the great reptiles and every moving living creature with which the waters swarm, *according* to *their* kinds, and every winged flier *according* to *its* kind. And God saw that *it was* good.**
> **1:22 And God blessed them saying, "Be fruitful and multiply and fill the waters in the seas and multiply the fliers on the land."**
> **1:23 And evening was, and morning was—fifth day.**

Genesis 1:20 begins the fifth day of the six-day Genesis chronicle. As proposed by *exemplar* creation, on this day God describes the animals that He created in past undefined time between Genesis 1:2 and 1:3. Swarming water creatures indicate the presence of giant schools of fish and other abundant marine life. Winged fliers represent all creatures that can fly, including birds, bats and insects. Both marine and flying creatures represent an intricate arrangement of elemental materials into extremely complex living machines that move, metabolize food, excrete waste, self-repair, reproduce and respond to their environment. An enormous variety of creatures with different shapes, sizes and capabilities have multiplied to fill the seas and skies of planet Earth.

The order of events described in Genesis 1:20-21 is significant. In verse twenty, a direct quotation by God describes the presence of marine life and flying creatures on Earth, and in verse twenty-one He is identified as their creator. These verses represent a sequence of events starting with (1) the creation of marine life and flying creatures in the past between Genesis 1:2 and 1:3, (2) the existence of these creatures in Genesis 1:20 and (3) identifying God as creator of these creatures in Genesis 1:21.

As usual, notice that in Genesis 1:20, the words "create, make, form, plant, grow and build" are not used. This verse describes what has already been created. If God describes marine life and winged fliers as present in Genesis 1:20, He cannot create them in Genesis 1:21 since they already exist.

The word "create" is mentioned in Genesis 1:1 and then not until the fifth day in

Genesis 1:21. From the beginning of the six-day Genesis chronicle in Genesis 1:3 through Genesis 1:20, create is not used. The complete absence of this word discredits the idea that God created things during this time. As proposed by *exemplar* creation, during the first five days and part of the sixth, God only reviews, describes and names what was created prior to Genesis 1:3.

Genesis 1:21 identifies God as the creator of all marine life and flying creatures on Earth in direct refutation of the polytheistic beliefs prevalent in Egyptian culture during the time of Moses. These animals were not created by Amon, the Egyptian king of gods, or Sobek, the crocodile god.

In the present, Genesis 1:21 is a refutation of philosophical naturalism fleshed out as chemical evolution and Darwinian evolution. According to these beliefs, microscopic life developed from non-life and then evolved into a multitude of species over time. These ideas are rejected, and God is identified as the creator of all marine life and flying creatures. He is the powerful intelligence who created the information, complex functional systems, mathematically improbable structures and fine-tuned arrangements found within these living things.

Genesis 1:21 mentions God as creator of the "great reptiles." The Hebrew word translated "reptiles" is used several times in the Old Testament to describe reptilian animals. This verse is unique because it mentions large animals previously unknown to science until the relatively recent discovery of their skeletal remains. Great marine reptiles include the lizard-like mosasaurs, measuring up to 15 meters (50 ft) long. Icthyosaurs had a fish-like body shape and could measure up to 15 meters long with an estimated live weight of 32,000 kg (70,000 lbs). Plesiosaurs were marine reptiles resembling a turtle with an extremely long neck and small head. Their fossils have been found up to 14 meters (46 ft) long. Crocodiles, such as kronosaurus, were great amphibious reptiles that could be up to 10 meters (33 ft) long with an estimated live weight of 9,000 kg (20,000 lbs). Pterosaurs, classified as flying reptiles, had a wing span up to 11 meters (36 ft) long. Great reptiles mentioned in this verse, and their confirmation by science, support the historical accuracy of the Bible.

In Genesis 1:21, God pronounces Earth's marine life and flying creatures as good. The reproduction process God designed was working. DNA contained in their eggs was faithfully passing on genetic information from parent to offspring. Both types of animals were reproducing profusely, supported by a carefully balanced biosphere and plant life. The intricate dependent, symbiotic and synergistic relationships within biological life were functioning as designed. In Genesis 1:22, God blesses the animals with the statement, "Be fruitful and multiply."

Genesis 1:24 And God said, "The land brings forth living creatures *according* to *their* kind: beasts and moving *things* and animals of the land, *according* to *their* kind." And *it* was so.
1:25 And God *is the one who* made the animals of the Earth *according* to *their* kind: the beasts *according* to *their* kind and every moving *thing* upon

the ground *according* to *its* kind. And God saw that *it was* good.
1:26 And God said, "*We shall* make man in our image, after our likeness. And *he shall* have dominion on fish of the sea and on fliers in the sky and on beasts and on all the Earth and on all the moving *things* that move over the land."
1:27 And God created man in His image, in God's image created him, male and female *God* created them.
1:28 And God blessed them, and God said to them, "Be fruitful and multiply and fill the Earth and subdue it and have dominion on the fish of the sea and on the fliers of the sky and on all living *things* that move over the land."
1:29 And God said, "Behold, *I* give to you every plant seeding seed which is over the surface of all the land, and every tree in which fruit is seeding seed. As food *they* are for you."
1:30 "And to every animal of the land and to every flier in the sky and to all *that* moves over the land—in which is the living creature—every green plant is food." And *it* was so.
1:31 And God saw everything that *He* did, and behold *it was* very good. And evening was, and morning was—the sixth day.

Genesis 1:24 begins the sixth day of the six-day Genesis chronicle. As proposed by *exemplar* creation, on this day God describes the land creatures He created in past undefined time between Genesis 1:2 and 1:3. Earth is filled with land-dwelling beasts, moving things and animals.

Genesis 1:24-25 narrates events structured in a style identical to the second, fourth and fifth days. In verse twenty-four, a direct quotation by God describes a multitude of land dwelling creatures, and in verse twenty-five He is identified as their maker. These verses represent a sequence of events starting with (1) the making of these creatures in the past between Genesis 1:2 and 1:3, (2) the existence of these land creatures in Genesis 1:24 and (3) identifying God as maker of these creatures in Genesis 1:25.

Notice that in Genesis 1:24, the words "create, make, form, plant, grow and build" are not used. This verse describes what has already been created. If God describes the land creatures as present in Genesis 1:24, He cannot create them in Genesis 1:25 since they already exist.

Genesis 1:25 and 27 identify God, in contrast to the many heathen gods, as the maker of all land creatures and the creator of man. This is a direct rejection of false Egyptian religious beliefs about the origin of land animals and humans. Cattle were not made by Hathor, the Egyptian cow goddess, nor were humans formed by Khnum, the Egyptian god who allegedly created man.[18]

In the present, Genesis 1:25-27 refutes philosophical naturalism, a belief in conflict with the creation story for more than twenty-five hundred years. God is identified as the creator of all land creatures and man.

As proposed by *exemplar* creation, within Genesis 1:3-25, nothing is created or made. God's actions are limited to reviewing, describing and naming what He created previous to Genesis 1:3, and their state in the present.

In Genesis 1:26, God proposes to make man in His image. Man will have intelligence and power. He will be a self-aware, self-directed entity with freewill. He will be a designer, builder and creator in his own right. Man—created in the image of God and endowed with His characteristics—will be capable of love and hate, mercy and wrath, patience and judgment, but in a reduced capacity. Humankind will have dominion over all marine life, flying creatures and land animals.

God purposefully created humans different from animals. As pointed out by John MacArthur, "Man's very posture, standing upright, distinguishes him from four-footed beasts and creeping things. The animals' natural posture directs their gaze downward toward the earth. Man, on the other hand, is naturally positioned to look upward, toward the heavens, where he can contemplate the glory of God displayed there."[19]

In Genesis 1:26 and 1:27, there is no ambiguity about God's intentions or actions. God first announces His intent to *make* man. God's actions are then described using the word "create" three separate times. God specifically *creates* man in His image. In His image God *creates* them. Humans are *created* as male and female.

Inspired scripture does not reveal in detail how God created Adam and Eve. Man was created within a fraction of one day by a powerful, intelligent being—an act completely beyond the understanding of present science. This event is a direct refutation of Darwinian evolution, which proposes man is a product of random mutation and natural selection over millions of years. God's creation of man is linked to other scriptures in the Old and New Testament as follows:

- Genesis 5:1-2 This is the book of the generations of Adam. When God created man, he made him in the likeness of God. Male and female he created them, and he blessed them and named them Man when they were created. (ESV)
- Genesis 9:6 "Whoever sheds the blood of man, by man shall his blood be shed, for God made man in his own image." (ESV)
- Deuteronomy 4:32 For ask now of the days that are past, which were before you, since the day that God created man on the earth, and ask from one end of heaven to the other, whether such a great thing as this has ever happened or was ever heard of. (ESV)
- Isaiah 45:12 I made the earth and created man on it; it was my hands that stretched out the heavens, and I commanded all their host. (ESV)
- Jeremiah 27:5 It is I who by my great power and my outstretched arm have made the earth, with the men and animals that are on the earth, and I give it to whomever it seems right to me. (ESV)
- Malachi 2:10 Have we not all one Father? Has not one God created us? Why

then are we faithless to one another, profaning the covenant of our fathers? (ESV)

• Matthew 19:4 He answered, "Have you not read that he who created them from the beginning made them male and female." (ESV)

• Mark 10:6 But from the beginning of *their* creation, He made them male and female. (ET)

• Luke 3:38 The son of Enos, the son of Seth, the son of Adam, the son of God. (ESV)

• Acts 17:28 In him we live and move and have our being; as even some of your own poets have said, "For we are indeed his offspring." (ESV)

• 1 Corinthians 11:9 Neither was man created for woman, but woman for man. (ESV)

• 1 Corinthians 15:45 Thus it is written, "The first man Adam became a living being;" the last Adam became a life-giving spirit. (ESV)

• Romans 9:20 But who are you, O man, to answer back to God? Will what is molded say to its molder, "Why have you made me like this?" (ESV)

• Colossians 3:10 And have put on the new self, which is being renewed in knowledge after the image of its creator. (ESV)

• 1 Timothy 2:13 For Adam was formed first, then Eve. (ESV)

• Hebrews 2:6-7 It has been testified somewhere, "What is man, that you are mindful of him, or the son of man, that you care for him? You made him for a little while lower than the angels; you have crowned him with glory and honor." (ESV)

Genesis 1:28 is often viewed as a cultural mandate. God blesses man and woman and tells them to multiply and fill the land, and to subdue and have dominion over the animals.

God specifically made man with unique features that permit him to dominate animals. He did not give him the speed, strength or fighting equipment given most wild creatures. Man has no strong claws or sharp teeth. Humans lack the keen sense of smell, extraordinary hearing and exceptional long range vision found in some animals. He is completely without a single strong defensive mechanism such as the hard shell of a turtle, the camouflage of a chameleon, the quills of a porcupine or the scented spray of a skunk. On the other hand, man's phenomenal brain, five senses, full-color vision, exceptional language ability, fully functional opposable thumb, dexterous hands and distinctive upright stature allow him to have superiority over the animal kingdom and fulfill God's mandate.

Genesis 1:29 points out that seed-bearing plants and fruit trees are food for man. Plants are a major part of man's food supply, including leaves such as spinach and roots typified by carrots. Generally, seeds are the basis for the majority of man's food and include grain seeds such as wheat, rye, barley and oats. Also included are corn, peas, lentils and beans in addition to seed pods such as grapes, plums, lemons, apples, pears, peaches, oranges and watermelons. God in His wisdom designed a variety of seeds and seed pods edible by humans.

Genesis 1:30 mentions that green plants are food for animals. The digestive systems of animals are made to handle a greater variety of leaves, roots and seeds than man's. This verse subtly points out a fundamental issue of biological life—plants are the bottom of the food chain. Ultimately all animal life, whether herbivorous, omnivorous or carnivorous, is dependent upon plants for food.

We can infer from Genesis 1:29-30 that within the Garden of Eden there were no significant carnivorous or predatory animals. Plants were the primary food source. Within the garden, man was a vegetarian and ate fruits and vegetables. Animals of the garden were essentially herbivorous. Birds fed on seeds, worms and harmless insects. No saber tooth tigers, lions, hyenas, alligators, wolves, venomous snakes or poisonous spiders lived in the garden. The garden was a protected place from the violence, predator-prey relationships and survival of the fittest that ruled the world outside.

Between Genesis 1:3 to 1:25, God reviews, describes and names what He created prior to Genesis 1:3. He also affirms its current state of operation. For example, light is proclaimed as "good" in Genesis 1:4. The collected waters and dry land are described as "good" in Genesis 1:10. Sprouting sprouts and growing sprouts are announced as "good" in Genesis 1:12. The Sun and Moon are recognized as "good" in Genesis 1:18. Swarms of marine life and flying creatures are labeled as "good" in Genesis 1:21. Land animals, beasts and moving things are identified as "good" in Genesis 1:25.

In Genesis 1:31, God looks upon His total creation—both past and present. Past things include all He created before Genesis 1:3. Present things include what He created after Genesis 1:25 on the sixth day—Adam and Eve, the Garden of Eden and animals within the garden. God shows pleasure in all He has created by summarizing it as "very good."

Genesis Chapter 2

Genesis 2:1 And the heavens and Earth *were* finished, and all their multitude *of miscellaneous things*.
2:2 And on the seventh day God finished His work that *He* did, and rested on the seventh day from all His work that *He* did.
2:3 And God blessed the seventh day and sanctified it, because on it *He* rested from all His work—that God did *when* creating.

Genesis 2:1 brings to a close the six-day Genesis chronicle and the completion of God's work. The host of miscellaneous things He did to bring the heavens and

Earth to their current state is now complete.

Genesis 2:2 mentions God resting on the seventh day from "work that He did." Work can be divided into physical and non-physical activities. Physical work God did on the sixth day includes creating, making, forming, planting, growing, building, placing and bringing things. Non-physical work He did during the entire six-day period includes reviewing, inspecting, describing, distinguishing, naming and blessing. Both physical and non-physical activities God did were work.

A comprehensive inspection of the Sun, Moon and Earth were no doubt a part of the six-day activities. God may have examined animals and plants for proper growth and reproduction. He may have checked Earth's air, water and soil as part of a comprehensive quality review. To pronounce all He observed as "good" and "very good," God may have conducted a thorough assessment of the entire planet.

Was God really tired and in need of rest? Of course not! God simply established a pattern of resting on the seventh day, later codified in the Ten Commandments. At Mount Sinai God formally instituted the Sabbath as a day of rest, and as pointed out in Mark 2:27, it was for man's benefit. God knew man needed a day of rest every week to refresh, renew and restore himself.

Genesis 2:4 The *following is the* account of the heavens and the Earth when created; on the *sixth* day *when* Yahweh God did *things to the* Earth and heavens.

Genesis 2:4-25 is best explained as a detailed description of events on the sixth day. During the sixth day God created man and woman, planted the Garden of Eden, grew trees in the garden and formed animals within the garden. This account continues through Chapter 3 and ends in Genesis 4:26. In Genesis 5:1, a new account begins with a short history of the descendants of Adam.

Genesis 2:4 starts with a general statement that an "account" about the creation of the heavens and Earth will be given. The specific account concerns the *day* God *did* the Earth and heavens. Within the six-day Genesis chronicle, described in Genesis 1:3-31, God only created on the sixth day.

Genesis 2:4 states that God created the "heavens and Earth." The heavens are mentioned first to emphasize the greatest of God's works. With an estimated 125 billion galaxies, each containing an estimated 400 billion stars, creation of the heavens was a magnificent feat. Following second in order is the Earth—a lesser creation and mentioned last.

The second half of Genesis 2:4 points out that God "did things to the Earth and heavens" on the sixth day. Notice the order is now reversed. The Earth is mentioned first in order of importance and then the heavens. On this day, the majority of God's effort applied to the Earth and less was expended on the heavens. This is one of only two places in the Old Testament where the expression "Earth and heavens" occurs.[20]

What did God do to the heavens on the sixth day? Scripture is silent on any specific details. Evidence from astronomy has identified several phenomena within our solar system which appear to be relatively recent and without a plausible natural explanation. These phenomena include one moon with a smooth ice-covered surface and the ice rings of Saturn. We can speculate these are among the things God did when He intentionally left signs in our solar system for modern man as evidence of His presence. (Chapter 13 discusses this evidence in more detail.)

> **Genesis 2:5 And each bush of the field was not yet on the land, and each plant of the field *did* not yet grow, because Yahweh God had not caused rain over the land and there was no man to cultivate the ground *of the garden.***
> **2:6 And mists rose up from the land and watered all the surface of the ground *of the garden.***
> **2:7 And Yahweh God formed the man *of* dust from the ground and breathed in his nostrils the breath *of* life. And the man existed as a living creature.**
> **2:8 And Yahweh God planted a garden from the east in Eden, and there *He* put the man whom *He had* formed.**
> **2:9 And Yahweh God grew from the ground every tree *that was* desirable to the sight and good for food, and the tree of life in the middle of the garden and the tree of knowledge of good and evil.**

Genesis 2:5 is a description of the ground chosen for the Garden of Eden, before God planted it. We can infer from this verse that the climate over the land selected for the garden was hot and arid, similar to the present-day Arabian Desert. Little or no rain fell, and there were no bushes or plants.

Genesis 2:6 describes how "mists" rose from the land and watered the ground. This verse refers to an effect within the Garden of Eden that was a substitute for rain. The Hebrew word translated "mist" is only mentioned one other place in the Bible and its exact meaning is unclear. We can attribute God as being responsible for the mists that watered the ground. In essence, He installed a watering system.

Genesis 2:5-6 does not teach that rain never fell on the entire Earth until the flood of Noah. Such an effect would have required the suspension of Earth's hydrologic cycle, one of the most fundamentally important systems on the planet. This cycle describes when water evaporates, condenses into clouds and then returns to the ground as rain, sleet and snow. Snowmelt and rain runoff find their way into springs, streams, rivers, lakes, seas and oceans. If water in the higher elevations were not replenished, all springs, streams and rivers would eventually dry up. If no rain fell for hundreds of years previous to Noah's flood, it would have created serious environmental problems.

Genesis 2:7 describes in more detail how Adam was formed on the sixth day out of "dust from the ground." Scientists have confirmed that the ground (Earth's crust) contains all the necessary elements to make the human body. The top ten elements and their use in the human body are listed in Table 1.2. Approximately 70% of body weight is water, composed of the elements oxygen and hydrogen. Other elements used in biological processes in decreasingly smaller amounts below those listed include magnesium, iron, zinc, copper, selenium, boron, iodine, chromium and cobalt.

Table 1.2 – Elements in the Human Body by Weight

Rank	Element	Weight	Percent	Location and Use
1	Oxygen	43 kg	61.43	Water, amino acids, DNA, RNA
2	Carbon	16 kg	22.86	Amino acids, DNA, RNA, sugar molecules
3	Hydrogen	7.0 kg	10.00	Water, amino acids, DNA, RNA
4	Nitrogen	1.8 kg	2.57	Amino acids, DNA, RNA, ammonia
5	Calcium	1.0 kg	1.43	Bones, teeth
6	Phosphorus	780 g	1.11	Bones, DNA, RNA, ATP, phosphate molecules
7	Potassium	140 g	0.20	Brain, nerve, muscle function
8	Sulfur	140 g	0.20	L-cysteine metabolism, sulfate molecules
9	Sodium	100 g	0.14	Salt, cellular tissues, blood, sweat
10	Chlorine	95 g	0.14	Salt, cellular tissues, blood, sweat

Notes:
(1) Weights are based on a 70 kg (154.3 lb) person.
(2) Hydrogen and oxygen are found primarily as water in the body.
(3) Sodium and chlorine are found primarily as salt in the body.
Reference: Emsley, John (1998). *The Elements*.

The primary difference between a man's physical body and a pile of dirt is the arrangement of the atoms. It is the sequencing and placement of atoms that create DNA, amino acids, proteins, enzymes, cells, nerves, arteries, muscles, bones, organs and other structures within the human body. During the creation of man, God rapidly assembled the required complex structures while protecting them from deterioration. When Adam was complete, God breathed life into him to start his bodily processes.

Genesis 2:8-9 describes God making the Garden of Eden where He plants and grows a variety of fruit trees good for man to eat. Food is available and abundant. Two special trees are placed in the center of the garden—the tree of life and the tree of knowledge of good and evil. Adam is placed inside as the first occupant.

God deliberately placed the tree of life and the tree of the knowledge of good and evil in the garden. Yahweh gave man free will, in addition to a means of expressing that will. Through these two trees, man could choose life or death. God granted humankind freedom of choice, with full power to obey or disobey.

Exemplar creation views the Garden of Eden as having firm boundaries. The garden was a secure place for Adam and Eve to grow and mature. Plants and animals were carefully selected to create a place of safety and harmony. There were no thorns or thistles. The garden was a carefully controlled, nurturing environment to ensure man survived in an otherwise hostile world. The garden offered protection from the predatory animals that lived outside, where tooth and claw reigned. Beyond its protecting boundary, wild animals roamed the land and survival of the fittest was the rule. Food edible by man was scarce and life was hard.

> **Genesis 2:10 And a river went out from Eden to water the garden and from there diverged and became four branches.**
> **2:11 The name of the first is Pishon; it winds through all the land of Havilah, where there is gold.**
> **2:12 And the gold of that land is good; bdellium and onyx stone are there.**
> **2:13 And the name of the second river is the Gihon. It winds through all the land of Cush.**
> **2:14 And the name of the third river is the Tigris; it goes east of Asshur. And the fourth river is the Euphrates.**

These verses give the geographical location of the Garden of Eden with respect to four rivers. Contrary to some claims, the location of the Garden of Eden is unknown. Two of the rivers cannot be positively identified with existing rivers today. This does not invalidate the garden's existence, but only illustrates that information is sometimes lost due to the passage of time even with a careful description.

> **Genesis 2:15 And Yahweh God took the man and put him in the Garden of Eden to cultivate and to keep it.**
> **2:16 And Yahweh God commanded to the man saying, "From any tree of the garden, you *may* eat-eat."**
> **2:17 "And from the tree of the knowledge of good and evil, *do* not eat from, because on the day you eat from it *you shall* die-die."**
> **2:18 And Yahweh God said, "*It is* not good that the man be alone. For him *I shall* make as his counterpart a helper."**

God placed Adam in the Garden of Eden and gave him the responsibility of cultivating and caring for the garden as part of his training. God in His foreknowledge knew humankind would sin and be cast out. God looked beyond Adam's sin and equipped him to survive in a hostile world beyond the garden.

God gave Adam specific instructions for eating from the trees of the garden. Adam was encouraged to "eat-eat" and could partake of any tree except the tree of knowledge of good and evil. If Adam ate from that tree, he would "die-die." In the original Hebrew, both eating and dying are emphasized through word repetition. (Throughout the Old Testament, word repetition is often used for emphasis although not evident in English translations.)

Genesis 2:17 contains a stern warning that if Adam ate of the tree of the knowledge of good and evil he would die. In Hebrew culture, a future event could be considered as completed in the present when commanded by one in authority—in this case God. The day Adam ate of the forbidden tree, he was condemned to die. The surety of his physical death was absolutely certain, although it occurred years later.

In Genesis 2:18, God recognizes that Adam should not live alone and purposes to make a helper for him. This is fulfilled in the creation of Eve in Genesis 2:21-22. However, before Eve was created, God formed the animals of the garden.

> **Genesis 2:19 And Yahweh God formed from the ground each animal of the field and each flier of the sky *within the garden*. And *God* brought *them* to the man to see what he *would* call them and whatever the man called the living creature, that was its name.**
> **2:20 And the man called names to all the beasts and to the fliers of the sky and to all animals of the field *within the garden*. And for Adam *was* not found as his counterpart a helper.**

Genesis 2:19 describes when God formed animals of the field and fliers of the sky for the garden. Notice that no marine life is mentioned. Since the garden was located on arid land (without any significant bodies of water), the lack of marine life is not unexpected. All animals were formed from the elements of the ground, as was man in Genesis 2:7. A small number of land animals and fliers (speculated at around fifty species) were created to populate the garden and completed this learning, nurturing environment.

Genesis 2:19-20 are often misunderstood. There is no need to believe God created all animals on the entire Earth at this time and brought them to Adam for naming in one day. There is no need to require Adam to hold a marathon session for naming thousands of animals on his first day alive. Such belief represents a misunderstanding of the six-day Genesis chronicle.

> **Genesis 2:21 And Yahweh God caused a deep sleep to fall over the man and *during* sleep took from one of his ribs and closed up under the flesh.**
> **2:22 And Yahweh God built—which *He* took *of* the rib from the man—into a woman, and brought her to the man.**
> **2:23 And the man said, "This is now bone from my bones and flesh from my flesh; for this *shall* be called woman, because this *was* taken from man."**
> **2:24 So over *this reason*, a man *will* leave his father and his mother and hold fast on his wife. And *they shall* be as one flesh.**
> **2:25 And the two were naked, the man and his wife, and were not ashamed.**

Genesis 2:21-22 describe the building of Eve. With our current knowledge of biotechnology, this description appears completely plausible. God put Adam to sleep and removed genetic material from one of his ribs. Advances in biology point out that DNA from one person can be used to make another, and stem cells inside bone are ideally suited for cloning. Males have one X and one Y chromosome, while females have two X chromosomes. Conceptually it is possible to make a woman from a man, but not vice versa. God could have duplicated Adam's one X chromosome. Two X chromosomes and Adam's remaining 44 chromosomes provided the basic genetic material for making the first female. Eve was similar to Adam except for being female, and as his sexual complement was truly a counterpart. God completed the job by closing up the incision under the flesh in Adam's side.

Genesis 2:23-25 describe Adam's acceptance of Eve and acknowledgement of their common flesh. The first wedding ceremony in history is not described in any detail, but the requirements for leaving father and mother and becoming one flesh in marriage are emphasized. Upon marriage, the bond between child and parent is weakened and replaced by that between husband and wife. This verse emphasizes a one man, one woman permanent bond and establishes the pattern of marriage God intended for humankind. A man is to hold fast to his wife so tightly they can be considered one, which becomes physical reality in their offspring. A child receives half of her chromosomes from her father and half from her mother.

Genesis 2:25 describes Adam and Eve living in a state of innocence—naked and not ashamed. The necessity of a warm climate is subtly pointed out, since humans without clothing are quite limited in cold temperatures they can continuously endure. The general environment of the garden was selected for their comfort.

Conclusion

Exemplar creation views the time between Genesis 1:1 and 1:3 as undefined by scripture. The time could be one day or billions of years. Genesis 1:1 begins with a statement about God creating the heavens and Earth. In Genesis 1:2, Earth is described as lifeless, dark and covered with deep water sometime after its creation. Between Genesis 1:2 and 1:3, God creates all of Earth's biosphere, the fossil record is laid down and significant geological change occurs. At Genesis 1:3, the transition to 24-hour days occurs.

The vast majority of Earth's history occurs prior to Genesis 1:3, and unfortunately little information is given about this period. The exact time, sequence and method God used to create the heavens and Earth are not revealed in scripture, and man is left to his own resources to fill in the details. A combination of special and general revelation helps us understand events during this era.

Between Genesis 1:3 and 1:31, the six-day Genesis chronicle occurs and time is measured in six literal days.

Between Genesis 1:3 and 1:25, the first five days and part of the sixth, is a time of reviewing, describing and naming. The sequence in which God describes things in Genesis 1:3-25 is not necessarily the same as the creation sequence prior to Genesis 1:3. Genesis 1:3, 6, 9, 11, 14, 20 and 24 are not commands of God, but are descriptive statements.

On the sixth day, God begins creating again. In Genesis 1:26, God proposes to *make* man. In Genesis 1:27, God *creates* man. In Genesis 2:7, God *forms* man of dust from the ground. In Genesis 2:8, God *plants* the Garden of Eden. In Genesis 2:9, God *grows* trees from the ground for the garden. In Genesis 2:19, God *forms* the animals of the garden. In Genesis 2:22, God *builds* a woman from Adam's rib. During the sixth day, God's physical work is described using the words "create, make, form, plant, grow and build."

Exemplar creation considers Genesis 2:4-25 as best explained in reference to the Garden of Eden and activities on the sixth day. A detailed account is given of God's actions when He created Adam and Eve, planted the garden and formed a small number of animals for the garden.

Genesis Chapter 3 continues the story of Adam and Eve in the Garden of Eden. Through Satan's influence, Adam and Eve disobey God and eat of the tree of knowledge of good and evil. They are punished and cast out of the garden, never to return. Adam and Eve give birth to two sons, Cain and Abel, and the elder kills the younger in the first recorded murder in history. Adam and Eve have another son, Seth, in addition to many more sons and daughters as mentioned in Genesis 5:4. Cain and Seth marry their sisters and begin populating the Earth. From this point onward, humanity continues to multiply until the flood of Noah.

The complete Genesis creation story has several levels of complexity capable of appealing to a wide reading audience, ranging from a small child to the scientist and theologian. On the surface, it is a simple story of an omnipotent God and how He created the heavens and Earth. On a deeper level, there are significant implications about the creation of life, the making of man in the image of God, the building of Eve from Adam's rib, the establishment of marriage, human free will, effects of the first sin, Satan's influence and man's punishment for sinning. On a still deeper level are implications concerning the trinity (Gen. 1:26), the dominion mandate (Gen. 1:28), a prelude of the Sabbath later codified in the Ten Commandments (Gen. 2:2-3), introduction of the name Yahweh for God (Gen. 2:4) and the messianic prophecy for the coming of Jesus Christ (Gen. 3:15). On all three levels, the Genesis story is true.

Exemplar creation is based on a careful study of the original Hebrew language of the Old Testament and presents the Genesis creation story as an abbreviated, non-technical, historical account of origins that is truthful and trustworthy. The Genesis creation account and empirical science do not conflict where they intersect. Special revelation and general revelation can be harmonized. The *exemplar* translation of Genesis Chapters 1 and 2 can be accepted by educated, intelligent, biblically knowledgeable and scientifically literate Christians as a true historical account. The Genesis story can be believed without apology.

Blessed is the one who finds wisdom, and the one who gets understanding. (Proverbs 3:13) ESV

Chapter 2

Understanding Science

A significant issue for Bible students and Christians in our present world is the relationship of science and the Bible. Can the Bible legitimately conflict with science? Does science contradict the biblical account of creation? Has science disproved the Bible? A failure to properly answer these questions has serious consequences. In England, France, Germany and other nations in Europe, belief in the Bible is equated with ignorance, mythology and superstition and plays a shrinking role in those societies. Government backed by science and its influence on education, law, social programs and public agendas has a skeptical, if not hostile, position in regard to biblical based beliefs.

To an extent, the decline in biblical influence on society is the fault of the Bible believing community. Many religious leaders, in their personal belief and teaching, have disconnected the Bible and science. In regard to origins, they believe the two are incompatible and can never be harmonized.

One position proposes that the Bible is our ultimate authority. Science should not be dismissed, but "must be kept in its proper ministerial role as a servant to the Bible and never placed in a magisterial role over and above the Bible."[1] If conflict occurs between the two, then science is wrong. In effect, science must be subservient to biblical theology. Proponents of traditional young-Earth creation theology believe that neither the Sun nor the stars can be billions of years old. Earth and its plants and animals were created roughly 6,000 years ago. The Earth was covered by a global flood around 2349 BC. There will never be agreement with science and the Bible regarding origins. Science is wrong.

A second position held by some members of the biblical community is that the Bible is a religious book written to meet the spiritual and emotional needs of humanity. According to this viewpoint, only in matters of faith and practice is the Bible relevant. Supporters say the Genesis creation account in the Bible is in irreconcilable conflict with science. The first eleven chapters of Genesis are myth and metaphor.[2] Adherents of this position point out that evolution is supported by an overwhelming consensus of scientific opinion. A literal interpretation of the Genesis account must be dismissed in preference to natural cause as God's method of creation. There will never be agreement with science and the Bible regarding origins. The Bible is wrong.

Adding fuel to the controversy is philosophical naturalism, a belief that the universe and all therein must be the result of strictly natural cause.[3] Explanations of origin are limited to natural, physical, mechanistic and material causes. All appeals to intelligent cause, such as God, are rejected. This philosophy is fleshed out in ideas of the Big Bang, evolutionary star formation, the solar nebular hypothesis, chemical evolution and biological evolution. Ideas supportive of natural cause give a complete explanation of origins, from the first atom to the human brain. Proponents of philosophical naturalism believe they are justified in rejecting all ideas not based on natural cause. God does not exist.

Which of the above three positions about origins is correct? This book purports to answer that question. As a prerequisite, the fundamentals of science, its limitations and terms must be understood. Knowledge of science adds clarity to any discussion about the complex issue of origins.

Any biblical view of origins should consider the impact of science on present-day thinking. Science is considered observable, repeatable, demonstrable, factual, truthful and the basis of our modern society. Technology based upon science has significantly reduced the effects of pestilence, disease, famine and natural catastrophe. Science has made it possible for mankind to have the highest standards of transportation, communication, housing, clothing, food, medical care, entertainment and general comfort the world has ever known. As Phillip Johnson has rightly observed, science is the only universally valid form of knowledge within our culture.[4] For these reasons, a serious attempt must be made to reconcile science and the Genesis creation account. To do anything less is to contribute to the decline of biblical belief in our present world.

Empirical Science

Science has been defined several different ways. According to *Webster's Third New International Dictionary*, the word "science" comes from the Latin word *scientia,* meaning "knowledge."[5] In a general sense, science is defined as the "possession of knowledge as distinguished from ignorance or misunderstanding."[6] More specifically, science is "knowledge concerned with the physical world and its phenomena."[7] The *McGraw-Hill Dictionary of Scientific and Technical Terms* defines science as: "A branch of study in which facts are observed and classified, and usually quantitative laws are formulated and verified. It involves the application of mathematical reasoning and data analysis to natural phenomena."[8] The *Academic Press Dictionary of Science and Technology* defines science as: "The systematic observation of material events and conditions in order to discover facts about them and to formulate laws and principles based on those facts."[9] Dr. Brent Dalrymple defines science as "the rational investigation of the physical world and its phenomena."[10] Dr. Michael Behe defines it as "an unsentimental, no-holds-barred struggle to understand nature."[11]

First and foremost, science is an empirical process. Empirical science is "based on actual measurement, observation or experience, rather than theory."[12] Empirical science is based on practical human experience with the physical world. Science has no strict methodology and may be conducted using different approaches, but ultimately relies on knowledge of phenomena and materials in the world around us. The vast majority of scientists—Archimedes, Galileo Galilei, Johannes Kepler, Nicolaus Copernicus, Isaac Newton, Gregor Mendel, Louis Pasteur, Marie Curie, John Dalton, Robert Boyle and Edwin Hubble—based their work on observation, measurement and experimentation. Significant empirical science cannot be performed in a sealed room, isolated from knowledge of the world outside, using only logic, mathematics or computers.

Table 2.1 – The Classes and Branches of Science
(Five areas of thought, study, research, information and practice)

Applied Science	Basic Science	Biological Science	Earth Science	Space Science
Medicine Agronomy Weaponry Engineering Pharmacology Transportation Communications	Physics Chemistry	Biology Botany Zoology Ecology Genetics Biochemistry Paleontology	Geology Hydrology Geophysics Meteorology Volcanology Oceanography	Astronomy Cosmology Astrobiology Astrophysics

Reference: Strahler, Arthur (1992). *Understanding Science: An Introduction to Concepts and Issues*. p76.

However, some scientists are theoreticians, and use primarily logic and mathematics to explain phenomena. For example, Albert Einstein did not conduct scientific research and was primarily a theoretician. Yet without knowledge of what other scientists had discovered about light, sound, motion, the Doppler effect and gravity, he could not have developed his theories. Without having personal knowledge of flashlights, sound, whistles, rockets, trains and elevators, he could not have explained his ideas. Einstein's theories have been validated to an extent by other scientists through observation, measurement and experimentation.

Secondly, yet of great significance, science is defined by philosophy. Philosophers of science define its boundaries, presuppositions and overarching ideals. If defined incorrectly, science will be misguided. Likewise, if the scientific community follows incorrect ideals and if the academic community teaches accordingly, then science has the potential for being horribly misused. For science to achieve its full potential, the acceptance of correct philosophical principles and faithfulness to those principles is required.

Science has traditionally been subdivided into applied science, social science and natural science as pointed out in *Understanding Science: An Introduction to Concepts and Issues.*[13] Five major classes of scientific thought, study, research, information and practice can be arranged as shown in Table 2.1.

Empirical science (from *scientia,* meaning knowledge) can be defined as a rational means of inquiry restricted to observation, measurement and experimentation to acquire dependable knowledge about the physical world in the form of verified empirical data. Generalizations of verified empirical data are developed as laws, principles and formulae. Empirical science relies solely on practical human experience and is inherently limited because of its method of inquiry. Lucid, logical thought is a basic requirement of this activity.

Definition 2.2 – Empirical Science

A rational means of inquiry restricted to observation, measurement and experimentation to acquire dependable knowledge about the physical world in the form of verified empirical data. Generalizations of verified empirical data are developed as laws, principles and formulae. Empirical science relies solely on practical human experience and is inherently limited because of its method of inquiry.

The above definition can be restated in broader terms as: Empirical science is a rational search for dependable knowledge about the world based strictly on physical evidence, and is inherently limited because of inaccessible and unavailable evidence.

The goal of empirical science is to obtain dependable knowledge using observation, measurement and experimentation as follows:

(1) Observation uses the five senses of sight, touch, smell, hearing and taste, often aided and enhanced through the use of instrumentation and other equipment. Telescopes, microscopes, weight scales, rulers, thermometers, decibel meters, spectrographs and other instruments have greatly enlarged the range and precision of human observation.

(2) Measurement refers to the way scientists quantify mass, length, volume, force, hardness, temperature, heat, pressure, time, sound, frequency, magnetism, voltage, amperage, light intensity, color and other phenomena. Measurement systems are comprised of precise reference standards, measurement units and accurate measurement equipment. Precise measurement is a significant, integral part of present-day science. Measurement permits observations to be accurately quantified, allows the exchange of information using standardized units and makes possible the identification of subtle physical characteristics that would otherwise be undetectable.

(3) Experimentation refers to a test under controlled conditions that is conducted to demonstrate a known truth, determine the validity of a hypothesis or investigate an idea.

Verified empirical data forms the foundation of all science. Observations, measurements and experiments that can be consistently repeated by other investigators (e.g. scientists, engineers and technicians) are the basis of empirical science. Scientific knowledge is cumulative[14] and has been built up over time. For example, the periodic table and knowledge of the elements have been accumulated over hundreds of years through the combined effort of many scientists.

Because of the complete dependence of empirical science on verified empirical data, corrections within empirical science throughout history have been minor. Probably the most notable corrections apply to the restatement of laws, principles and formulae to restrict their application within specific boundaries. For example, the recent theories of relativity and quantum mechanics have not replaced or overthrown Newtonian mechanics[15] founded upon the work of Isaac Newton (1642-1727). Newton's laws of motion are accurate generalizations when properly bounded; however, they do not apply to relativistic velocities or behavior at the atomic level.

By definition, empirical science is solidly based on physical evidence. Empirical science relies solely on practical human experience, without regard to theory or significant inference. Doctors, pharmacists and agronomists use the discoveries of empirical science to improve our standard of living. Scientists, engineers and technicians use empirically derived information to develop, design and manufacture technological products for our modern world. Mass-produced products are based on the consistent operation of physical phenomena. The cause, operation and structure of some phenomena may never be known, but a lack of that information does not inhibit application through technology. Empirical science is dependable knowledge.

Verified Empirical Data

Empirical science is based on verified empirical data which includes quantifiable numerical data in addition to descriptive information. Verified empirical data can be defined as dependable information about the physical world (i.e. universe) based upon repeated independent human observation, measurement and experimentation over time with consistent results. Verified empirical data is without significant inference and is not theory, hypothesis, conjecture, speculation, unverified data, estimation, extrapolation or consensus of scientific opinion. Verified empirical data is not appreciably tentative, but is subject to correction.

Verified empirical data has a small probability of being false. The repeated confirmation of results using different equipment for observation, measurement and experimentation at different locations by neutral or skeptical scientists over a period of time makes this type of information reliable. This is not to say verified empirical data can never be wrong, but that is has a small probability of being wrong. The precision, understanding and completeness of empirical data are constantly improving.

The book *Science and Creationism: A View from the National Academy of Sci-*

ences is misleading in its discussion of empirical data. Rather than verified data, the book uses the term "fact" defined as: "In science, an observation that has been repeatedly confirmed and for all practical purposes is accepted as 'true.' Truth in science, however, is never final, and what is accepted as a fact today may be modified or even discarded tomorrow."[17] The first sentence roughly corresponds to Definition 2.3 about verified empirical data. Facts are observations that have been repeatedly confirmed and accepted as true. Facts make up the dependable knowledge base of science. However, the second sentence in the NAS definition undermines the first. The second sentence actually refers to the inferences of science (theories and hypotheses) which are subject to being modified or discarded tomorrow. The NAS statement is confusing and misleading because it mixes fact and inference.

Definition 2.3 – Verified Empirical Data

Dependable information about the physical world based upon repeated independent human observation, measurement and experimentation over time with consistent results. Verified empirical data is without significant inference and is not theory, hypothesis, conjecture, speculation, unverified data, estimation, extrapolation or consensus of scientific opinion. Verified empirical data is not appreciably tentative, but is subject to correction.

Unfortunately, other science books hold a similar view of scientific facts. The textbook *Life-The Science of Biology* says, "Statements of biological 'fact' are mixtures of observations, predictions and interpretations."[18] Predictions and interpretations are not the equivalent of fact and it is misleading for educators to treat them as such, especially in a science textbook. Predictions and interpretations made by scientists are based on inference and judgment, and may be completely wrong.

Contrast the above definitions of "fact" with a standard dictionary definition. *The American Heritage College Dictionary* defines fact as: "(1) knowledge or information based on real occurrences and (2) something demonstrated to exist or known to have existed."[19]

In science, facts should represent verified empirical data. With little effort, scientists who are knowledgeable of a particular area can separate verified empirical data from extrapolations, estimates, unverified data, theories, hypotheses, conjecture and speculation. When factual and non-factual information are commingled together without being identified, the potential exists for deception and manipulation in science education and public policy based on science. Science can be misused by unscrupulous people.

Limitations of Empirical Science

Empirical science is limited and in some instances inferior to other means of knowledge. In regard to past events, written history is significantly superior to empirical science. Credible eyewitness accounts of historical events often give details totally beyond the ability of science. What can science tell us about Benjamin Franklin? What can science tell us about the San Francisco earthquake of 1906? About the American Revolutionary War? If a written history of these events did not exist, most information about them would be unknown or unknowable. Without written history, the times, causes, effects and other significant details about major past events would be lost forever.

The vast majority of history is beyond the reach of empirical science. In most instances, past events cannot be observed, measured or experimented upon because of the one-way flow of time. Events cannot be replayed. Most history cannot be verified by scientific means, but nevertheless can be true.

Although history is not science, it is important to humanity. History tells us where we came from, how we got here and who we are. History contains information about past cultures, nations, governments, wars, great people, significant events and technological developments. History fleshes out the progress of political, religious, philosophical and scientific ideas. Our knowledge of the past has a strong influence on current beliefs and behaviors. History gives us perspective.

Non-repeating historical events of an unusual nature are often viewed with skepticism. For example, suppose there is an isolated group of primitive natives in a remote jungle who have never seen a gun. One day a small party meets a big game hunter and is intrigued by his rifle. The hunter puts on a demonstration by shooting several coconuts and small animals. When the group returns to their village, they excitedly tell the other natives about the strange booming stick which killed animals and made holes in coconuts. Village elders reject their story along with the evidence. The dead animals could have been killed by other means and the hole in the coconuts could have been made by a crude drill. The big game hunter cannot be found to repeat the event. Village elders conclude the members of the hunting party are lying.

In a similar manner, it is difficult for scientists to evaluate non-repeating historical events. Any unusual event not examined and described by an objective, trained observer is viewed with skepticism. Evidence in support of the event can often be explained due to other means. One-time events in the form of miracles by God, displays of advanced technology or natural occurrences have little value to science. A single nonrecurring event has little worth other than for historical, emotional, religious or philosophical purposes. Generally, an unrepeatable event has no technological importance. Without lasting effects, one-time events can neither be proved nor disproved by science.

Scientists often debate how to treat historical information. Should all historical accounts be rejected? Should only "modern" historical accounts be accepted? Should lasting physical evidence be required before accepting a science related historical account? There is no clear, crisp answer to these questions. There is no easy answer for how to treat one-time, non-repeating historical events. For example, on November 11, 1572, Tycho Brahe, an astronomer, noticed a new, very bright star in the heavens. To the naked eye, it was visible in daylight for about two weeks and at night for about sixteen months. In his notes, Tycho left a written description of this event and a sketch of the star's location. Today astronomers know he saw a supernova. The Hubble Space Telescope and Chandra X-ray Observatory have identified remnants of Tyco's supernova at the location described by him. An expanding cloud of debris has been clearly observed. If scientists only had Tyco Brahe's written historical account, it is doubtful much significance would be given to his observation.

People have observed hundreds of similar unusual phenomena and left written accounts. Scientific acceptance of these accounts varies and is based on the credibility of the eyewitnesses and their descriptions. Similar events repeated at other times in history or lasting physical effects available for study today often support these written descriptions.

Even in the present, many phenomena are beyond understanding by empirical science. Information has no plausible natural explanation. Science cannot explain moral values, evil or purpose. Science cannot explain the human emotions of love, hate, greed, sorrow, joy, happiness, friendship, passion, guilt, awareness, pleasure, faith and distrust. Other than crudely attempting to explain these as chemical reactions within the human brain, empirical science has no satisfactory answer.[16]

Empirical science is quite limited in understanding the origin, ultimate cause, operation and structure of many phenomena. Nuclear forces, positive proton charge, negative electron charge and gravity are basic forces that exist, yet where did they come from? How do they operate? Why are there atoms? How did matter and energy originate? These questions may never be answered. In regard to many phenomena, all empirical science can say is they exist and are consistent within the known universe. Many of the basic questions about nature and the universe at large may never be answered by empirical science because of its inherent limitations.

Philosophies of Truth-Seeking through Science

Philosophy is a search for wisdom using logic and reasoning. The term philosophy is derived from two Greek words: *philein* meaning "to love" and *sophia* meaning "wisdom." Philosophy is "the love of wisdom" and the philosopher is a "lover of wisdom."[20] Philosophy may be defined as an intellectual pursuit of wisdom in regard to human belief and behavior.

Philosophy is a search for wisdom while science is a search for dependable knowledge. Even though the two endeavors are related, philosophy does not have the same limitations, objective standards and verification processes of science.

Philosophically, science may be considered a search for truth about the world based on physical evidence and its logical inferences. Certain fundamental philosophical ideas accompany this truth-seeking process and include the following:

(1) Preconceived requirements – Conceptually, science has no preconceived requirements based on philosophy, ideology or religion that restrict a search for truth founded on physical evidence. There are no predetermined demands that only certain information will be found or not found. There are no stipulations to ignore certain evidence. There are no commitments to force data to agree with a particular belief system. Inferred explanations (theories and hypotheses) which best explain applicable physical evidence in a logical manner ultimately prevail.

(2) Artificial limits – A philosophy of truth-seeking requires that artificial limits placed on scientific thought, regardless of the source, be identified and neutralized. For example, philosophical naturalism requires that all explanations must be based on natural, physical, mechanistic and material causes. By inference, philosophical naturalism requires that natural cause can plausibly explain all that is observed in nature and the universe at large. In truth, natural cause cannot remotely explain the origin of life and therefore should be rejected as an absolute constraint.

(3) Unrestricted sources – The search for truth and the development of good ideas in science are not restricted to a particular group of people. Conceptually, science is not racially, culturally, sexually, politically, philosophically, ideologically or religiously biased. Good ideas can originate from anyone. Verified empirical data in support of an idea is more important than the source of the idea. Preferably, theories and hypotheses should be evaluated based on their fit with physical evidence rather than rejected for coming from an undesirable source.

(4) Self-correcting – Conceptually, science is a self-correcting process. However, for correction to occur a search for truth based upon verified empirical data and its logical inferences must remain the foundation of science. Evidence that is ignored or suppressed inhibits the correction process. The acceptance of implausible and illogical inferences hinders self-correction. When the search for truth is abandoned in favor of popular ideas, philosophies and beliefs, then correction is stifled.

Some popular theories of science have been terribly wrong. Scientific ideas accepted at one time and now known to be false include explanations for the cause of disease (disproved by our knowledge of germs, viruses and genetic mutations), the phlogiston theory of combustion, the caloric theory of heat, maternal impression as the cause for birth defects and the inheritance of acquired characteristics. The popular theories of spontaneous generation and a geocentric universe were believed for more than two thousand years. A relentless pursuit of truth based on physical evidence ultimately discredited all of these ideas.

The Continuum of Science

Most people tend to think of science in terms of empirical science. Actually, science as accepted today is a combination of empirical science and *related tentative information.* Most discussions do not seriously distinguish between the two. However, to make sense of science in the twenty-first century, the difference between empirical science and tentative information must be fully understood.

Table 2.4 – Science and Related Tentative Information					
Physical world	Verified empirical data	Generalizations Laws Principles Formulae	Estimation Extrapolation Unverified data Computer modeling	Theory Hypothesis Conjecture Speculation	Unknown phenomena Metaphysical cause
Factual	Factual	Factual	Tentative Data	Ideas	Appeals
Empirical Science			Tentative Information		
The Divide					

The layman (non-professional educator) tends to think of science education as a collection of facts and the study of facts. In reality, science education is more than just the study of facts (verified empirical data). Science as taught in our schools, universities, scientific journals and popular science publications is a continuum from verified empirical data at one extreme to metaphysical cause at the other. Table 2.4 illustrates the range of science and related tentative information.

Empirical science, on the left side of the divide in Table 2.4, is based on the physical world around us. Study of the physical world has resulted in the development of verified empirical data and generalizations of that data in the form of laws, principles and formulae. Empirical science represents dependable knowledge about nature and the universe at large.

Tentative information related to science is listed to the right of the divide in Table 2.4 and may be partitioned into three major categories.

(1) Tentative data – *Estimation* represents calculated data based on incomplete information (such as the total number of bacteria in a cubic mile of sea water). *Extrapolation* represents information about a known effect that is extended into the unknown (such as future population growth). *Unverified data* may represent new information on the cutting edge of research and exploration. *Computer modeling* attempts to duplicate real-world phenomena using a computer program.

Some tentative information is quite credible, although not equivalent to verified empirical data. For example, when engineers design an airplane wing they often use a computer program based on finite element analysis to determine the stress in critical structural components. When the program is properly applied, the predicted stresses can be quite accurate.

*(2) **Ideas*** – Thought about scientific issues is expressed in the form of theory, hypothesis, conjecture and speculation. *Theory* and *hypothesis* are inferences based on an interpretation of empirical data. Conceptually, theory is a well-accepted idea while a hypothesis can be proposed by a single individual. *Conjecture* and *speculation* are unsupported ideas. Thought attempts to answer questions beyond the reach of empirical science related to origin, ultimate cause, operation and structure.

*(3) **Appeals*** – When scientists attempt to explain things beyond empirical science, they often appeal to unknown phenomena and metaphysical cause. Appeals may be in the form of *natural* metaphysical cause in contrast to *intelligent* metaphysical cause. For example, chemical evolution is the idea that life evolved from non-life; however, it has no plausible mechanism and the severe limitations of natural cause preclude it from occurring. As a result, proponents of chemical evolution often appeal to unknown phenomena described as "natural, self-emergent, complex, functional systems." On the other hand, opponents of chemical evolution appeal to intelligent cause—evidenced by information and complex function.

As a whole, tentative information represents the growing edge of science. Some things that are tentative today may become confirmed or disproved tomorrow. While tentative information is a legitimate part of science and related activity, it is definitely subject to change.

The dividing line between empirical science and tentative information is not always clear. Conflict among scientists is normally not due to the evidence of empirical science but rather interpretations of the evidence. For example, in 1996 news broke that a rock from Mars had been found in Antarctica containing evidence of Martian life in the form of tiny sausage shaped objects reminiscent of some species of terrestrial bacteria. This announcement was strongly disputed by other scientists skeptical of the claim.[21] The issue was not the molecules, minerals and structures found within the rock. The evidence was carefully documented and agreed upon by all concerned. The issue was the *interpretation* of the evidence and subsequent disagreement over the origin and cause of the sausage shaped objects.

Scientific Theory

Facts are boring, while theory is exciting because it appeals to the imagination. Theories are inferences and a means of explaining things beyond the immediate reach of empirical science. Theory is allowed to go beyond observation, measurement and experimentation. Theory goes where man has never gone before and may not be capable of going. For example, scientists have theorized different temperatures and structures within the interior of the Sun. It is inconceivable that man will ever be able to study the Sun's interior directly.

Inferred explanations (theories and hypotheses) are an important and legitimate part of scientific thought and science education. Scientists develop theories in an attempt to explain questions about (1) origin, (2) ultimate cause, (3) operation and (4) structure. Theory attempts to explain the *origin* of energy, matter and life in the universe. Theory attempts to explain the *ultimate cause* of disease, birth defects and cancer. Theory attempts to explain in detail the *operation* of gravity, electricity and nuclear fusion. Theory attempts to explain the *structure* of atoms and the interior of the Earth and Sun.

The National Academy of Sciences in *Teaching About Evolution and the Nature of Science* defines scientific theory as: "A well substantiated explanation of some aspect of the natural world that can incorporate facts, laws, inferences and tested hypotheses."[22] This definition is flawed in several respects. First of all, theory should be identified as an inferred explanation since all theories are inferences. There is a huge difference between a *correct* explanation of how maggots originate in dead meat and an *inferred* explanation. An incorrect inference led to the false theory of spontaneous generation which dominated scientific thinking for more than two thousand years. Secondly, the statement that theory "can incorporate facts, laws, inferences and tested hypotheses" is similar to saying, "people can be nice." People can also be rude, dishonest, thieves and murderers. *Can* is equivalent to "may." With current usage, theory *may* have the ability to incorporate facts, laws and tested hypotheses, rather than being a requirement. Thirdly, the NAS definition uses the term "natural world," no more objective than "supernatural world." Use of the word "natural" shows a deliberate bias and presumes the world is due to natural cause. The preferred term is "physical world." With the NAS definition as a pattern, two improved versions are:

- Theory is an inferred explanation about some aspect of the physical world, and its *key features are directly supported* by verified empirical data, laws, principles or experiments. (This definition excludes an idea from being called a theory whose key features are not firmly supported.)

- Theory is an inferred explanation accepted by a consensus of scientists about some aspect of the physical world and *may be supported* by verified empirical data, laws, principles or experiments. (This definition is more inclusive and represents the broad scope of ideas currently called theory.)

An alternate explanation of scientific theory given in Definition 2.5 emphasizes the components which make up theory. Scientific theory is defined as a tentative, inferred explanation of a physical phenomenon based on incomplete information, whose components are empirical data, logic and belief. Empirical data related to theory may be supportive, conflicting, incomplete, inaccurate, extrapolated or misinterpreted. Four forms of logic which may be used to support theory are deductive reasoning, inductive reasoning, mathematics and computer modeling. Belief may originate from a philosophical idea, religious idea, personal conviction, gut feel, intuitive thought, hunch or inspiration.

Definition 2.5 – Scientific Theory

A tentative, inferred explanation of a physical phenomenon based on incomplete information, whose components are empirical data, logic and belief. Empirical data related to theory may be supportive, conflicting, incomplete, inaccurate, extrapolated or misinterpreted. Four forms of logic used to support theory are deductive reasoning, inductive reasoning, mathematics and computer modeling. Belief may originate from a philosophical idea, religious idea, personal conviction, gut feel, intuitive thought, hunch or inspiration.

Supporting verified empirical data is the most important component of theory. A strong theory will have large amounts of supporting evidence unique to that explanation. Sometimes evidence is not uniquely supportive and can be claimed by opposing theory. For example, Darwinian evolution uses similarity among species as support for common descent. Similarity can also be explained by design theory as evidence of a common designer or a common design philosophy.

A second important aspect of theory regards the amount of contradictory data. Strong theories have little or no discrediting evidence. Conflicting evidence may discredit a weak theory to the point it is no longer believed. For example, mutations within animal and plant life are an absolute fact. Biological evolution proposes that beneficial mutations occur and natural selection spreads them throughout a species genome. Allegedly, beneficial mutations accumulate over many generations to produce new features, functions and species. The actual cataloging of thousands of protein mutations in the human genome shows they are either neutral or deleterious, rather than beneficial.[23] Beneficial human protein mutations are either nonexistent or extremely rare, information that strongly discredits the theory of biological evolution.

Every theory begins as a hypothesis developed by a single individual, in turn originating from a philosophical idea, religious idea, personal conviction, gut feel, intuitive thought, hunch or inspiration within that person's mind. Through the accumulation of evidence in support of the original intuitive thought and its acceptance by other scientists, a hypothesis can develop into a theory. However, we should be aware that philosophy or religion (especially in the study of origins) can be a powerful motivator to sustain the acceptance of a theory.

Strong theories have explanatory power. For example, the theory of continental drift is based on significant evidence, including the measurement of present day continental drift, ocean floor spreading, the identification of separate continental plates, volcanic mountain chains, the fit of the eastern edge of South America to western

Africa, similar plant and animal fossils found on different continents and evidence of dramatic climate change in Antarctica based on ice cores, coal and fossils. The theory of continental drift gives a powerful, unifying explanation for this related collection of evidence.

Theory can mislead by incorrectly explaining empirical evidence. For example, early in the history of science it was theorized that gravity attracted heavier objects more strongly than light objects. Today we know this explanation is false. Early experimenters noticed that when simultaneously dropped from the top of a building, a rock always fell faster than a feather. Today we know that gravity attracts both objects equally, but air resistance acting on the feather makes it fall slower (in a vacuum they will fall at the same rate). When this experiment is repeated today, the results are identical to that hundreds of years ago but the explanation is totally different. When all factors are not understood, theory can incorrectly explain evidence, and the two should be clearly separated.

Strong theories have predictive power and are able to envision specific things that will or will not be found. They are able to foresee possible arrangements, relationships and effects. Correct predictions give plausibility to the truth of theory, while failures discredit it.

In the teaching of science, all theory should be critically analyzed. Competing theories can be evaluated based on (1) the amount of specifically supportive evidence, (2) the amount of discrediting evidence, (3) failed or successful experiments, (4) the ability or inability to plausibly explain important phenomena within the realm of applicable theory and (5) failed or successful predictions. Within current usage of the term, some theories are highly credible while others have major faults and should be weeded out through critical analysis. For example, Big Bang theory attempts to explain the universe due to natural cause. Expansion of the universe is strongly supported by redshifted light from distant galaxies, and current expansion of the universe can be logically extrapolated backward in time to when the universe was much smaller. However, the extrapolation eventually breaks down and the naturalistic origin of the universe from an infinitesimal small quantum fluctuation is completely speculative. Allegedly, a quantum fluctuation created an enormous amount of high temperature energy, in turn spontaneously becoming basic matter and eventually self-organizing into hydrogen and helium atoms.

Defining Science

Science and related tentative information illustrated in Table 2.4 encompasses the continuum of science. Considering the range of information, thought and practice, how should science generally be defined?

The National Academy of Science in their book *Science, Evolution and Creationism* defines science as: "The use of evidence to construct testable explanations and predictions of natural phenomena, as well as the knowledge generated through this process."[24] This definition falls short in several respects. First, there is no mention of the work scientists do in discovering, quantifying and categorizing empirical evidence—the basis of scientific knowledge. Secondly, this definition over emphasizes theory. While explanations and predictions are a part of science, the majority of effort is in expended in other pursuits. Thirdly, the NAS definition uses the term "natural phenomena," no more objective than the term "supernatural phenomena." The preferred term is "physical phenomena."

As an improvement to the NAS definition, science can be defined in light of the continuum illustrated in Table 2.4. Keep in mind that science is both an activity (what scientists do) and a body of knowledge (what scientists know).[25] A good definition must include empirical science and related tentative information.

Science can be defined as a means of inquiry (an activity) using observation, measurement and experimentation to acquire knowledge about the physical world (i.e. universe). Scientists develop generalizations of dependable knowledge in the form of laws, principles and formulae. Inferred explanations of a tentative nature are developed as theories and hypotheses. Science has no preconceived demands for explanations or new discoveries and is open to correction in the light of new evidence. Verified empirical data forms the basis of science.

Definition 2.6 – Science

Science can be defined as a means of inquiry (an activity) using observation, measurement and experimentation to acquire knowledge about the physical world. Scientists develop generalizations of dependable knowledge in the form of laws, principles and formulae. Inferred explanations of a tentative nature are developed as theories and hypotheses. Science has no preconceived demands for explanations or new discoveries and is open to correction in the light of new evidence. Verified empirical data forms the basis of science.

The above definition can be restated in broader terms as: Science is a rational search for dependable knowledge about the world based on physical evidence and its logical inferences, and is inherently limited because of inaccessible evidence, unavailable data and unknown information.

Definition 2.6 highlights the significant difference between dependable knowledge and tentative explanations. Inferences in science create most of the controversy and are the biggest source of error. Both theory and hypothesis are tentative, inferred explanations.

Scientists develop generalizations of dependable knowledge in the form of laws, principles and formulae as follows:

(1) Scientific laws – A law of science in a formal sense is a carefully worded statement that describes an observed phenomenon within specific boundaries, based on consistent verified empirical data. The National Academy of Sciences in *Teaching About Evolution and the Nature of Science* defines scientific law as: "A descriptive generalization about how some aspect of the natural world behaves under stated circumstances."[26] Several important scientific laws have been identified, such as the first and second laws of thermodynamics and Newton's laws of motion and universal gravitation. Other laws include Kepler's laws of planetary motion, Boyle's Law, Coulomb's Law, Ohm's Law, Dalton's Law of Partial Pressures, Faraday's Law of Electromagnetic Induction, Mendel's laws of heredity and laws of light reflection and refraction. The operation of some laws can be precisely described by mathematical formulae.

Scientific laws are often referred to in a colloquial sense. The term "laws of chemistry and physics" is often used in an informal manner in reference to all phenomena in the universe considered true.

(2) Scientific principles – A principle of science is a factual statement of a more general nature than a law. For example, Avogadro's principle states that all gases at the same pressure and temperature contain almost the same numbers of molecules. Archimedes' principle states that a solid body immersed in a fluid is buoyed up by a force equal to the weight of the displaced fluid. Steno's principle of superposition states that layers of rock are arranged in a time sequence, with the oldest on the bottom and the youngest on the top. The principle of the uniformity of nature states that the laws of chemistry and physics discovered on Earth apply throughout the universe.

(3) Scientific formulae – A formula of science generalizes a phenomenon in precise terms. *Chemical* formulae can describe the atoms within a molecule or the reactants and products of a chemical reaction. *Mathematical* formulae can describe effects discovered by physics such as the acceleration of an object due to an applied force. *Geometric* formulae reflect relationships observed in geometric objects such as triangles, squares, circles, cubes and spheres. Some laws, such as Newton's laws of motion and universal gravity, can be expressed by formulae.

Calling an idea a law, principle or formula does not make it true. Consistent verified empirical data in support of the law, principle or formula is what makes it true. Without reliable empirical support, these terms become meaningless. A law, principle or formula should be consistent without exception within its defined boundaries.

The Nature of Science by James Trefil, PhD and professor of physics at George Mason University, discusses many of the laws, principles, formulae, effects, rules, constants, relationships and phenomena that define twenty-first century science. As pointed out by Dr. Trefil, science does not always use these terms consistently.[27]

In summary, it is important to understand the two basic divisions of science—empirical science and related tentative information. Science is a continuum from verified empirical data to metaphysical appeals. A substantial difference exists between the two. Empirical science to the left of the divide in Table 2.4 can be accepted without reservation. Information to the right of the divide should not be automatically accepted as true or false, but should be carefully evaluated against verified empirical data. In general, tentative information will be controversial, debatable and subject to change. A proper understanding of science permits a more objective analysis of many contentious issues.

Methodological Naturalism

Science is a way of acquiring dependable knowledge about the world. In the not too distant past, humanity emerged from the Dark Ages into a time of enlightenment and scientific discovery. During this emergence, empirical science became the basis for understanding the physical world. Although difficult for individuals in the twenty-first century to realize, at one time mankind explained a host of phenomena through a combination of superstition and religion. Fever, infection, sickness, disease, mental illness, birth defects, crop failures and pestilence were often attributed to superstitious causes. Meteors, comets, supernovae, eclipses, earthquakes, volcanic eruptions and storms were often attributed to religious causes. Virtually everything not understood could be attributed to these beliefs.

As science developed, philosophical ideas emerged to define how science should be conducted. Practitioners, philosophers and thinkers came to recognize the importance of these ideas. Four current ideas which define how science should operate are as follows:

(1) Empirical – Dependable knowledge about the physical world is based upon observation, measurement or experimentation. Verified empirical data forms the basis of science.

(2) Rational – Logical thinking is a tool of science and includes deductive and inductive reasoning. Mathematics is a form of logical thinking based upon numbers. A computer program can represent logical thinking.

(3) Skeptical – Scientists are doubtful of observations, measurements or experiments that cannot be repeated by other investigators. Inferences are tentative and evaluated on the amount of supporting or discrediting evidence, failed or successful experiments, failed or successful predictions and competing theory.

(4) Objective – The theories and hypotheses of science are open to critical analysis, debate and change. Scientists are not dogmatic about their unsupported ideas and appeals.

One of the controversial philosophies of science is now referred to as "methodological naturalism." The term methodological naturalism appears to have originated with philosopher Paul de Vries of Wheaton College in his 1983 paper, "Naturalism in the Natural Sciences."[28] His article was subsequently published in the *Christian Scholar's Review* in 1986.

Long before the term methodological naturalism was coined, science was considered a rational search for dependable knowledge based on physical evidence and its logical inferences. Experiment, measurement and observation have been a part of science since the time of Sir Francis Bacon (1561-1626) and Roger Bacon (1220-1292).[29] When properly understood, methodological naturalism can be readily accepted as an inherent part of science. Science must be based on physical evidence. The real debate is over attempts to constrain *inferences* and *appeals* to natural cause alone—to the total exclusion of intelligent cause.

Methodological naturalism may be defined as a method of science used to acquire knowledge about the physical world (i.e. universe) based upon observation, measurement and experimentation. By this method, inferred explanations (theories and hypotheses) and appeals are prejudiced toward natural cause during preliminary investigation. Natural cause is a biased assumption of science rather than a dogmatic requirement. This method has no position regarding the existence or non-existence of intelligent powers, theistic figures or God. This methodology is limited and phenomena beyond direct observation, measurement and experimentation (1) may or may not be true, (2) are inferred or unknowable and (3) beyond the limits of current empirical science.

Definition 2.7 – Methodological Naturalism

A method of science used to acquire knowledge about the physical world based upon observation, measurement and experimentation. By this method, inferred explanations and appeals are prejudiced toward natural cause during preliminary investigation. Natural cause is a biased assumption of science rather than a dogmatic requirement. This method has no position regarding the existence or non-existence of intelligent powers, theistic figures or God.

Three key points about methodological naturalism are: (1) The inferences and appeals of science are biased toward natural cause, rather than being constrained to only natural cause. (2) Methodological naturalism has no position regarding the existence of God in contrast to *philosophical naturalism* which denies the existence of God. (3) This methodology of knowledge seeking is specifically limited.

The controversy over methodological naturalism applies primarily to the level of bias toward inferences of natural cause. How strongly should scientists favor natural cause? The point at which theories based on natural cause are abandoned and other ideas are considered is a critical issue—especially in regard to origins. For example, present scientific research has brought to light compelling information that strongly discredits chemical evolution. At what point do scientists say that evolution as an explanation for the origin of life is a dead end and other ideas such as design theory should be considered? There is no simple answer to this question. To a large extent, the answer depends upon how well scientists understand what is being studied. Are single-celled organisms a "black box" or are their structures, mechanisms and operation well understood? Single-celled life is functionally complex and information rich. Based on current knowledge about microscopic life, the severe limitations of undirected change preclude a plausible explanation due to natural cause.

Despite the rhetoric of being objective truth seekers, scientists have strong emotional ties to their beliefs (e.g. philosophies, ideologies and religions) and can be resistant to change. Scientists are often guilty of making faith-based appeals in the name of science, and when their ideas are challenged, find them difficult to abandon. Conceptually, science is self-correcting and errant ideas are extinguished through contradictory conclusions based on evidence, open debate and critical review.

When the biased assumption of natural cause becomes a dogmatic requirement, several problems immediately surface. Methodological naturalism changes from being a method of science to a controlling philosophy of cause within scientific thought. Methodological naturalism becomes indistinguishable from philosophical naturalism. It becomes a predetermined requirement for all inferences and appeals of science. An absolute requirement for natural cause undermines the empirical, rational, skeptical and objective nature of science, and distorts it in the following ways:

(1) Non-empirical – Regardless of its observed limitations, natural cause is the only explanation allowed. Critical evidence allegedly in support of natural theory is permitted to be missing or unknown.

(2) Non-rational – Essentially any reasoning in support of a natural explanation is accepted. Inferences need not be logical or plausible. Speculative ideas easily trump negative evidence.

(3) Non-skeptical – Only theories of natural cause are accepted, regardless of missing supporting evidence, specifically discrediting data, failed experiments, failed predictions, lacking mechanisms and competing credible ideas.

(4) Non-objective – Scientists are dogmatic in their support of natural cause. Theories of natural cause are not subject to critical analysis, open debate or refutation. Appeals to natural metaphysical cause are inherently superior to intelligent cause.

In summary, empirical science is a search for dependable knowledge based on observation, measurement and experimentation. Scientific thought easily moves beyond those limitations in the form of inferences, unsupported ideas and appeals. Methodological naturalism biases scientific thought towards natural cause; however, it is not an absolute requirement. Our understanding of phenomena beyond the reach of empirical science may or may not be true, are inferred or unknowable.

Philosophical Naturalism in Science

Philosophical naturalism is essentially an all-encompassing belief in natural cause. *Webster's New World College Dictionary* defines philosophical naturalism as: "The belief that the natural world, as explained by scientific laws, is all that exists and that there is no supernatural or spiritual creation, control, or significance."[30] Other references say philosophical naturalism is a belief that nature is all there is[31, 32] and rejects belief in supernatural beings and its entities.[33] Philosophical naturalism proposes that undirected change has produced everything that exists in time and space.

Definition 2.8 – Philosophical Naturalism

As applied to science, philosophical naturalism is a belief that the quest for knowledge is limited to explanations that are natural, physical, mechanistic and material. In regard to origins, this belief proposes that random change, undirected phenomena and extensive time have resulted in nature and the universe at large. All explanations of origins not based strictly upon natural cause are denied. God, theistic figures or intelligent powers are rejected as creative causes.

As applied to science, philosophical naturalism is a belief that the quest for knowledge is limited to explanations that are natural, physical, mechanistic and material. In regard to origins, this belief proposes that random change, undirected phenomena and extensive time have resulted in nature and the universe at large. All explanations of origins not based strictly upon natural cause are denied. God, theistic figures or intelligent powers are rejected as creative causes. By inference, philosophical naturalism requires that natural cause plausibly explain all things considered natural in the universe.

The difference between philosophical naturalism and methodological naturalism is quite subtle, yet very important. In regard to inferred explanations (theory and hypothesis), methodological naturalism makes natural cause a *biased assumption*. On the other hand, philosophical naturalism makes natural cause an *absolute requirement* for all inferred explanations, unsupported ideas and appeals.

Proponents of philosophical naturalism insist that only ideas based upon natural cause be allowed within scientific thought such as the Big Bang, evolutionary star formation, the solar nebular hypothesis, chemical evolution and biological evolution. All ideas of origin based on intelligent cause are placed outside the realm of scientific thought and categorically rejected.

The Achilles' heel of philosophical naturalism is the severe limitations of natural cause. Science has uncovered a world significantly more complex than philosophers twenty-five hundred years ago realized. Philosophical naturalism conflicts with twenty-first century scientific knowledge. For example, undirected phenomena cannot remotely account for the information and functional complexity found in a single-celled organism, representative of the first life. In an attempt to overcome this problem, proponents of natural cause often appeal to missing information, unknown mechanisms and undiscovered laws—a completely faith-based exercise.

Philosophical naturalism and its ancillary ideas have no good answer for the severe limitations of natural cause. Random change has not produced anything comparable to the music, art, literature and technology man has created. Likewise, undirected phenomena have not demonstrated the ability to create information, complex functional systems, mathematically improbable structures and fine-tuned arrangements in any significant amount. Evolution as observed in the real world is severely limited. This raises serious doubt about philosophical naturalism and its claim that natural cause gave rise to galaxies, stars, a habitable planet, life and man. These doubts invalidate the rigid application of philosophical naturalism to control scientific thought regarding theories of origin.

Defining Evolution

Evolution is often defined as undirected, random, purposeless, mindless change over time. Undirected change readily explains rusting iron, crumbling rocks, water erosion, hardening sediments, decaying plants, meteor impacts, erupting volcanoes, earthquakes, fires, tsunamis, hurricanes and floods. Undirected change is responsible for dying stars and the ultimate heat death of the universe. Within biology, undirected change explains naturally occurring amino acids, simple molecules, chromosome abnormalities, gene defects, protein isoforms, point mutations, genetic disease, cancer, aging, genetic drift, limited species divergence, new viral strains and the rise of antibiotic resistant bacterial populations. Undirected change is capable of forming snowflakes, mineral crystals, concretions, freeze-thaw rock circles, sand dune formations, dried mud crack patterns, eroded structures, clouds, sand particles and salt grains. This kind of evolution is well supported by science and an established fact.

Definition 2.9 – Evolution

Evolution is undirected change over time and has a factual and theoretical component. Evolution as fact is the ability of undirected phenomena to produce simple effects, elementary arrangements and mathematically probable structures in addition to wearing away, deteriorating and randomizing. Evolution as theory uses natural cause over time to explain things that are clearly not of intelligent origin or influence. Some believe evolution is a powerful constructive mechanism when given significant time.

Stephen J. Gould, a leading spokesman for evolution, states that evolution is both a fact and a theory.[34] His basic assessment is correct. Evolution has a factual component and a theoretical component. Evolution may be viewed as both fact and theory regarding its constructive and destructive abilities.

Evolution as fact is undirected change as described above. As a weak constructive mechanism, undirected phenomena are capable of producing simple effects, elementary arrangements and mathematically probable structures. As a destructive mechanism, undirected phenomena can wear away, deteriorate and randomize. This type of change has been observed in the present, is an ongoing process and is true.

Evolution as theory uses natural cause over time to explain things that are clearly not of intelligent origin or influence. For example, natural cause can rationally explain stellar ageing, stellar novae, meteor impacts and planetary cooling. Undirected change can logically explain genetic mutations, limited species divergence and new viral strains. Natural processes can credibly explain continental drift, ocean floor spreading, mountain erosion, volcanic mountain building, river delta formation and sedimentary strata.

Some scientists consider evolution a *powerful constructive mechanism* when given significant time. Philosophically, they believe undirected change has created all of the information, complex functional systems, mathematically improbable structures and fine-tuned arrangements observed in nature and the universe at large. Nonliving molecules somehow became living things. Random molecular combinations somehow created the DNA molecule and information contained therein. Single-celled organisms somehow developed into complex mammals. Random molecules and gases somehow condensed into stars. In general, ideas of this kind are driven by a belief in philosophical naturalism and are challenged by the severe limitations of natural cause.

Ideas of evolution can be divided into their factual and theoretical components. Ideas include the Big Bang, evolutionary star formation, the solar nebular hypothesis, chemical evolution, biological evolution and geological evolution. Each of these can be divided into their factual and non-factual components, with the ratio of fact to theory varying widely. For example, chemical evolution is essentially unsupported by evidence and is all "theory." This term is used in a colloquial sense because no accepted theory for the chemical evolution of life exists and no plausible mechanism supports such an explanation. At the other extreme, geological evolution is supported by a substantial amount of evidence.

Competing Ideas to Evolution

Natural theology at one time held influence within the scientific community. Based on the study of nature, God could be inferred as its creator. In 1802 William Paley wrote *Natural Theology* and used the example of a watch to support his argument for God. By examining the fit and function of a watch, intelligent design could be inferred as its cause.[35]

Within scientific circles, natural theology was extinguished because of the popularity of Darwinian evolution. A watch was not alive and, although a complex functional object, could not evolve into another watch. Animals were alive and through variation and natural selection a microscopic organism could allegedly change into a larger, more complex species. Darwin did not address the origin of either watches or microscopic life. To Darwin, the first single-celled life just happened and was nothing more than simple material.

Today, intelligent design has reemerged as a counter to Darwinian evolution. Phillip Johnson, an emeritus professor of law from the University of California at Berkeley, is one of the leading figures in its rebirth. He has written several popular books about the evolution controversy, including *Darwin on Trial* and *Reason in the Balance*. Unlike natural theology, intelligent design does not begin with belief in God, does not presuppose a creator or miracles, and is theologically minimalist.[36] Intelligent design has theological implications, but is not a theological enterprise. On the other hand, there is no presupposition that an intelligent power does not exist or could not have been involved in making the universe. Information and functional complexity found in nature are best explained as a product of intelligence. A universe attributed to a combination of natural cause and intelligent cause is allowed.

Design theory is introduced in this book and is based upon (1) knowledge of phenomena in the physical world about us, (2) acknowledging the limitations of undirected change and (3) human design. Design theory begins with an understanding of the universe in which we live and appeals to the accumulated knowledge of science. Based on what we know, the undirected laws of chemistry and physics are severely limited as a constructive mechanism. Undirected change is incapable of creating all of the information, complex functional systems, mathematically improbable structures and fine-tuned arrangements that exist in nature and the universe at large. Using human design as a basis, design theory is an alternate rational explanation to natural cause. A design agent or agents originating from outside the limitations of the known universe is theorized, without any presupposition about the agent's origin, identity, character or nature. Design theory is a scientific alternative to philosophical naturalism and its flawed ideas.

Terminology Traps

The origins debate is primarily about whether the universe is due to natural cause or intelligent cause. Use of the word "natural" shows a strong bias. For example, the term *natural* universe implies the universe is the result of natural cause—no more objective than the term *supernatural* universe.

The word natural is firmly embedded in scientific thinking and nomenclature, used by both evolutionists and creationists. Scientists tend to talk about natural phenomena rather than phenomena. Books generally use the term natural world rather than physical world. Teachers discuss the natural laws of chemistry and physics rather than the laws of chemistry and physics. We need to be acutely aware of the implications of the words natural and supernatural. For example, nature may be defined as "Everything on Earth not of human origin or influence." However, nature may not be due to natural cause.

The term "natural science" should be abandoned, being no more objective that the term "supernatural science." *Science* and *empirical science* are the preferred terms. Use of the term natural science creates a terminology trap and tends to promote only natural explanations and natural theories, eliminating intelligent cause as a consideration.

On the other hand, empirical science is based on observation, measurement and experimentation. Science has no absolute requirements about what will or will not be found. Science is a search for dependable knowledge based upon physical evidence and its logical inferences. Natural cause is a biased assumption of science rather than a dogmatic requirement. Scientists have the ability to discern between natural and intelligent cause.

Conclusion

Science has revolutionized our world. Science and technology continue to produce products that improve transportation, communications, housing, clothing, food production and medical care. Science is slowly but surely reducing the effects of pestilence, disease and famine. Science is improving our standard of living while discovering new ways to ensure a clean, safe environment. Biology is an especially exciting frontier with huge potential, promising new cures for disease, improved food production and better living. Without question, science has dramatically improved the human condition.

One criticism, applied primarily to science education, is the commingling of empirical science and related tentative information. Their commingling in education, textbooks and reference works has corrupted science to an extent. For example, history in textbooks is based upon real past events and significant theory is not taught. While historians may speculate why certain events happened and postulate different scenarios, in general these are clearly identified. Whereas in science, significant theory is often taught in textbooks without being separated from factual information, and students are frequently unable to distinguish between the two. While a legitimate part of scientific thought, all theories, hypotheses, conjecture and speculation in science education should be plainly identified.

No theory of science taught in public education should be given a free ride. To allow this to occur is to promote narrow minded thinking about tentative information. Theory should be taught in the light of (1) the amount of specifically supportive evidence, (2) the amount of discrediting evidence, (3) failed or successful experiments, (4) the ability or inability to plausibly explain important phenomena within the realm of applicable theory, (5) failed or successful predictions and (6) competing theory.

Theory is best taught when accompanied by strong critical analysis. To prevent theory from becoming dogma, it should be accompanied by a discussion of its weaknesses and flaws to foster healthy skepticism. This approach does not require the teaching of opposing theory—sometimes no viable competing theory exists. When no plausible theory exists, it is acceptable to say, "We don't know." It is better to admit ignorance that to misguide young minds by teaching flawed belief.

Another criticism, directed primarily at science education and science institution management, is the misunderstanding of philosophical naturalism and methodological naturalism. The difference between them is subtle, but significant. Some educators and managers often confuse the two and incorrectly believe scientifically minded people must be philosophical naturalists. Since philosophical naturalism is antithetical to biblical origins or any form of intelligent cause, they feel compelled to stamp out the opposition. This misguided zeal has led to activities in our public schools, colleges, museums and science institutions reminiscent of the persecution of Galileo. Science is a search for dependable knowledge based on physical evidence and its logical inferences rather than an absolute commitment to philosophical naturalism. Natural cause is a biased assumption of science rather than a dogmatic requirement.

The goal of this chapter has been to explore the meaning, terms and concepts of science. Like any endeavor, the more one practices a given profession and the longer one studies a subject, the more one knows about the issues. Knowledgeable people are helpful in pointing out inconsistencies, assumptions, alternate explanations, arguments and critical facts. This is true regardless of the subject. A good understanding of science does not solve the origins issue, but provides the basis for a more meaningful discussion. Armed with a proper view of science, we can achieve a better insight into our beginnings than might otherwise be possible.

All scripture is given by inspiration of God, and is profitable for doctrine, for reproof, for correction, for instruction in righteousness. (2 Timothy 3:16) KJV

Chapter 3

Biblical Creation Theology

Bible students and Christians need to clearly understand the basics of science, philosophical naturalism, evolution, design and creation theology. A failure to understand these subjects and their implications perpetuates past mistakes. Inferior knowledge about these issues continues the legal, political and theological errors of the biblical community. Conversely, a proper knowledge of this material builds personal faith in the integrity of scripture and permits the development of correct origins theology. A firm understanding permits the proper evaluation of competing ideas, theological or otherwise, which attempt to explain our origins.

Origins theology, bounded by Genesis Chapters 1 to 11, intersects with science and in this respect is different from other theology. Biblical teachings about God, the trinity, the incarnation of Christ, angels, heaven, hell, sin, salvation and moral issues are separate from science. Essentially all the miracles performed in the Old and New Testament are without any lasting physical effects and can neither be confirmed nor disproved by science. Simply because we cannot duplicate biblical events does not disprove God's miracle performing ability. On the other hand, the Genesis creation account describes a major event. Logical inferences of science based on physical evidence from the past strongly intersect with the historical account recorded in Genesis. Traditionally, empirical science and the biblical creation story have strongly clashed in regard to time. Although many scientists and religious leaders consider harmonizing the two an impossibility, a careful examination of the evidence clearly shows this is not the case. The *exemplar* translation of Genesis Chapters 1 and 2 permit the Bible and science to agree, while strongly upholding the truthfulness and trustworthiness of God inspired scripture.

However, harmonizing the two is not a frivolous enterprise. It requires an understanding of science, the Bible, knowledge of creation theology, a history of interpretation about Genesis and a strict appeal to the inspired biblical Hebrew text as our authority. In addition to these fundamentals, a strong desire for truth is an absolute requirement.

Biblical Authority

A powerful, intelligent Creator inspired men to record His message to humanity. Copies of the original manuscripts have been preserved and passed down through the centuries to establish the basis for the Bible we have today. Over thousands of years, God's providence has protected the Bible against corruption and concerted attempts to destroy it.

The Bible is the final authority for Christian belief, worship and practice. Church traditions, ecclesiastical authorities, elected or appointed spiritual leaders, church councils, creeds, religious declarations, philosophies, extra-biblical writings, mystical beliefs or practices, personal revelations and self-proclaimed prophets do not replace biblical authority. While a part of our religious heritage and perhaps enlightening, they must be subservient to scripture. From its inception, the Bible has remained steadfast throughout cultural, moral, religious, philosophical, political, scientific and technological change. Centuries of history have established God's word as the only steadfast guide for the church and its theology. The Bible was written to be relevant throughout humankind's past, present and future.

The Bible stands above all other secular and religious books in the world for its authorship, content, claims and antiquity. *Evidence That Demands a Verdict* by Josh McDowell is a timeless commentary about the Bible. He discusses its authors, their professions and the times in which they wrote. McDowell details the manuscripts upon which the Bible is based and convincingly points out that more ancient copies of the Bible exist than any other document of antiquity. By careful comparison, scholars have been able to verify the faithful reproduction of the original writings. Copies of manuscripts in the original Hebrew language of the Old Testament and Greek of the New Testament serve as the basis for today's Bible.

The Bible is the only true and dependable source for the Christian church's theology and belief. As Dr. Harold Lindsell points out in *The Battle for the Bible*, establishing the Bible as the ultimate authority for religious information is of paramount importance for the person who believes in God. Church doctrine should support the Bible rather than destroy its credibility.

The Genesis creation story has a huge impact on our acceptance of the Bible as true. As John MacArthur says, "If the biblical creation account is in any degree unreliable, the rest of Scripture stands on a shaky foundation."[1] If the Genesis creation story is rejected, it becomes easy to discard other acts of God in the Old Testament. If the miracles of the Old Testament are discarded, it becomes easy to disallow the miracles of the New Testament. If we disallow the miracles of Jesus Christ in the New Testament, can we logically accept His power to forgive sins? Can we trust the Bible in matters of faith and practice, yet disbelieve it in matters of history and science? To a large extent, our belief in the Bible stands or falls on our view of God as a powerful, intelligent creator who made the heavens, Earth and all therein. When God is firmly established as creator, other acts of God in the Bible are readily acceptable. The Bible is its own testament to its truthfulness, supported by the following verses:

- 2 Timothy 3:16 All Scripture *is* given by inspiration of God, and *is* profitable for doctrine, for reproof, for correction, for instruction in righteousness. (NKJV)
- Titus 1:2 In hope of eternal life which God, who cannot lie, promised before time began. (NKJV)
- Hebrews 6:18 That by two immutable things, in which it *is* impossible for God to lie, we might have strong consolation, who have fled for refuge to lay hold of the hope set before *us*. (NKJV)
- 2 Timothy 2:15 Be diligent to present yourself approved to God, a worker who does not need to be ashamed, rightly dividing the word of truth. (NKJV)

The above New Testament verses clearly point out that: (1) All scripture, both Old and New Testament, is inspired of God. (2) God cannot lie, so writings He inspired are true. (3) The word of God is true and should be rightly divided.

Today's Bible is based on copies of the original God inspired manuscripts passed down through the centuries. Minor errors have occurred through repetitive copying and in some instances intentional change. Textual criticism and painstaking comparison among ancient manuscripts have been used to correct these errors and return to the original text of the Bible.[2]

In addition to textual errors, biblical error has resulted from a general loss of information over time. The exact meaning of some words in reference to animals, plants, minerals, objects, geographical features, cities and personal names has been lost in history. The meaning of some ancient verbs, nouns and other words is less than clear. Some colloquial expressions, poetical statements and metaphorical sayings are not understood with any certainty. Because of these effects, certain words, terms and verses are translated on a best effort basis. Scholars familiar with the Bible are fully aware of these issues.

The mature Bible student and Christian can easily accept minor textual uncertainties and interpretative issues without having a disparaging view of the Bible. God inspired the writing of the Bible in its original inerrant form. Beginning at inception, its copying, understanding and translation have deteriorated to a minuscule extent because of the influence of fallible men and the passing of time. Today, through careful study of the Bible we can be confident of the message God intended us to have.

Philosophical Naturalism

"Nature is all there is and all basic truths are truths of nature," summarizes philosophical naturalism.[3] As mentioned in Chapter 2, philosophical naturalism proposes that the quest for knowledge is limited to explanations that are natural, physical, mechanistic and material. This philosophy denies the existence of God as a powerful intelligence who created the universe. Philosophical naturalism is atheistic and its adherents either reject or are skeptical of all biblical teachings.

Philosophical naturalism is similar in some ways to Christianity. Christianity has clergy who preach its beliefs, sacred writings (the Bible), doctrinal teachings, regular meetings and buildings of worship. Philosophical naturalism boasts recognized leaders who teach its beliefs, books that champion its cause, basic ideas upon which it is founded and periodic gatherings of its believers. Both philosophical naturalism and Christianity are significantly faith-based.

Philosophical naturalism permits one to believe in a personal god, but only as a construct of the human mind. The idea of a warm, loving, caring god is a comforting thought, and one is permitted to acknowledge a universal god similar to the tooth fairy or Santa Claus. God is allowed, but relegated to the role of a senile, impotent, emasculated, eviscerated and powerless figure. God is the product of wishful thinking and a self-deluded mind. He has no physical influence on the real world.

Ideas in support of philosophical naturalism about origins developed over a period of years and can be attributed to several individuals. Anaximander (610-546 BC), a Greek philosopher, is credited to having "envisaged a formless mass that was both the source and destination of all material things." He thought life originated from water.[4]

Empedocles of Acagras in Sicily (492-432 BC), a Greek philosopher, developed the four-element theory of matter. He theorized that everything consisted of various combinations of air, fire, water and earth,[5] an idea accepted by Western scientific thought for more than two thousand years. He believed animal life originally grew out of the earth, similar to growing plants.

Aristotle (384-322 BC) taught natural philosophy. His writings indicate he believed some animals spontaneously developed from putrefying earth, vegetable matter and mud. Some insects grew from dew falling on leaves, in decaying dung, in the hair or flesh of animals, or in excrement. Some species of mullet (fish) grew spontaneously out of mud or sand.[6]

Epicurus (341-270 BC), a Greek philosopher of Athens, was a key figure in promoting empirical science. He insisted that nothing should be accepted unless tested through direct observation and logical deduction. Philosophically, he believed the universe was infinite and eternal and that events in the world are ultimately due to the random motions and interactions of atoms.

Lucretius (95-55 BC), a Roman philosopher, published the book *On the Nature of Things* in 56 BC. He "envisaged an evolutionary universe, one that developed slowly to its present state physically, biologically and sociologically."[7] As Isaac Asimov points out, "Lucretius was the boldest spokesman in the ancient world of anti-religious views. His book was specifically intended to lift what he considered the burden of religious fears from the backs of mankind."[8]

From these beginnings, a number of other thinkers, observers and experimenters promoted a strictly natural explanation for life, nature and the universe. By the time of Christ, Greek philosophers had established naturalism as an acceptable worldview. From that point onward, philosophical naturalism was an integral part of Western culture. In 1859, Darwin's theory of evolution popularized philosophical naturalism, which has since grown to dominate scientific thought.

Proponents of philosophical naturalism readily explain important beliefs about origins, moral values and ultimate purpose. In these three areas, naturalism and the Bible are absolutely opposed.

In regard to *origins*, philosophical naturalism has been fleshed out in several forms. According to Big Bang theory, the universe started from a natural singularity. Evolutionary star formation proposes that all initial stars formed due to self-gravity from hydrogen and helium gases. The solar nebular hypothesis explains Earth and our solar system due to a series of natural events. Chemical evolution attributes the first life to random chance, elemental materials, simple molecules and vast amounts of time. Darwinian evolution builds upon the first single-celled organism using small changes and natural selection to create Earth's diverse species. In short, philosophical naturalism and its supporting ancillary ideas purport an all-sufficient explanation of origins.

Considering *moral values*, the end result of believing in philosophical naturalism is the absence of moral absolutes. There is no supreme standard of right and wrong. Unrestrained immorality, rather than moral restraint, is more enjoyable and preferred. Proponents believe moral values are nothing more than a consensus of opinion. Social convention is useful in giving structure to society, but guilt should not be attached to breaking its rules. Remorse from immoral activity is only the echo of an inhibited conscience. Adherents believe all morals are subjective and should be defined from a historical, cultural and situational perspective.

Philosophical naturalism erases the guilt of sinful activities and gives intellectual credibility to moral relativism. Proponents say man is not responsible to God, but answers only to his own conscience. Appetites of the flesh are made to be satisfied. Survival of the fittest and our animal nature explains man's cruelty to man. According to this belief, Adolf Hitler is to be commended for his vision of a master race superior to all others. He is to be applauded for putting theory into practice and eliminating the weak, infirm, retarded, old and racially impure. Advocates of philosophical naturalism deny the existence of God, any dread of judgment and a reckoning of life's deeds. People need not fear an accounting to a holy, powerful, omniscient God knowledgeable of their most secret actions and personal thoughts. Death and the grave end all. As the most intelligent animal, man becomes his own god and philosophical naturalism his religion. With natural cause as his creator, God is a superstitious distraction and is neither needed nor desired.

As for *ultimate purpose*, according to proponents of philosophical naturalism, there is none. Human life is nothing more than a random, mindless accident in a vast cosmos. Man has no more purpose than a speck of dust floating in the emptiness of space. According to proponents, if there is any reason for living, it is to fulfill personal ambitions and physical desires. Hedonism and selfishness reign. Success is defined by the fulfillment of sexual fantasies and fleshly impulses. Rather than serving God, man has every right to be self-serving. This philosophy allows man to go from the cradle to the grave without any purpose in life other than the pursuit of self-satisfaction.

Philosophical naturalism has tremendous appeal. Fleshed out as evolutionary theory, it is an intellectual sounding alternative to the Genesis creation story. Charles Darwin was a mediocre scientist, but was brilliant when he gave scientific credibility to philosophical naturalism in the form of species evolution. Darwin's theory allowed man to explain his existence based on natural cause, without a need for God. Evolution intellectually freed man from dependence on a creator. Proponents say it loosed humanity from the tyranny of biblical oppression and guilt. In essence, philosophical naturalism is a substitute for Christianity, capable of telling man where he came from, how to live along the way and where he is going.

Philosophical naturalism is in direct, irreconcilable conflict with biblical Christianity. They are antithetical beliefs. Conflict occurs on both a scientific and non-scientific level. On one level, disagreement occurs over the origin of the universe, Earth, life and humankind, while on the other level, disagreement occurs over moral values and ultimate purpose. It is impossible to concede the scientific side of this argument to philosophical naturalism, yet retain a worldview based on biblical values and purpose. No single individual can embrace both, as one will eventually destroy the other. In light of this, the discrediting and rejection of philosophical naturalism in regard to origins is of paramount importance.

Trustworthiness of the Bible

Thinking Bible students and Christians find it difficult to accept the Old Testament as truthful and trustworthy when the Genesis creation story is perceived to conflict with science. This tempts discarding the first two chapters of Genesis. However, the creation account cannot be cleanly removed from the rest of the Bible. If considered myth, it becomes impossible to determine where Genesis makes a transition from mythology to historical reality. Outside of Genesis many passages in the Bible refer to the creation account and become problematic.

Because of the supposed conflict of the Genesis creation story with science, theologians skeptical of the Bible have developed a clever way around the issue. The Bible becomes an authority in matters of *faith and practice,* but may be wrong in matters of a *historical or scientific nature,* such as the creation account. In other words, proponents of this view believe we should trust the Bible when it says to love our neighbor, be forgiving, pray, abstain from sin and tithe. We should trust in Jesus

Christ for the forgiveness of sins, salvation and His promise of heaven. On the downside, the Bible need not be considered historically accurate. This theology permits outright rejection of the Genesis creation story, miracles of the Old Testament and miracles of Jesus Christ himself. Subsequently, Jesus' virgin birth, sinless life, crucifixion, resurrection and ascension become questionable and open to denial.

Theology that disconnects faith and practice from physical reality is opposite to what Jesus lived and taught. Jesus performed many miracles during His ministry and used physical power to prove spiritual authority. This is illustrated in Matthew 9:5-6 when a paralytic man was healed. Jesus posed the question, "Which is easier to say, your sins are forgiven or to say get up and walk?" Anyone could claim power to forgive sins, since forgiveness is an invisible act and easy to say. On the other hand, healing is highly visible and hard to do. Jesus' miracles of healing were undeniable physical evidence of supernatural power that authenticated His spiritual authority to forgive sins.

Outside the confines of the theological community, accepting the Bible in faith and practice but rejecting it in matters of a historical or scientific nature does not make sense. It is bad theology and bad logic. Because of this issue, some members of the clergy have quietly abandoned the source of their faith. They continue to hold positions within their church. They preach, teach, counsel, advise and write, but the flame has gone out. They go through the motions to keep a roof over their heads and bread on the table, but their hearts are cold. Their spiritual calling has become a ritual without hope, vision or purpose.

Charles Templeton, a close friend and contemporary of Billy Graham, epitomized this struggle. Templeton was a rising star in the Christian evangelical world in the 1940s. While attending seminary, he became acutely aware of scientific evidence supporting an old age for the Earth, Darwinian evolution and related theological issues. He wrestled with the apparent clash of science and the Bible. Finally Templeton's doubts prevailed, and in 1957 he abandoned his promising ministry for secular pursuits.

Near the end of his productive life in 1996, at the onset of Alzheimer's, Templeton wrote *Farewell to God: My Reasons for Rejecting the Christian Faith.* His basis for rejecting God began with the perceived conflict between scripture and science. He chose to trust the physicist, anthropologist, geneticist and geologist rather than the theologian.[9] As Templeton said in his own words, "Whom should you believe, the Christian church or your own common sense?"[10] He was unable to reconcile evolution, with the supposed support of science, and the Bible.

Without question, Templeton made the wrong decision. He failed to separate empirical science from the inferences and appeals of science. He failed to recognize the influence of philosophical naturalism on scientific thought. Templeton failed to understand that the battle is between the Bible and philosophical naturalism, rather than the Bible and science. He failed to realize that if a cause is just and blessed by God, eventually ideas, strategies and support can be marshaled to achieve victory.

Perhaps if *Darwin on Trial* by Phillip Johnson, *Darwin's Black Box* by Michael Behe and *Evolution: A Theory in Crisis* by Michael Denton had been available in Templeton's time, his decision might have been different.

Templeton was not an average preacher—he was a great preacher. He had a brilliant mind and could captivate audiences with the powerful, clear message of the gospel. His leadership abilities were extraordinary and had he remained in the ministry, might have eclipsed Billy Graham as a national figure. After leaving the ministry in 1957, Templeton began a successful career in news and broadcasting. He wrote a total of twelve books, with several being best sellers in Canada. One book was adapted for a movie. From a secular perspective, Templeton's achievements easily put him in the top one percent of the population.

Templeton died on June 7, 2001 due to a combination of failing health and Alzheimer's. He had a successful career and was eulogized by several newspapers and news services. Yet Templeton's death was tragic in many respects. Tragic in the sense his mind was destroyed by Alzheimer's and his Godly faith by philosophical naturalism. Tragic because he abandoned God based on a combination of bad theology and bad scientific information.

Many consider Templeton's leaving the ministry an anomaly, while in reality it denotes a systemic problem. How many promising Bible students, ministers, educators and Christian leaders must lose their vision because of a perceived conflict between the Bible and science? Templeton's life is proof that the Genesis creation story is a significant issue for biblical Christianity in the twenty-first century.

Intelligent, talented, educated and motivated Christians of sound judgment deserve a solution to the Genesis problem. Failure will result in the ongoing erosion of church leadership from the loss of capable people who require credible answers about biblical based, origins theology. Failure to find a solution will allow philosophical naturalism and its workhorse evolution to continue their domination of intellectual high ground within the church, religious education and personal Christian belief.

Empirical Science and the Bible

The Old Testament contains a significant amount of historical information about people, cities and events not found in other ancient manuscripts and stands alone above all other books for its antiquity, content, style and believability. Among significant teachings is the Genesis creation story where God is identified as creator of the heavens, Earth, living things and humankind.

Theologians often call the Bible God's *special revelation* while His creation is called *general revelation*. The Bible and the physical world may be considered separate entities, each containing truth capable of being revealed through study. Since truth cannot conflict with truth, the two should harmonize.

The Bible is not a book of science; however, it discusses historical issues which have scientific implications. When the historical record of the Bible and empirical science intersect, they should not conflict. Biblical events and resultant lasting physical evidence should be capable of complementing each other where they overlap.

As illustrated in Table 2.4, science today can be divided between empirical science and tentative information. To make sense of science, the two must be separated. Verified empirical data gained through repeated observation, measurement and experimentation can be trusted; however, tentative data, inferences, unsupported ideas and appeals should be viewed with skepticism.

The Bible will legitimately conflict with scientific thought driven by philosophical naturalism. The Genesis account of God as creator opposes natural cause as an explanation for the heavens and Earth. The Bible clashes with evolutionary thought about the origin of life. Inspired scripture disagrees with Darwinian evolution as an explanation for Earth's many species. Appeals to biblical theistic cause directly conflict with appeals to natural metaphysical cause.

In most instances, empirical science can neither prove nor disprove historical events. Most past events, whether secular or biblical, cannot be verified as true or false by science. Any lasting physical effects have disappeared. It cannot be proved Benjamin Franklin flew a kite. It cannot be proved George Washington ever fought and won any battles. It cannot be proved Moses had a brother or lived in Egypt. The majority of written history is unsupported by physical evidence. However, when there are lasting effects from a major event, evidence and historical accounts should not conflict. This applies to both secular and biblical history. Lasting effects from biblical events such as the Genesis creation story and Noah's flood should not conflict with empirical science.

There is no expectation for empirical science to support miracles of the Bible with no lasting impact, such as the burning bush which spoke to Moses. Science is not asked to confirm the appearance of a cloud that guided the nation of Israel in the wilderness. These were one-time events which can neither be proved nor disproved. One-time events with no lasting effects fall into the same category as the story in Chapter 2 of the primitive natives and the big game hunter. The hunting party cannot prove the existence of the booming stick, nor can the village elders disprove its existence.

Fundamental Principles of Biblical Creation Belief

Exemplar creation is based on four fundamental principles that are foundational and interdependent.

- God physically created the heavens and Earth
- God inspired the written account of His creation.
- The written account will not conflict with the physical creation.
- Any perceived disagreement is the responsibility of humankind.

Exemplar creation proposes that the first two chapters of Genesis contain the essence of Christian belief about biblical origins. The creation story is an abbreviated, non-technical, historical account of real events and is not mythological or metaphorical. Words used in Genesis Chapters 1 and 2 have no special meanings applicable only to the creation story, and their usage is consistent with the remainder of the Old Testament. The account of Genesis Chapters 1 and 2 in the Hebrew Bible is an accurate copy of the original God inspired text.

Table 3.1 – A Basis for Origins Belief

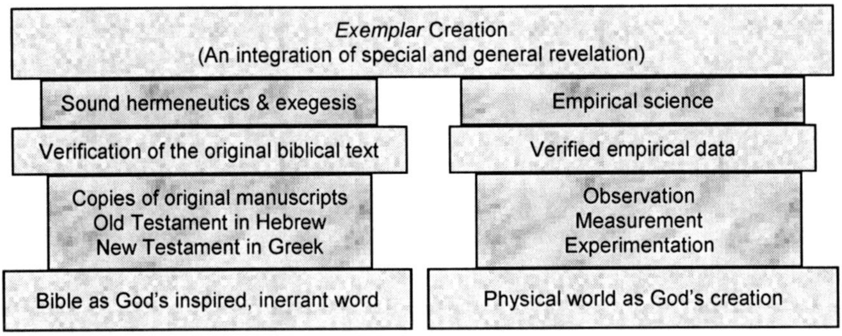

Truth cannot conflict with truth. Special revelation and general revelation should not clash. In reality, traditional origins theology strongly disagrees with general revelation in regard to time. For example, traditional young-Earth creation theology interprets and translates the Genesis account to arrive at a creation date for the universe of roughly six thousand years ago. On the other hand, scientific information strongly supports an old Earth, Moon and stars. Evidence compellingly attests to the existence of animals and plants on Earth substantially before a 4004 BC.

To believe God's written account of creation can conflict with His material creation is to admit that God is either deceitful or inept. For the Bible to say one thing and physical reality to be entirely different forces one to question God's honesty or competence. This is equivalent to an engineer designing a laptop computer and then having it described in the owner's manual as a toaster. Customer service can rationalize, apologize, patronize and grovel as needed to soothe irate customers. The fact is, an inexcusable mistake has occurred.

God is neither dishonest nor incompetent. Apparent conflict between the Genesis creation story and general revelation cannot be blamed on God. He did not give us false information in the Bible nor falsify physical evidence to mislead us. He is not a deceiver. The blame for any error lies solely upon humankind—not God.

Exemplar creation proposes that a correct understanding of Genesis and applicable physical evidence easily harmonize. A strong, biblically based, interpretative framework readily clarifies previously misunderstood words. The resulting *exemplar* translation supports the Genesis creation story as historically truthful and trustworthy. One of the most challenged accounts in scripture becomes a solid foundation upon which Bible believers can go from defense to offense.

Table 3.1 illustrates the basis upon which Bible students and Christians can form their beliefs about origins. The foundation consists of the Bible as God's inspired word and the physical universe as His creation. Parallel lines of information and reasoning build upwards and merge at the top as an integrated explanation of origins based on special and general revelation.

The Origin of Incorrect Creation Theology

Some individuals, after studying ancient literature, have taken the position that the Genesis creation story is nothing more than a compilation of Sumerian, Egyptian, Babylonian, Assyrian and Canaanite beliefs. Poetical and metaphorical expressions in Job, Psalms, Proverbs and Isaiah are given to support the argument that the Bible is full of incorrect scientific information. Historical accounts and descriptive words in scripture are tenuously connected to pagan mythology, and used by critics to attack the Bible as untrue.

The reasoning of critics is similar to taking present-day factual events and then discounting their credibility by connecting common words, expressions and themes with fictional works. With this type of logic, the account of man's walk on the Moon can be accused of being false because of its connection to hundreds of science fiction books, short stories and movies.

The Bible makes it clear that the Hebrews repeatedly worshipped idols, adopted false beliefs, embraced pagan practices and married outside their faith in disobedience to God. The nation of Israel over and over again fell away from God, and scripture clearly documents their unfaithfulness. So while it is true many Israelites were influenced by the false beliefs of the people around them, this does not require the Genesis creation story to be in error. The Old Testament was written by faithful men who rejected the pagan ideas and practices of the surrounding cultures.

Moses was inspired by God to write the first five books of the Old Testament, often called the Pentateuch (or Torah by the Jews). He probably based most of his work on written historical accounts handed down through Abraham and his descendants, although some of the Genesis account was no doubt given by direct revelation from God. When Moses wrote the Pentateuch, what he penned in Hebrew on the original manuscripts was without error.

Exemplar creation proposes that interpretation of the Genesis account changed over time. Sometime after the writing of the Pentateuch around 1491 BC,[11] but before 300 BC, Hebrew explanations of the Genesis creation story changed. Young-Earth creation theology developed during this time, coinciding with the rise of errant views about the resurrection, the state of the dead, forgiveness, divorce, observing the Sabbath, personal cleansing and other beliefs. By the time of Christ's ministry, a host of extra-biblical and false beliefs were actively promoted by several Jewish religious sects. Fortunately, these beliefs were kept out of the New Testament. For example, a number of verses attest to God as creator of the heavens, Earth and humankind; however, traditional young-Earth creation theology is not taught. No New Testament scripture specifically mentions that God created the heavens and Earth in *six days* or *instantly* spoke them into existence.

The writings of Flavius Josephus (37-100 AD), a Jewish historian, show that young-Earth creation theology was the dominant Jewish belief about origins immediately following Christ's ministry. Josephus wrote extensively about Jewish history and commented on the Old Testament. In regard to the Genesis creation story, Josephus explained that creation occurred in six days. Heaven and earth were created on the first day. Earth was in a thick darkness and a wind moved across the surface. God created light and separated it from darkness. He named night and day. On the second day, God placed the heaven over the whole world and determined that it should stand by itself. He placed a firmament around it and made it agreeable to the earth, as a source of moisture, rain and dew. On the third day, God appointed dry land to appear with the sea round about it. He made the plants and seeds spring up. On the fourth day, He adorned the heaven with the Sun, Moon and the other stars, and appointed their motions and courses. On the fifth day, God produced the living creatures that swim in the sea and fly in the air. On the sixth day, God made the four-footed beasts, both male and female. He also formed man on this day. In just six days, the world and all therein was made. The seventh day was a time of rest called the Sabbath.[12] Josephus' account of creation essentially agrees with traditional young-Earth theology as taught today.

Philo the Jew, or Philo of Alexandria (20 BC-50 AD), was also a Jewish historian. He lived in Alexandria, Egypt which had a large Jewish population at the time. Philo gave a detailed commentary on the Genesis creation story generally supportive of traditional young-Earth creation theology. According to Philo, God separated light from darkness. The whole body of water in existence was spread out over all the earth and penetrated its parts like a sponge soaked with moisture. Earth was swampy land and deep mud. The elements of earth and water were mixed up and combined together in one confused mass of an undistinguishable and shapeless nature. God caused the salt water to be gathered together.[13] God began to adorn the land and bade it bring forth grass, corn, herbs, fodder for cattle and food for man. He commanded every kind of tree to spring up, both wild and cultivated. The earth produced full grown trees with fruit. On the fourth day, He adorned the heavens. The earth brought forth all kinds of plants and fruits before the Sun and Moon existed.[14] God created stars perceptible by the external senses.[15] On the fifth day, God proceeded to create the races of mortal creatures. He created the aquatic animals, including the races of fish and sea monsters, followed by the races of birds. On the sixth day, God had the earth bring forth cattle and beasts and creeping things of each kind. Man was created on the sixth day, the last of God's created creatures.[16] Philo's rambling commentary of the creation account appears to be a mixture of scripture, bad theology and an active imagination.

When and why young-Earth creation theology developed is unknown and any explanation is speculative. We can hypothesize the original creation story remained intact during the life of Moses, Joshua, the time of the judges and the first three kings of Israel (Saul, David and Solomon). Following the reign of Solomon, the kingdom of Israel was divided into Judah and Israel. The kings of the resulting two kingdoms repeatedly led the people away from God into idolatry until eventually both Judah and Israel were carried away into captivity. Perhaps it was during this time that incorrect theology emerged. We can surmise that some charismatic religious leader embellished the creation story to compete with the pagan gods of the cultures around them. The new theology portrayed God as essentially a giant wizard who instantly spoke the world and all therein into existence during six days. Based on their limited knowledge, this was a viable interpretation in light of the miracles of the Old Testament when God killed the Egyptian first born, parted the Red Sea, supplied manna for food and collapsed the walls of Jericho.

Regardless of when or why young-Earth creation theology developed, once established it became the theological paradigm of biblical origins. This theology first surfaced when the Pentateuch was translated from Hebrew to Greek in the Septuagint around 275 BC. The Jews subsequently passed their creation beliefs to the early New Testament Christian church. In time, young-Earth ideas became a part of Roman Catholic Church theology, Muslim writings and belief, and Protestant doctrine.

Young-Earth creation theology has had a significant influence on translations of the Old Testament from Hebrew, evidenced in the Greek Septuagint (275 BC) and Latin Vulgate (405 AD). This theology is found in English translations starting with the hand copied Wycliffe Bible (1382-1388). Young-Earth creation theology influenced William Tyndale's translation of the Pentateuch into English published in 1530, the Myles Coverdale Bible of 1535, the Matthew's Bible of 1537, the Great Bible of 1539, the Geneva Bible of 1560, the Bishop's Bible of 1568 and the King James Version of 1611.[17] All subsequent English translations of the Bible have been affected by this theology.

A Historical Analysis of Creation Theology

Two major lines of reasoning can be used to evaluate the Genesis creation story. One line of reasoning accepts traditional young-Earth creation theology as true and traces it acceptance from the *past to the present.* A second line of reasoning is skeptical of traditional young-Earth theology and goes from the *present to the past* to show

(1) how this theology conflicts with empirical science, (2) how this errant belief was added to the Bible and (3) how we can appeal directly to the original Hebrew text for our understanding of Genesis.

Supporters of traditional young-Earth creation theology typically start with the past and work forward to the present to show it has been the accepted belief throughout history. Proponents say God spoke the heavens, Earth and all therein into existence during six days around 4004 BC according to Ussher.[18] The Hebrews were God's chosen people who preserved and copied the Old Testament. Around 275 BC, the Septuagint translation was begun when Jewish scribes translated the first five books of the Old Testament (the Pentateuch) from Hebrew into Greek. This translation is a major work and strongly supports traditional young-Earth creation theology. Further along in history, the writings of Philo (20 BC-50 AD) and Josephus (37-100 AD) show that young-Earth creation theology was the accepted belief of the time. Moving on to 500 AD, the Masoretic scribes began adding vowel points and accent marks to the Hebrew text. These support young-Earth creation theology. Early translations of the Bible into Latin support this theology, followed by German, French and English versions. Proponents view this historical continuum as overwhelming confirmation of the truth of their position. From their perspective, traditional young-Earth theology is supported by more than twenty-two hundred years of translation, commentary and belief.

A second line of reasoning, counter to the above, works from the present backward to the past. The four fundamental principles of *exemplar* creation previously mentioned apply: (1) God created the heavens and Earth. (2) Genesis contains the God inspired written account of creation. (3) God's word and work will not conflict. (4) Any perceived disagreement between His word and work is humankind's responsibility.

Starting around 1700, scientists began to seriously question a young age for the Earth. Up until that time, traditional young-Earth creation theology had been the reigning biblical paradigm and had never been forcefully challenged. As scientific knowledge grew, so did physical evidence in support of an old age for the Earth. Newly discovered evidence was in irreconcilable conflict with young-Earth creation theology. Over the last 300 years scientific data supporting an old age for living things, Earth and the heavens has accumulated to the point it will never be overturned. Something is terribly wrong with traditional interpretations of the Genesis account in regard to time.

As we go back in time, the Masoretic Text appears as a key piece of evidence. The Old Testament we have today is translated from the Masoretic Text, produced by the Masoretes (a group of Hebrew scribes) who faithfully copied and preserved the Hebrew Bible after the time of Christ.

The Old Testament was written primarily in Hebrew with a small amount in Aramaic using twenty-two alphabetic characters. There are no vowels in biblical Hebrew—all letters are consonants. During the time of Christ, when Hebrew was read aloud, the speaker put vowels into the words. To standardize the pronunciation of words, Masoretic scribes added vowel points and accent marks (small dots, dashes and other symbols above, below or within the Hebrew letters) to the text.

Several sources help identify when vocalization marks were added. According to *The Anchor Bible Dictionary*, between 500 and 700 AD Hebrew scribes developed vowel points and accent marks to aid in the reading of Hebrew text.[19] Another source, *The New Encyclopedia of Britannica,* says that the Masoretic text was meticulously assembled and codified and supplied with diacritical marks to ensure correct pronunciation. This work began around the 6th century and was completed in the 10th century AD by scholars at Talmudic academies in Babylonia and Palestine.[20] A third source, *The Text of the Old Testament* by Ernst Würthwein says vowel points and accent marks originated between 650 and 750 AD.[21]

The work of the Masoretes raises several important issues. First, vocalization marks were added 500 years or more after Christ's ministry, proven by the Dead Sea Scrolls and other archaeological discoveries. Vocal marks were not in the original text and they are uninspired additions to the Bible. Secondly, traditional young-Earth creation theology was the dominant origins belief of the day, and the Masoretes fully accepted it. Vocal marks they added support their theology. Thirdly, vowel points and accent marks control the interpretation of some key verbs in the Genesis creation account. By the seemingly innocuous addition of vocal marks, Masoretic scribes embedded young-Earth creation theology into the biblical text.

As we go further backward in time, the Septuagint becomes the next significant document. The first five books of the Old Testament (the Pentateuch) were translated from Hebrew into Greek by Jewish scribes in Alexandria, Egypt around 275 BC.[22] *The Anchor Bible Dictionary* gives the third century BC as a general date for translation of the Torah, possibly in Egypt.[23] The remaining books of the Old Testament were translated at various times during the next one to three hundred years. Translation of the Old Testament from Hebrew into Greek was a response to the spread of the Greek language due to the conquests of Alexander the Great. The Septuagint allowed Jewish proselytes who did not speak Hebrew to study the Old Testament in their native Greek tongue.

The Septuagint raises several significant issues. First, the Septuagint is not a dependable translation as pointed out in the introductory remarks in *A New English Translation of the Septuagint.*[24] Some words are not translated, poorly translated, intentionally mistranslated or added to the original text. Secondly, the Septuagint is not inspired. There is no inspired translation of the Old Testament into any language—Greek, Aramaic, Syrian, Latin, German, English or otherwise. Thirdly, the Hebrew scribes who translated the Septuagint believed in traditional young-Earth creation theology and used their work to embed their beliefs into the Bible for all time. These scribes were familiar with the native Hebrew and Greek languages of the day and were conceptually capable of making a good translation; however, their poor scholarship and theological beliefs produced a less than accurate work.

Present-day scholars and theologians rely heavily on the Septuagint translation and to a lesser extent the Masoretic vocalization marks for their understanding of the Genesis creation story. Hebrew scholars use these two works as the basis for their interpretation of the Hebrew text. Even though the Septuagint and Masoretic works are not inspired, they are accepted as absolutely authoritative in regard to the creation account. Traditional logic is that the Hebrews were God's chosen people, they spoke the native languages and were the copiers and keepers of the Old Testament; therefore, what they believed must be correct.

As we continue back in time, the next document is the original biblical text. Genesis Chapters 1 and 2 are identical in all reliable Hebrew manuscripts. With absolute certainty, we have the words penned by Moses thirty-five hundred years ago. When the Septuagint translation and Masoretic vocalization marks are laid aside as a controlling influence, the original Genesis creation story is allowed to surface. When the language of the creation story is connected with the rest of the Old Testament, misinterpreted words and phrases soon come to light. A Genesis creation story emerges that is fully truthful and trustworthy. Special and general revelation harmonize.

Prominent Theologies of Biblical Origins

Any explanation of the Genesis creation story should be evaluated based on how well it accurately reflects the Hebrew text and how it respects the Bible. The Genesis creation account and empirical science should not conflict where they intersect and should be capable of harmonization. This is the benchmark by which all biblical creation theology should be measured.

Four major theological explanations of the Genesis creation story are (1) traditional young-Earth creation theology, (2) gap theology, (3) day-age theology and (4) the framework view and analogical day position. All four of these beliefs can be evaluated based on the above benchmark.

Traditional young-Earth creation theology is the oldest and most well known. Its proponents steadfastly believe in the inspiration and inerrancy of the Bible—the clear teaching of scripture must not be compromised. Unfortunately, they have inherited an incorrect interpretation that strongly conflicts with empirical science *in regard to time*. Traditional young-Earth creation theology proposes that the Sun, Moon and stars were created on the fourth day of the Genesis story. The universe, Earth and all life were created in six 24-hour days in the recent past around 4004 BC (Ussher's chronology). Henry Morris allows extending the creation date back to around 10,000 BC.[25] The recent book *Refuting Compromise* by Jonathan Sarfati, PhD, gives a defense of traditional young-Earth creation belief. There are irreconcilable problems in trying to harmonize this theology with the discoveries of empirical science within the last three hundred years. Some of the main issues are:

(1) Conflict with empirical science regarding time. Overwhelming physical evidence supports the heavens, Earth and living things as being significantly older than six thousand years. Evidence includes starlight, ice core annual layers, Moon impact craters, tree ring dating, radiocarbon calibration curves, lake sediment correlation curves and radiometric dating.

(2) Inability to explain vast coal, fossil and sedimentary deposits within a six thousand year time frame. Noah's flood is not a plausible explanation for these deposits.

(3) Reliance on the addition, mistranslation and poor translation of key words in Genesis Chapters 1 and 2. Most significantly, Genesis 1:3, 6, 9, 11, 14, 20 and 24 are misinterpreted as commands rather than descriptive statements.

Because of the conflict of traditional young-Earth creation theology with empirical science, several additional explanations of the Genesis creation story have been developed.

Gap theology was among the first alternate interpretations to force the presumed teaching of the Bible (at the expense of scripture) to agree with science. Thomas Chalmers (1780-1847), founder of the Free Church of Scotland, brought gap theology into prominence through his preaching.[26] A defense of gap theology was given by George H. Pember in *Earth's Earliest Ages* first published in 1884. This theology was promoted by the 1917 Scofield Reference Bible in its explanatory notes for Genesis. In 1970 Arthur C. Custance published *Without Form and Void* where he promoted gap theology through an in-depth analysis of the word "was" in Genesis 1:2.

Gap theology proposes that millions of years occurred between Genesis 1:1 and 1:2. During this time, dinosaurs and other prehistoric animals lived, hydrocarbon reserves were formed, the fossil record was laid down and geologic change occurred. Genesis 1:2 is translated to say, "But the earth had become a ruin and a desolation."[27] The Earth was subsequently reformed and repopulated during the six-day Genesis chronicle. Several issues which discredit this theology include:

(1) Geology and paleontology do not support an event within the last 100,000 years or more which destroyed the Earth's surface (causing the extinction of all life) to make it without form and void. Science does not support the recent reappearance of all life on Earth.

(2) The creation of the Sun, Moon and stars on the fourth day is not satisfactorily explained.

(3) This theology requires mistranslation of the stative verb "was" as "had be-

come" in Genesis 1:2 to show change. This theology does not acknowledge the addition, mistranslation and poor translation of certain key words in Genesis Chapters 1 and 2. This theology fails to recognize the expression, "And God said" in Genesis 1:3, 6, 9, 11, 14, 20 and 24 is not a prelude to a command but rather an introduction to a descriptive statement.

Day-age theology is a second idea developed to force the presumed teaching of the Bible (again at the expense of scripture) to agree with science. Each day of the creation account is considered millions or billions of years long. No good explanation has ever been given for treating the six days of Genesis, specifically defined by morning and evening, as millions of years of time. When each day is considered to be millions of years long, significant problems arise owing to the sequencing of events. *A Matter of Days* by Hugh Ross, PhD, discusses day-age theology and its connection with natural cause and evolution. Serious issues with this theology include:

(1) Ross believes the Bible teaches Big Bang cosmology.[28] He appears to believe that fine-tuned expanding gases (ashes) from the Big Bang eventually coalesced into gas and dust clouds to develop galaxies, stars, planets and moons.[29] Strictly natural cause cannot remotely explain these heavenly bodies.

(2) Accepting each day as millions of years relegates significant portions of the Genesis creation story to a metaphorical explanation.

(3) This theology promotes misinterpretation of the word "day," clearly defined by evening and morning. This theology does not acknowledge the addition, mistranslation and poor translation of certain key words in Genesis Chapters 1 and 2. This theology fails to recognize the expression, "And God said" in Genesis 1:3, 6, 9, 11, 14, 20 and 24 is not a prelude to a command but rather an introduction to a descriptive statement.

A third group of theologies developed to force scripture to agree with science are the framework view and analogical day position. Proponents of these ideas consider the Genesis account as significantly metaphorical. Meredith G. Kline is a well-known advocate of the first view which is "a literary framework for us to understand the work of creation, without committing anyone to see the days as either sequential or normal."[30] Analogous days defined by C. John Collins "are God's workdays, their length is neither specified nor important, and not everything in the account needs to be taken as historically sequential."[31] Dr. Collins discusses these ideas in *Genesis 1-4: A Linguistic, Literary and Theological Commentary*. Problems with these theologies include:

(1) These ideas do not resolve the conflict between a straightforward reading of the six-day Genesis chronicle (Genesis 1:3-31) and empirical science. The clear language of six consecutive days and the sequence of events for each day are not satisfactorily explained.

(2) Both theologies treat a large part of the Genesis creation story as metaphorical and are another attempt to explain away the presumed clear teaching of scripture. Neither skeptics nor advocates of the Bible are swayed by this interpretation.

(3) These theologies do not acknowledge the addition, mistranslation and poor translation of certain key words in Genesis Chapters 1 and 2. They fail to recognize the expression, "And God said" in Genesis 1:3, 6, 9, 11, 14, 20 and 24 is not a prelude to a command but rather an introduction to a descriptive statement.

There is a tendency to group together all theologies that propose an old age for the Earth. While *exemplar* creation teaches an old age for the Earth and universe, it cannot be equated with gap theology, day-age theology and the analogical day position. The scriptural basis is significantly different. *Exemplar* creation does not require the metaphorical treatment of any words in the creation account, does not attempt to explain away the clear teaching of scripture and relies on a correct interpretation of previously mistranslated key words and phrases.

Although not a biblically based theology, theistic evolution is the belief that God used natural cause as His method of creation. William Dembski points out that the Darwinist establishment despises theistic evolution.[32] Philosophical naturalism absolutely rejects any theistic influence—there is no God. Natural evolution by definition is undirected change without any intelligent guidance. Theistic evolution represents muddled thinking by people who firmly believe in evolution, yet want to believe in some kind of god.

One prominent believer in theistic evolution is Francis S. Collins, head of the Human Genome Project and author of *The Language of God.* In his book Dr. Collins defines six basic points of theistic evolution paraphrased as follows.[33]

(1) The universe came into being out of nothingness approximately 14 billion years ago. [No mention is made of God.]

(2) The universe appears to have been precisely fine-tuned for life despite the extremely small probability. [God is not referenced as the cause.]

(3) Life originated on Earth, although the precise mechanism is unknown. [No appeal is made to God and no plausible natural mechanism is identified.]

> (4) Once life arose and biological evolution got under way, no special supernatural intervention was required. [Collins never says where God was required.] The process of random change and natural selection resulted in the development of Earth's many diverse and complex species over very long periods of time. [In a leap of faith, Collins appeals to these limited mechanisms.]
>
> (5) Humans are a result of the biological evolution process and share a common ancestry with the great apes. [God is not identified as man's creator.]
>
> (6) Humans are unique in ways that defy an evolutionary explanation. Our spiritual nature is shown by the existence of Moral Law—the knowledge of right and wrong—and the search for God that characterizes all human cultures throughout history. [No appeal is made to the Bible as the basis for our belief in God or as an ethical, moral or religious authority.]

Theistic evolutionists consider science and religious faith as two completely separate realms that never intersect. The debate over origins is typically framed between *science and faith*, between *science and the Bible* or between *science and religion*. By implication, the opposing position is not science and therefore false. As previously pointed out, science is a rational search for dependable knowledge based on physical evidence and its logical inferences, and does not have an inherent position about origins. A much better way to frame the debate is between *natural cause and intelligent cause* or between *evolution theory and design theory*.

Theistic evolutionists accept natural cause as a total explanation of origins, yet claim to be people of faith. They typically consider the Genesis story and related verses as myth. Their belief requires ignoring the severe limitations of natural cause to create information, complex functional systems, mathematically improbable structures and fine-tuned arrangements. Advocates of theistic evolution readily appeal to natural metaphysical cause, which is somehow more logical and satisfying than an appeal to intelligent cause. Collins condemns "God of the gaps" type thinking[34] even though he apparently has no issue with appeals to unknown natural phenomena to fill in the gaps.

Theistic evolutionists may be people of faith; however, the god they believe in has little relation to the God of the Bible. Their god is distant and uncaring, lacking the desire, power or intelligence to create anything. He is a fuzzy, ill-defined, universal, cosmic kind of god. In fact, he may be dead or simply a product of wishful thinking.

All of the above theologies and ideas have major biblical and scientific issues, and give a small indication of the absolutely chaotic condition of current Bible based origins belief. They illustrate the need for a correct interpretation and translation of the Genesis creation story.

In summary, biblical theology is similar to scientific theory in some respects.

Scientists use their study of the physical world as the basis for scientific theory. This is a legitimate enterprise; however, they can become enamored with theory and ignore conflicting evidence. In a similar manner, theologians who use the Bible as their authority can fall in love with theology and disregard contradictory information found in special and general revelation. In the debate over origins, scientific theory must be held accountable to the physical evidence of empirical science. In like manner, biblical based creation theology must be held accountable to scripture and physical evidence. This is the standard by which all viewpoints of the Genesis story should be measured.

Conclusion

Exemplar creation establishes a framework for resolving the conflict between biblical creation and empirical science—where they intersect. *Exemplar* creation supports the Bible as God inspired and the doctrine of inerrancy. On the other hand, it provides an interpretation and translation that permits special and general revelation to harmonize. The Genesis creation story is an abbreviated, non-technical, historical account that is truthful and trustworthy.

Key issues in this chapter include:

(1) The Bible in the original manuscripts is inspired of God, inerrant and the basis of authority for Christian belief.

(2) While perhaps informative and enlightening, church traditions, ecclesiastical authorities, elected or appointed spiritual leaders, church councils, creeds, religious declarations, philosophies, extra-biblical writings, mystical beliefs or practices, personal revelations and self-proclaimed prophets do not serve as a final authority for Christian religious belief.

(3) The Bible is in irreconcilable conflict with philosophical naturalism in regard to origins, moral values and ultimate purpose. While not a religion, philosophical naturalism is a faith-based belief in direct opposition to significant biblical teachings, which include God as creator

(4) The Genesis creation account and empirical science should not conflict where they intersect. Failure to harmonize the two by a rational method will ultimately undermine the credibility and authority of the Bible.

(5) Special revelation and general revelation should harmonize where they overlap. God's written word, defined by the Bible, should not conflict with His physical work, defined by empirical science. Truth does not conflict with truth.

(6) Educated, intelligent, talented, motivated Christians of wisdom and influence deserve a credible solution to the perceived conflict between the Genesis creation story and empirical science.

(7) Explanations and translations of the Genesis creation story have been in error for more than twenty-two hundred years because of the influence of traditional young-Earth creation theology developed prior to 275 BC.

People want the truth about origins. Individuals with a desire for sound theology and a willingness to change can accept *exemplar* creation as an accurate interpretation of the Genesis creation story. Bible students and Christians can accept a translation of the creation account that respects the Bible as the inspired word of God, yet harmonizes with the evidence of empirical science. With history as our guide, the true story of biblical origins will prevail over centuries of theological error. In a free society, truth percolates to the top and ultimately prevails.

As iron sharpens iron, so a man sharpens the countenance of his friend. (Proverbs 27:17) NKJV

Chapter 4

Evidence of Age

Scientists in the eighteenth century began disputing a recent creation date for the Earth as they discovered evidence to the contrary. As science matured, an analysis of starlight, ice core data, radiometric isotope decay, tree-ring growth, carbon-14 dating and Moon craters uncovered compelling evidence discrediting a six thousand year old universe. Traditional young-Earth creation theology *in regard to time* was not correct. This chapter examines the scientific evidence supporting an old creation date for the Earth, Sun and stars.

The method and amount of time God took to create the universe have been debated for centuries and is an arguable question when based strictly on information found in Bible. Traditional young-Earth creation theology teaches *creatio ex nihilo* (creation out of nothing). By speaking, God instantly brought the universe into existence out of absolute emptiness. The following two verses are often quoted in support of this belief:

• Psalm 33:6 By the word of the LORD were the heavens made; and all the host of them by the breath of his mouth. (KJV)

• Hebrews 11:3 Through faith we understand that the worlds were framed by the word of God, so that things which are seen were not made of things which do appear. (KJV)

Psalm 33:6 and other poetical verses should not be used to explain God's method of creation. For example, pillars are described as supporting the Earth in 1 Samuel 2:8; Job 9:6 and Job 26:11. We know the Earth is not supported by pillars; therefore, these verses are speaking metaphorically. Likewise, Psalm 33:6 extols the majesty and power of God as maker of the heavens in a metaphorical sense and is not a detailed account of how the heavens were created.

According to Hebrews 11:3, the worlds were framed by the word of God. This is substantially different than saying God spoke and commanded the Earth to be created instantly from nothing. Keep in mind the Greek word translated "world" is not equivalent to "Earth." According to this verse, things seen were made of things invisible (things which do not appear). This can be explained in reference to subatomic particles such as protons, neutrons and electrons. God made the world out of basic matter that cannot be seen.

In reality, the Bible gives few details of how God created the universe. Chapters 1 and 2 of Genesis present an abbreviated, non-technical, historical account of what God did, but little about how He actually did it. The exact time, sequence and method God used to create the heavens and Earth are not revealed in scripture, and man is left to his own resources to fill in the details. If God chose to create in a series of major events over time, that is His prerogative. If He took billions of years to make the heavens, Earth and all therein, that is within His purview. Theology has no grounds to demand that God act in a specific manner or time scale in the absence of explicit biblical teaching. This is doubly true when there is compelling physical evidence to the contrary. God can act instantaneously or over billions of years. As an ageless creator, time has little significance.

Exemplar creation proposes that a combination of special and general revelation best explain origins. The Bible clearly teaches that God created the heavens and Earth. On the other hand, science strongly supports intelligent cause over billions of years as responsible for the galaxies, stars, our solar system, Earth and life. When combined together, the Bible identifies God as the designer and empirical science clarifies the time involved.

Measuring Distance to Stars and Galaxies

Stars have always intrigued humankind. How did they originate? How do they work? How far away are they? Present-day astronomy can now answer some of these questions. For example, astronomers have developed techniques and technology that allow accurate distance measurements to some stars and galaxies. This capability has been developed through intensive effort involving thousands of scientists and technicians during the last one hundred years.

Trigonometric methods can be used to measure the distance to stars within roughly 300 light years of Earth. Using trigonometry, if the length of one leg and the acute angle opposite that leg are known, the size of the resultant triangle can be determined. For measuring the distance to a star, the Earth's average orbital diameter about the Sun of 298 million kilometers (186 million miles) is used as one leg of the triangle.

Measuring the acute angle to a star (target star) from Earth is slightly more involved. The angle to a target star is measured in relation to a distant reference star that is so far away it appears to be stationary, regardless of how it is viewed from Earth. At a specified point in the Earth's orbit, the angle between the target star and reference star is measured. Six months later when the Earth is on the opposite side of its orbit, the angle between the target star and reference star is measured again. The smaller of the two angles is subtracted from the larger to obtain the acute angle (parallax angle) to the target star. The distance to the target star can be calculated by dividing Earth's orbit diameter of 298 million km by the sine of the parallax angle. This is called the *trigonometric parallax* method.[1]

A precisely measured parallax angle is key to accurately calculating the distance to a star. Technology has been developed to permit the measurement of parallax angles in milliarc seconds. One thousand milliarc seconds equal one arc second. In turn, 60 arc seconds equal one arc minute, 60 arc minutes equal one arc degree and 360 arc degrees equals a full circle.

In 1989, the Hipparcos Mission satellite was launched by the European Space Agency and remained active until communications were terminated in 1993. From space, Hipparcos measured 118,218 stars for their parallax angles.[2] (This data is available to the public in the Hipparcos Data Catalog on its website.) The angles measured are very small, with the smallest listed at around 10 milliarc seconds. For example, star number 1340 has a parallax angle of 10 milliarc seconds (mas), which calculates to a distance of 326 light years.

Another technique used for measuring the distance to stars millions of light years from Earth is based on absolute brightness and the inverse square law. This method uses the three characteristics of (1) apparent brightness, (2) absolute brightness and (3) color index. The first characteristic, *apparent brightness*, is the brightness of a star as viewed from Earth. The farther away a star is from Earth, the less its apparent brightness. The second characteristic, *absolute brightness*, is a measure of the total light produced by a star, similar to a 100 watt light bulb. Regardless of how near or far away the light bulb, it always gives off the same amount of light. The third characteristic is the *color index* of a star. Stars range in color from blue, to blue-white, white, yellow-white, orange and red. Blue stars are the hottest in temperature and red stars are the coolest. Star color can be precisely measured and assigned a color index.

Astronomers have discovered that the color index of a star is approximately proportional to its absolute brightness. This relationship applies to stars identified as *main sequence stars*. The correlation of color index to absolute brightness was discovered independently by astronomers Ejnar Hertzsprung and Henry Norris Russell around 1912.[3]

The principle behind measuring the distance to a main sequence star is based on the inverse square law. As a main sequence star gets farther from Earth its apparent brightness decreases as the square of the distance. For example, suppose a star one light year from Earth has an apparent brightness of one. The same star two light years from Earth will have one-fourth the brightness. The same star three light years from Earth will have one-ninth the brightness. This relationship continues until the star can no longer be seen. The inverse square relationship of apparent brightness to distance is used to determine how far away a star is from Earth.

For main sequence stars, the color index is approximately proportional to its absolute brightness. Regardless of the distance from Earth, if a star is visible, its color index can conceptually be measured and thus its absolute brightness determined. By comparing the absolute brightness of a main sequence star (determined by its color index) against its apparent brightness (determined by direct measurement), the distance to a star can be calculated. This technique can be used to measure individual stars out to one million light years or more from Earth. No extrapolation of data is required and the physics behind this procedure have been verified on Earth.

Stars in the Andromeda Galaxy (M31), approximately 2.5 million light years away, have been measured using the inverse square technique. In 2005, astronomer Ignasi Ribas of the Catalonia Institute for Space Studies in Bellaterra, Spain and his colleagues announced they had used this technique to precisely determine the distance to two binary stars in the Andromeda Galaxy. The stars were calculated to be 2.52 million light years away from Earth (meaning it takes their light 2.52 million years to reach Earth) within an accuracy of six percent.[4] The greatest difficultly was accurately determining the color index and apparent brightness of the stars.

Measuring the distance to stars by the inverse square law is not related to popular cosmological theories. Measurements based on this technique are unrelated to ideas about the ultimate size of the universe, the age of the universe or Big Bang theory, and are independent of spectral redshift explanations. Astronomers who are both Bible believers and non-believers accept the inverse square method and acknowledge that stars exist more than one million light years from Earth.

Starlight and Time

Old starlight from distant stars challenges the credibility of traditional young-Earth creation theology in regard to time. Light from distant stars is just now reaching Earth. For example, light from a star two million light years away must travel two million years to reach Earth. Upon reaching Earth, events recorded in the starlight two million years ago can be observed. The distance starlight travels and the information it contains conflict with the belief that all stars were created in 4004 BC.

What do young-Earth creationists say about this evidence? At first, they questioned the techniques used to measure the distance to stars such as the trigonometric parallax and inverse square method. Their objections fell by the wayside as astronomers who believed in traditional young-Earth creation theology came forward and supported these measurement methods.

The argument then developed that God instantly created starlight between the Earth and distant stars. The problem with this idea is that starlight contains unique information about specific events such as novae, luminosity variations and regular cyclic fluctuations. For example, in 1987 a supernova was discovered in the Tarantula Nebula at an estimated distance of 170,000 light years away, meaning the actual event occurred 170,000 years ago. Many supernovae occur every year and in 2006 alone 551 were identified.[5] All of them represent stored information within starlight from past events just now reaching Earth.

God did not instantly create starlight. If God created starlight and falsified the information it contains, then He is a great deceiver and has put Bible believers in an indefensible position. Ken Ham, a leading young-Earth creationist, has rejected the idea that God instantly created starlight between the Earth and distant stars.[6]

Time dilatation is another argument developed by those who believe God created the stars 6,000 years ago. The book *Starlight and Time* by D. Russell Humphreys, PhD, hypothesizes that massive time dilatation occurred because of a white hole.[7] While the first four creation days ticked away on Earth, billions of years passed out in the universe among the stars.

Humphreys' hypothesis is based on the observation that a clock within a high gravity field will run slower than one exposed to lower gravity. For example, a clock on the surface of the Earth will run more than 38 milliseconds per day slower than one in a geosynchronous orbiting satellite where gravity is weaker. This difference is very small, but nevertheless real. The time dilation hypothesis proposes that our planet was subject to an immense gravity field, causing time to pass slowly, while time among the stars occurred at a normal rate. The universe aged billions of years while only a few days passed on Earth. This enormous amount of time dilation is completely speculative.

Dr. Humphreys uses the time dilation hypothesis in an attempt to harmonize starlight, time and a 6,000 year old universe. His explanation of Genesis 1:1-19 ranks among the most bizarre yet conceived.[8] Humphreys' idea is repeatedly trotted out as the answer to the starlight problem[9] in a desperate effort to salvage traditional young-Earth creation theology.

Another idea proposed to solve the starlight problem involves changes in the speed of light, hypothesized to have decreased over time (it was allegedly faster in the past).[10, 11] There is nothing to support this belief other than inaccurate measurements made by previous investigators. More recent, precise experiments show that the speed of light in a vacuum is constant. In addition, for light to reach Earth from distant stars within the first four days of creation, its speed would have been millions and billions of times what it is today—not just a few percent. No significant data directly supports this speculative idea.

In the final analysis, traditional young-Earth creation theology is in direct conflict with starlight and an old age for the stars. This theology has a significant problem with time. Extensive time is needed for starlight to reach Earth from stars created in the remote regions of the universe. A 6,000 year old universe does not allow adequate time. In contrast, *exemplar* creation allows time in the undefined past prior to Genesis 1:3 for starlight to reach Earth. *Exemplar* creation easily harmonizes with the evidence of empirical science.

Moon Impact Craters

Impact craters present a serious challenge to the credibility of traditional young-Earth creation theology in regard to time. Since the Moon and Earth are linked together, their history in some respects is similar. Impact craters on the Moon are a reflection of impact events on Earth. Six thousand years is not adequate time for impact craters to occur on Earth and then be removed by weathering.

Close examination of the Moon shows its surface is pockmarked with craters as a result of meteor impacts over time. Craters range from less than a foot in size to the Ptolemaeus impact basin, 164 km (102 miles) in diameter. In many areas there are craters within craters. Owing to the absence of an atmosphere on the Moon, craters formed in the distant past are still present today. There is no wind or water to erode them away or fill them in with sediment; therefore, they provide a clear record of past impact events. The *Encyclopedia of Earth Sciences* hypothesizes that most craters were formed very early in the Moon's history more than 3 billion years ago.[12] The *Age of the Earth* proposes that the majority of Moon craters occurred from 3.5 to 4.6 billion years ago.[13]

Extensive observations have been made of Moon craters. A merge of the NASA-RP-1097 and USGS lunar crater data lists 8,596 lunar craters more than 1 km (0.62 miles) in diameter, by latitude, longitude, name, size and identification number. This includes 1,088 lunar craters more than 50 km (31 miles) in diameter.[14] Table 4.1 shows the distribution by size of Moon craters from 1 km (0.62 mile) to 600 km (373 miles) in diameter.

When compared with the Moon, the Earth has few impact craters. The Earth Impact Database lists a total of 174 confirmed meteor craters in the world.[15] Of that number, 155 are above one kilometer in diameter. The most well-recognized crater in America is the Barringer Meteor Crater, located twenty miles west of Winslow, Arizona. This crater has a classic shape and is approximately 1.25 km (4,100 ft) in diameter and 174 meters (571 ft) deep. The edge of the crater rises approximately 46 meters (150 ft) above the surrounding terrain.

Based on their relationship, the Earth and Moon have been subjected to the same meteor activity in the past. Earth was hit by proportionally more meteors because of its larger size. The Moon has 8,596 meteor craters more than one kilometer in diameter which corresponds to approximately 115,500 meteor impacts on Earth, based on the ratio of their surface areas. The 1,088 meteor craters on the Moon more than 50 km in diameter equate to approximately 14,620 meteor impacts on Earth. Weathering and other natural processes have erased the remains of all meteor impacts on Earth except for 174 known craters.

Traditional young-Earth creation theology has no time to fit this number of meteor impacts into 6,000 years of Earth's history. For example, for 115,500 impacts to occur before Noah's flood (between 4004 and 2349 BC according to Ussher[16]) would have required 70 meteor impacts per year over one kilometer in diameter and 8.8

impacts per year over 50 km in diameter. It is doubtful civilization could survive this level of meteor activity. To have the impact rate then drop dramatically after Noah's flood to 0.040 per year (represents 174 impacts over 4,350 years) is without a plausible explanation.

Table 4.1 – Moon Impact Crater Size Distribution
(Arranged in decreasing order by diameter size)

Diameter Size Range (kilometers)	Quantity	Diameter Size Range (kilometers)	Quantity	Diameter Size Range (kilometers)	Quantity
501 - 600	2	151 - 160	10	71 - 80	134
401 - 500	1	141 - 150	13	61 - 70	234
301 - 400	6	131 - 140	22	51 - 60	287
201 - 300	19	121 - 130	31	41 - 50	468
191 - 200	5	111 - 120	31	31 - 40	711
181 - 190	8	101 - 110	52	21 - 30	1211
171 - 180	8	91 - 100	84	11 - 20	2027
161 - 170	10	81 - 90	93	1 - 10	3129

Reference: McDowell, Jonathan (2007). A merge of NASA-RP-1097 (Anderson and Whitaker 1982) and USGS lunar crater data.

The time Noah was in the ark represents a second possible period for meteor impacts. Any hypothesis that 115,500 meteor impacts happened during Noah's flood presents several problems. This number of meteor strikes in 375 days (the time Noah was in the ark) would have been devastating and represents 308 meteor strikes per day, including 39 strikes more than 50 km in diameter. It is doubtful Noah's ark could have survived the devastation caused by this level of meteor activity.

In addition to the above problems are the meteors themselves. For all meteors to hit Earth during the time Noah was in the ark requires a continuous swarm of meteors for at least 28 days. This is the time required for one complete revolution of the Moon as it orbits the Earth (to expose its entire circumference). A minimum of one orbit is required to achieve a distribution of meteor impacts over the Moon's surface. A swarm of meteors extending the diameter of the Moon's orbital plane, or 812,000 km (504,577 miles), would have been needed for a minimum of 28 days. In addition, two other swarms would have been required. Craters on the Moon are the result of meteors from at least three different directions: the equatorial plane, in addition to the north and south poles. From what scientists know about our solar system, it would be implausible for three meteor swarms to occur and converge on the Earth from three different directions within a 375 day period.

A third possible period for meteor impacts is after Noah's flood. If this happened, most impact sites would be visible today since not enough time has elapsed for erosion and weathering to remove them. There would be many more than the 174 meteor craters currently identified.

To explain Moon craters, some proponents of traditional young-Earth creation theology use the argument of *apparent age*. The Moon is new, but God created impact basins on its surface to make it appear old. New things can have an apparent age.

While the argument of apparent age seems plausible, close examination shows the idea is flawed. For example, when God created Adam he appeared to be a young adult with a normal growth history. In reality he was only a few hours old. The creation of Adam as an adult is accepted; however, in viewing Adam more closely something is amiss. The bottoms of his feet have no thick calluses from walking barefooted. He has no cracked or broken toenails. There are no scratches, cuts or scars on his legs, arms and hands from normal living. His hands are not callused from working nor are his fingernails chipped or dirty. Sunlight has not aged his skin. He has no facial wrinkles and his teeth are perfect without cavities, decay, stains, chips or wear. An examination of his arteries shows they are in perfect condition without any cholesterol build-up. While God could have made Adam appear old due to the effects of deterioration over time, it is doubtful He did.

The idea of apparent age should not be used to solve problems traditional young-Earth creation theology has with lunar impact craters. It is a mistake to say God created the Moon with craters to give it an appearance of age. This makes God a great deceiver and puts the Bible believer in an indefensible position. If the Moon were truly 6,000 years old, it should have virtually no meteor impact craters.

In summary, evidence of 8,596 Moon impact craters is compelling proof that a proportional number of impacts also occurred on Earth. Within the past, approximately 115,500 meteors with the capability of forming a one kilometer diameter crater or larger have struck Earth. Six thousand years does not allow enough time for this number of meteor impacts to occur and then be removed by natural processes. In contrast, *exemplar* creation allows time in the undefined past prior to Genesis 1:3 for these impact events. *Exemplar* creation easily harmonizes with the evidence of empirical science.

Antarctica and Greenland Ice Cores

Ice core data present a serious challenge to the credibility of traditional young-Earth creation theology in regard to time. In certain parts of the world, snow piles up on a continual basis. Summers are cold enough to keep winter snowfall from melting, resulting in accumulation to enormous depths over time. Two of the most notable areas on Earth where this has occurred are Greenland and Antarctica. Snow has collected in some areas on these two continents to more than 3,000 meters (9,843 ft) thick. As snow accumulates, pressure turns it into ice. A count of annual layers within this ice gives a physical record going back more than 100,000 years.

Annual layers are formed by snow that falls throughout the year in Antarctica and Greenland. Most snow accumulates during winter, with less during summer.

Dust builds up on the snow during summer and provides a dirty film so that annual layers can be visually distinguished. Non-visual effects also occur based on the temperature difference between winter and summer snow. Non-visual effects that vary include the amount of deuterium (heavy hydrogen), oxygen-18 isotopes, CH_4 and carbon dioxide contained in the snow. Non-visual effects can be measured to confirm visual effects when counting ice layers.

Annual ice layers represent physical evidence of past climatic change. Motivations for examining the ice record include a desire to understand past planetary heating and cooling fluctuations. Other motivations include an attempt to accurately date past volcanic eruptions, such as the Minoan eruption of Thera (an island in the eastern Mediterranean Sea) that occurred roughly around 1600 BC.

Drilling an ice core is a huge project accomplished under challenging conditions. Ice cores are typically drilled at locations where the average temperature ranges between -25° and -65° F. A combination of temperature, wind and isolation make living conditions extremely difficult. All crew members, food supplies, living accommodations and drilling equipment must be transported either by air or overland to the core site. Drilling can take several years to complete and may be limited to the summer seasons. A core drilling project is an expensive, large undertaking and normally requires government funding, United States Air Force logistics support and personnel from several large scientific organizations and universities.

Ice cores are drilled, handled, stored and analyzed with specialized equipment. Drilling is done with a hollow core drill, in some instances thermally heated. Ice core sections range from 3 to 6 meters (9.8 to 19.7 ft) in length and have a typical diameter of 10 cm (4 in). Each section is carefully stored in a custom made container and identified. Containers are kept refrigerated at a point well below freezing during ground and air transport over thousands of miles. At the research center (such as The National Ice Core Laboratory in Lakewood, Colorado) they are stored in freezers awaiting examination.

The examination process for ice cores is extensive. Cores are sliced lengthwise using a paper thin saw blade into different sections, similar to cutting a log into boards. One thin, lengthwise piece is used to visually count annual layers in the core. Pieces are mounted on a special viewing table with backlighting to aid in counting the dust bands. Other complete lengthwise pieces are analyzed for electrical conductivity. Lengthwise pieces may be further cut into small crosswise pieces for isotope, gas, ion and dust analysis. Most work is accomplished in a clean room atmosphere under conditions approximating the inside of a deep freeze.

As depth increases in an ice pack, annual layers become thinner. Pressure causes ice to flow horizontally to the sides of the glacier, resulting in thinning of the lower layers. Increasing depth is accompanied by progressively thinner layers that are more difficult to count.

Table 4.2 lists the major ice cores drilled in the world. The year completed, ice depth and core depth are a matter of public record. The two far right columns list the number of counted or estimated annual ice layers at a specific depth in the ice core.

The first major ice core drilled and the first to reach bedrock in Antarctica was completed on January 29, 1968 at Byrd Station. The core was drilled with a cable-suspended electromechanical hollow rotary drill. Byrd station is located at 80° 01′ S, 119° 31′ W near the southwest corner of Antarctica at an altitude of 1,530 meters (5,020 ft) above sea level. Vertical thickness of the ice at this location was 2,164 meters (7,100 ft). Although no specific annual layers were counted for the Byrd core, the project obtained important data about the ice—it was indeed thick. It would take approximately 14,430 years or more to build an ice thickness of 2,164 meters at this site based on present annual moisture accumulation equivalent to 15 cm (5.9 in) of water per year (with no thinning of annual ice layers).

Dome C (old) ice core was drilled during 1977-1978 using a thermally heated core drill.[17] Full core depth was 906 meters (2,973 ft). Dome C is located at 74° 39′ S, 124° 10′ E at an elevation of 3,240 meters (10,630 ft). Average air temperature at this site is -64.3° F and is among the coldest drilling sites in Antarctica. The core was analyzed for deuterium, oxygen-18 isotopes, CO_2 and CH_4. Non-visual indicators were used to count 39,515 annual layers at a depth of 903.4 meters (2,964 ft). It would take approximately 113,000 years or more to build an ice thickness of 3,400 meters at this site based on present annual moisture accumulation (with no thinning of annual layers).

The Vostok ice core drilling project was completed in January of 1998 in Antarctica, and was a collaborative effort among Russia, France and the United States.[18] The Vostok core is the deepest ice core ever recovered with a length of 3,623 meters (11,887 ft). Drilling stopped about 120 yards short of a subterranean lake the size of Lake Ontario, as scientists did not want to contaminate the water with drilling fluids. No visual layer counting was performed on the Vostok ice core. Non-visual isotope measurements, glacial modeling and other methods were used to estimate 422,766 annual layers at a depth of 3,310 meters (10,860 ft). It would take approximately 123,000 years or more to build an ice thickness of 3,700 meters at this site based on present annual moisture accumulation (with no thinning of annual layers).

Drilling at Dome C (EDC3) was started in 1999 and completed in 2003.[19, 20] Dome C is located at 75° 06′ S, 123° 21′ E in Antarctica at an altitude of 3,233 meters (10,607 ft). The ice core was 3,260 meters (10,696 ft) deep, representing an estimated 800,000 annual layers. No visual layer counting was performed on the Dome C core. Isotope measurements, modeling and other methods were used to estimate the number of annual layers. It would take approximately 108,000 years or more to build an ice thickness of 3,260 meters at this site based on present annual moisture accumulation (with no thinning of annual layers).

The Greenland Ice Core Project (GRIP) completed its drilling phase in 1992.[21] The drill site was located at 72° 35′ N, 37° 38′ W at an altitude of 3,230 meters (10,598 ft). The core was 3,029 meters (9,938 ft) deep and ended at bedrock. This was the first major core to be extensively analyzed for annual layers. Manual counting identified 60,000 annual layers in the first 2,495 meters (8,186 ft) of the core. The manual counting of visual layers was cross checked against annual isotope cycles and other methods. The average layer thickness at 2,495 meters was 0.684 cm (0.269 in). At depths below 2,495 meters, the ice layers became progressively thinner and manual counting became difficult. The GRIP ice core is 2,983.2 meters (9,788 ft) deep and contains an estimated 248,763 annual layers.

Table 4.2 – Major Greenland and Antarctica Ice Cores

Continent	Ice Core Location	Date Completed (year)	Total Ice Thickness (meters)	Annual Moisture (cm/yr)	Annual Ice Layers	At Core Depth (meters)
Antarctica	Byrd Station	1968	2,164	15.0	---	---
Antarctica	Dome C (old)	1978	3,400	3.0	39,515	903.4
Greenland	GRIP	1992	3,029	25.0	248,763	2,983.2
Greenland	GISP2	1993	3,053	25.0	108,857	2,792.0
Antarctica	Taylor Dome	1994	1,811	2.5	214,760	554.0
Antarctica	Vostok	1998	3,700	3.0	422,766	3,310.0
Antarctica	Dome C	2003	3,300	3.0	800,000	3,260.0
Greenland	NorthGRIP	2003	3,085	19.0	123,000	3,080.0

Notes:
(1) The annual moisture accumulation is in centimeters of water per year.
(2) Dome C completed in 2003 is also known as EDC3.
(3) The annual ice layers listed occur up to the core depth in meters listed. Beyond that point layers become uncertain. GRIP has 248,763 layers at a core depth of 2,983.2 meters.

The Greenland Ice Sheet Project Two (GISP2) completed its drilling phase on July 1, 1993, by drilling 3,053 meters (10,017 ft) through the Greenland Ice Sheet into bedrock.[22] GISP2 is located at 72° 36′ N, 38°30′ W in central Greenland at an altitude of 3,208 meters (10,525 ft), approximately 30 km (18.6 miles) from the GRIP site. Using annual dust layers, the GISP2 core was manually counted for the first 37,900 layers. The GISP2 ice core is 2,792 meters (9,161 ft) long and contains approximately 108,857 annual layers when counted using annual isotope cycles.

The first 100,000 annual layers of electrical conductivity and oxygen isotope data for the Greenland GISP2 and GRIP cores have been compared, and their curves match very closely. For the same depth in meters, the GRIP core has more than twice as many layers as the GRISP2 core because of ice thinning.

The North Greenland Ice Core Project (NorthGRIP) completed its drilling phase in July 2003, after six years.[23] The core was drilled at 75.1° N, 42.3° W (north of the GRIP and GISP2 sites) to a depth of 3,085 meters (10,122 ft) where it hit bedrock. The core drill itself was 11 meters (36 ft) long and hung from a cable during drilling. Each ice core section was 3.5 meters (11.5 ft) long and 9.8 cm (3.86 in) in diameter. The clear, regular, annual layers in the core were visible down to 2,600 meters (8,531 ft), representing approximately 80,000 annual layers.[24] The NorthGRIP core has been cross-dated to the GRIP core back to 105,000 years before present using isotope data.

Scientists have analyzed the NorthGRIP core using several isotope and ion measurements to precisely count the number of annual layers. As mentioned, the concentration of several isotopes and ions in snow fluctuate on an annual basis because of the temperature difference between winter and summer. In the NorthGRIP core, oxygen isotopes, deuterium, calcium ions and sodium ions were cross checked against visually counted annual layers. When completely analyzed, this core is expected to provide a highly accurate count of annual layers back to more than 100,000 years before present.

At 3,085 meters (10,122 ft) the NorthGRIP drill bottomed out on bedrock, and when the drill was retracted some organic material had stuck to the end of the drill. According to James White, professor of Geological Sciences at the University of Colorado at Boulder, the pieces looked like blades of grass or pine needles.[25] This is physical evidence that plants existed on the Earth more than 100,000 years ago.

Empirical data from ice cores show that Antarctica and Greenland have been covered with ice for more than 100,000 years. GRIP, GISP2, Taylor Dome, Vostok, Dome C and NorthGRIP are solid evidence that the Earth is substantially older than 6,000 years. The cross correlation of data among different cores adds weight to the validity of this evidence.

What do supporters of traditional young-Earth creation theology say about this evidence? Critics of ice core data say annual layers are actually sub-annual layers.[26] Critics hypothesize that multiple snow falls during the winter season show up as sub-annual layers, indistinguishable from annual layers. The fallacy of this argument is that oxygen isotopes, deuterium, calcium ions and sodium ions in snow vary according to temperature. Winter snow has a different concentration of molecules than summer snow because of temperature. Annual snowfall accumulation based on non-visual measurements can be matched with the visual counting of dust layers. For example, the number of manually counted layers compared against the number of oxygen-18 isotope annual temperature cycles easily agree within five percent. This is conclusive evidence these are annual layers representing more than 100,000 years of Earth history.

Michael Oard, a young-Earth creationist, hypothesizes that the huge accumulations of ice in Greenland and Antarctica are due to heavy snows that started immediately after the flood of Noah approximately 4,350 years ago.[27] He says present accumulations are the result of 500 years of heavy snows and cold temperatures due to post flood effects, and represent the start of the ice age.[28] There is absolutely no evidence in the ice core record to support this belief.

In summary, more than 100,000 annual layers observed in ice cores is physical evidence of an old Earth. Multiple drilling sites and different personnel on two different continents at different times confirm the validity of this evidence. *Exemplar* creation allows time in the undefined past prior to Genesis 1:3 for the accumulation of extensive ice in Antarctica and Greenland. *Exemplar* creation easily harmonizes with the evidence of empirical science.

Antarctica Coal and Fossils

Ernest Shackleton was the first to discover coal in Antarctica. During the 1907-1909 expedition, Shackleton took his ship, *Nimrod,* into the Antarctic where he set up base camp at Cape Royds on Ross Island located in McMurdo Sound. In an attempt to reach the South Pole, his group traveled south across the Ross Ice Sheet and through the Transantarctic Mountains on the Beardmore Glacier. They were within 97 nautical miles (180 km) of being the first to reach the South Pole, when they stopped and began retreating.[29]

On the return trip from their historic attempt, Shackleton and his group made an interesting discovery 350 miles from the South Pole. As they left the Polar Plateau and were crossing the Beardmore Glacier, they discovered coal at an altitude of 3,100 meters (10,200 ft). Shackleton's party brought back samples and mapped the vein, located near Mount Buckley.

The Transantarctic Mountains in western Antarctica, immediately east of the Ross Ice Sheet, form one of the world's great mountain chains, extending for more than 2,200 km (1,370 miles) in length. Many mountain peaks exceed 4,000 meters (13,000 ft) in height[30] and contain numerous exposed coal seams.[31]

Coal has also been found in the Prince Charles Mountains located in eastern Antarctica at approximately 70° S, 70° E. Within these mountains, the Amery Oasis (near the junction of the Lambert Glacier and the Amery Ice Shelf) lies at an altitude ranging from 400 to 1,300 meters (1,300 to 4,300 ft) above sea level. The Bainmedart Coal Measures at the Amery Oasis contain extensive deposits of coal, with some seams more than 3 meters (10 ft) thick.[32]

Coal in Antarctica is physical evidence of past abundant plant life, before the climate turned deadly cold. Coal may be formed in swamps where vegetation accumulates over time and becomes compacted. Coal can also be formed when vegetation is stripped from the land, collected at settling points, covered with sediment and then compacted. Regardless of the formation process, coal is compelling evidence of climatic conditions favorable to plant growth.

Several notable animal fossils have been found in Antarctica. During a 1990-1991 expedition, William R. Hammer, a paleontologist at Augustana College, and William J. Hickerson discovered a new dinosaur. The skeleton was found at an altitude of 3,800 meters (12,450 ft) on Mount Kirkpatrick located in the Transantarctic Mountains near the Beardmore Glacier. Remains of three other dinosaurs—a very large prosauropod (a long-necked, small-headed, semi-bipedal, herbivorous dinosaur), a small pterosaur (a flying reptile) and another unknown theropod (a bipedal, carnivorous dinosaur with a three toed foot)—were also found. Professor Hammer's dinosaur, named Cryolophosaurus (meaning "frozen crested reptile"), is a large crested meat-eating theropod with two large hind legs for walking and two small fore limbs for grasping prey. Cryolophosaurus is the first carnivorous dinosaur found in Antarctica.[33]

On December 7, 2003, while on another fossil hunting expedition in Antarctica, Professor Hammer found a sauropod pelvis at an altitude of 3,900 meters (12,800 ft) on Mount Kirkpatrick.[34] The pelvis measured about a meter across and belonged to a sauropod estimated to be between 1.8 and 2.1 meters (5.9 to 6.9 ft) tall and up to 9 meters (29.5 ft) long. Sauropods were quadruped, herbivorous dinosaurs with a long neck and tail, and a small head. They are the largest dinosaurs to have roamed the Earth.

A variety of plant remains have been found in the Ohio Range of the Transantarctic Mountains, including fossilized tree trunks and ferns. The Ohio Range is known for its highly picturesque, thick layers of massively eroded sedimentary rock.

Animal and plant fossils are firm evidence that at one time abundant life existed in Antarctica. Today, other than sea life, virtually nothing lives there owing to the extreme cold. In the past, Antarctica's climate was substantially different and was warm enough to allow dinosaurs, trees, ferns and other vegetation to grow. Temperatures somehow dropped from an estimated average of +60° F in the past to an average of -50° F today. When this happened, animal and plant life disappeared and ice began to accumulate.

How did the Antarctic climate change from warm to bitter cold? This is an intriguing question. The leading theory proposes that at one time Antarctica existed closer to the equator where it received more sunlight. Due to continental drift, Antarctica moved to its present location over vast time.

What do proponents of traditional young-Earth creation theology say about this information? Some hypothesize that originally the Earth had a different atmosphere, resulting in a mild climate over the entire Earth. Proponents say that after Noah's flood, approximately 4,350 years ago, the global climate changed to its present state. This hypothesis conflicts with the ice cores of Greenland and Antarctica which show conclusively these two continents have been very cold for more than 100,000 years.

In summary, extensive time is necessary to allow the warm climate animals of Antarctica to grow, mature, die and become fossilized. Extensive time is needed for warm climate plants to grow, accumulate and form thick coal beds. Extensive time is needed for the continent of Antarctica to drift to its present location and become drastically colder. Extensive time is required for ice to accumulate to great thickness in Greenland and Antarctica. *Exemplar* creation allows time in the undefined past prior to Genesis 1:3 for these events to occur. *Exemplar* creation easily harmonizes with the evidence of empirical science.

Radiometric Dating Igneous Rocks

A precise method of determining the Earth's age has been a pursuit of scientists for many years. Radiometric dating is the result of that quest and indicates the Earth is billions of years old. This dating technique presents a serious challenge to the credibility of traditional young-Earth creation theology in regard to time.

Before radiometric dating was developed, several methods were used to estimate the Earth's age. Processes extrapolated to determine age include planetary cooling, sedimentation, salt and mineral accumulation in the oceans, helium diffusion and the decay of Earth's magnetic field. Extrapolations of these phenomena begin with certain assumptions, potential causes for error and known inconsistencies. Dates range from a few thousand to millions of years.[35]

To better understand extrapolation, the accumulation of salt in the Dead Sea serves as a good example. Scientists can calculate the amount of water and salt currently in the Dead Sea. In addition, they can measure the amount of water it receives annually from the Jordon River, and the amount of salt in each cubic foot of incoming water. From this information, the total amount of salt added to the Dead Sea in one year can be calculated. Using this information, scientists can compute the sea's saltiness 1,000 years in the future.

The problem with using extrapolation to determine the future or past saltiness of the Dead Sea is that conditions may change. The climate may change and subsequently affect the amount of rain and salt flowing into the Jordan River. The type of vegetation and amount of cleared land may vary over time and influence water runoff. The quantity of water removed from the Jordan River for irrigation may be altered. The inflow of salt from subterranean springs that flow directly into the Dead Sea may fluctuate. In the long term, many variables can affect the amount of water and salt flowing into the Dead Sea to make this an unreliable extrapolation over thousands of years.

On the other hand, radiometric dating, based on the extrapolation of isotope decay, has proven to be reliable. An atomic isotope is one form of the same element. Isotopes have the same number of protons and electrons, but a different number of neutrons. An unstable isotope may decay into another isotope of the same element or a completely different element. Isotope decay is the most consistent type of natural change known and is unaffected by non-nuclear processes acting near the surface of the Earth such as heat, cold, pressure, magnetism, centrifugal force, chemicals and shock.

Isotope decay rates have been carefully measured and are stated in half-life, the length of time required for one-half of the isotope atoms within a sample to decay into another form. By precisely measuring the original isotopes and decay products in an igneous rock, its age since solidification can be determined. In essence, when an igneous rock solidifies it becomes a ticking clock that can be read at some future date.

Radiometric dating applies to the dating of igneous rocks (rocks that were once completely molten). Magma from within the Earth and volcanoes are the most common source. In contrast, sedimentary rocks have never been molten and cannot be dated; however, it is possible to date individual igneous rocks contained within sediments. Radiometric dating cannot be used to directly date plant and animal fossils, but a lava flow on top of a fossil, or beneath it, can be dated to give an upper or lower age limit.

Five of the better known radiometric techniques for dating igneous rocks are rubidium-strontium, samarium-neodymium, potassium-argon, argon-argon and uranium-lead. The rubidium-strontium, uranium-lead and samarium-neodymium techniques are preferred because no knowledge of the original rock content is needed. (For a short discussion of these dating techniques, an Internet article by Dr. Roger C. Wiens is recommended.[36])

One of the most significant books written about our planet's age is *Age of the Earth* by G. Brent Dalrymple, PhD, research geologist at the U.S. Geological Survey and retired professor from Oregon State University. In his book, Dr. Dalrymple discusses radiometric dating and presents an extensive compilation of information from different sources. Table 4.3 gives a summary of the oldest igneous rocks provided in his book.

Dalrymple gives a total of 1,159 sample dates from five major locations on Earth with an age spread of 2.32 to 3.81 billion years. Since publication of his book, rocks dated up to 4 billion years old have been found in Greenland, the Canadian Shield near Lake Superior in Canada, Transvaal Province of South Africa and Western Australia.[37]

Age of the Earth also lists information collected from dating Moon rocks and meteors. Dalrymple gives a number of dates from lava flows (basalt rocks) on the Moon between 3.09 to 3.95 billion years old.[38] Some of the oldest Moon rocks date from 3.90 to 4.51 billion years old.[39] He gives 67 dates from 42 different meteors between 4.29 and 4.60 billion years old.[40]

Dalrymple discusses lead isotopes and their use in dating the Earth. Based on the Holmes-Houtermans model, lead isotope decay gives a calculated Earth age of 4.54 billion years.[41] He states that naturally occurring long-lived radioactive isotopes lead to the incontrovertible conclusion that the solid bodies of the Solar System formed 4.54 billion years ago.[42] Based on this and direct radiometric measurements, an age of 4.6 billion years is often quoted for the Earth.

Prevalent theory proposes that meteors and the Earth were formed at the same time. Molten meteors cooled first because of their smaller size and started the clock ticking. It took Earth (as a molten ball) roughly 0.6 billion years to cool enough to retain the first rocks. The earliest rocks found on our planet's surface are approximately 3.8 to 4.0 billion years old. Rocks older than this are theorized to have been recycled by subduction when they slid back into the molten mantle and remelted.

The controversy over radiometric dating has been fueled to an extent by occasional inconsistent measurements, often used by skeptics to attack its credibility. Inaccurate results have occurred due to the initial technical immaturity of the processes, poor laboratory techniques, lack of process knowledge, improper sample preparation and poor sample selection. As the technology for radiometric measuring has matured, measurements have become more reliable.

Table 4.3 – Radiometric Ages of Early Rocks
(A compilation of radiometric dates from *The Age of the Earth*)

Book Table	Description	No. of Sample Dates	Dating Details	Age Range (billions of yrs)
4.1	West Greenland	476	43 avg. ages	2.52 to 3.81
4.1	Labrador	131	12 avg. ages	2.52 to 3.76
4.2	Superior Province, North America	130	15 avg. ages	2.32 to 3.68
4.4	Pilbara Block, Western Australia.	175	19 avg. ages	2.61 to 3.57
4.6	Barberton Mountain Land, S. Africa	247	33 avg. ages	2.73 to 3.56
5.5	Lunar Mare basalt rocks	62	Dated 14 rocks	3.09 to 3.95
5.6	Oldest lunar rocks	30	Dated 13 rocks	3.90 to 4.51
6.3	Old meteors	67	Dated 42 rocks	4.29 to 4.60
Reference: Dalrymple, G. Brent (1991). *The Age of the Earth.*				

Accepting radiometric dating as a credible process does not mean it is an infallible process. Several factors can affect the accuracy of dates obtained from igneous rocks. Exacting laboratory procedures are a basic requirement. Igneous rocks to be dated must be carefully selected, and rocks not fully melted or contaminated by their surroundings can give incorrect dates. In his book, Dalrymple does not discuss the accuracy of dating volcanic ash nor the accuracy of dating rocks modified by meteor impact shock. The dating of reheated or partially melted igneous rocks may give inaccurate results, since the clock is incompletely reset. Both the original formation date of the rock and the reheating event date may be wrong.

In summary, radiometric dating is not based on visual evidence equivalent to starlight, Moon craters or ice cores. It is the measurement of isotope decay, undetectable by direct human senses, extrapolated backward in time. However, based on precise measurements using sensitive instrumentation and a consistent decay process, its extrapolation is highly credible. Igneous rocks that have been fully melted, uncontaminated while melted, solidified and not reheated are excellent candidates for dating. Trustworthy dates can be obtained as evidenced by consistent dates among specific rock outcrops using different dating techniques. Radiometric dating is compelling evidence that the Earth is billions of years old. *Exemplar* creation allows time in the undefined past prior to Genesis 1:3 for significant radioisotope decay to occur. *Exemplar* creation easily harmonizes with the evidence of empirical science.

Carbon-14 Dating Organic Material

Radiocarbon dating presents a serious challenge to the credibility of traditional young-Earth creation theology in regard to time. Carbon-14 dating is a method of determining the age of once living plants and animals (organic materials). This technique cannot date rocks. Completely fossilized plant or animal remains are not good candidates for dating. Evidence from research about radiocarbon dating indicates life has existed on Earth back to 30,000 years and beyond.

The radiocarbon dating of organic materials is based on decay of the carbon-14 isotope extrapolated backward in time (similar to the radiometric dating of igneous rocks). Accuracy depends on the consistency of all variables affecting the process.

The carbon family has three naturally occurring isotopes: carbon-12, carbon-13 and carbon-14. Carbon-12 comprises 98.892% of the total carbon in the atmosphere while carbon-13 accounts for 1.108% of the total.[43] Carbon-14 comprises about 0.0000000010% of the carbon in the atmosphere.[44] Carbon-12 and carbon-13 are stable and used as a reference, while carbon-14 is unstable and is the basis for radiocarbon dating.

The origin and decay of carbon-14 begins when nitrogen-14 atoms in the upper atmosphere are struck by cosmic rays. When a nitrogen atom has one proton removed and one neutron added, one carbon-14 atom is formed. Starting with a given amount of carbon-14, it takes approximately 5,730 years for one-half of the starting amount to decay and turn into another form (this determines its half-life).[45]

Once carbon-14 is formed, it combines with oxygen in the atmosphere to make carbon dioxide, which descends to the Earth's surface and is picked up by plants. During photosynthesis, the carbon atom is separated from the two oxygen atoms. Carbon becomes a permanent part of the plant's structure, while oxygen is released back into the atmosphere. When animals eat plants, the carbon contained therein becomes a part of their bodies, residing in their bones, teeth, horns, claws and other tissues.

When a plant or animal dies, intake of all carbon ceases and their remains become a ticking clock as carbon-14 slowly decays. Conceptually, by measuring the ratio of carbon-12 (or carbon-13) to the remaining carbon-14, the time since death can be determined.

Three major techniques have been developed for carbon-14 dating: the gas counting method, the liquid scintillation method and the accelerator mass spectrometry method. The newest method, accelerator mass spectrometry (AMS), has an advantage over the others because small samples from one gram to a few milligrams can be analyzed. The AMS method is able to count all carbon-14 atoms in a specimen at one time rather than counting their decay rate (in small samples, the decay of individual carbon-14 atoms can be irregular).

The basics of radiocarbon dating are easy to understand; however, taking a concept and turning it into a reliable dating method is more difficult. Many factors in radiocarbon dating can significantly affect its accuracy. The precision of radiocarbon dating becomes increasingly uncertain above 10,000 years because of the rapidly decreasing percentage of carbon-14 in a sample. As shown in Table 4.4, after five half-lives or 28,650 years only 3.12% of the original carbon-14 remains. The basic physics limit the process. In addition, several other known variables can affect the precision of radiocarbon dating.

(1) Atmospheric carbon – The first major variable in radiocarbon dating is the amount of carbon-14 in the Earth's atmosphere. Since carbon-14 in the air is the source of carbon for plants and animals, if the amount in the atmosphere varies over time, then the amount retained by a living organism will also vary. It is theorized that any change in Earth's magnetic field affects carbon-14 production by offering more or less protection from cosmic rays. Changes in the Sun can also vary the intensity of cosmic radiation bombarding Earth and influence carbon-14 production. Regardless of the cause, scientists know that the percentage of carbon-14 in the Earth's atmosphere has changed over time. The amount has been higher in the past, which causes radiocarbon dated objects to appear younger than they really are.

Table 4.4 – Carbon-14 Half-Life and its Decay in a Sample over Time
(Based on a half-life of 5,730 years)

Number of C-14 Half-Lives	Percent of Original Carbon-14	Percent of Original Carbon-13	Percent of Original Carbon-12	Number of Years Before Present
0	100.000%	100%	100%	0
1	50.000%	100%	100%	5,730
2	25.000%	100%	100%	11,460
3	12.500%	100%	100%	17,190
4	6.250%	100%	100%	22,920
5	3.120%	100%	100%	28,650
6	1.563%	100%	100%	34,380
7	0.781%	100%	100%	40,110
8	0.391%	100%	100%	45,840
9	0.195%	100%	100%	51,570
10	0.098%	100%	100%	57,300

Tree-ring dating (dendrochronology) has been used for many years to correlate radiocarbon dates. Scientists have counted tree rings and measured their radiocarbon ages to develop correlation curves. Using this technique, Dr. C. W. Ferguson at the University of Arizona, Tucson built a tree-ring chronology from bristlecone pine trees in the White Mountains of California going back almost 8,000 years.[46]

Scientists have built standardized calibration curves which permit correcting radiocarbon dates affected by carbon-14 variation in the atmosphere. In 1998, the IntCal98 radiocarbon calibration curve was released based on a partial tree-ring series. An updated version, IntCal04 was released in 2004 based on a complete tree-ring chronology of German pine, German oak, Irish oak and Pacific Northwest Douglas fir going back 12,410 calendar years before present (BP).[47] Beyond this, corals were used to extend the curve back to 26,000 calendar years BP. The complete IntCal04 calibration curve permits the correction of carbon-14 dates to a maximum of 26,000 years. Based on this calibration curve, 22,000 radiocarbon years are equivalent to 26,000 calendar years.[48]

A radiocarbon correlation curve back to 50,000 years was built by Dr. Richard Fairbanks at Columbia University and his associates based on coral. Corals were obtained from Barbados in the western tropical Atlantic, the Kiritimati Atoll in the central equatorial Pacific and the uplifted reefs of Araki Island in the western Pacific. This correlation curve is based on $^{230}Th/^{234}U/^{238}U$ radiometric measurements of coral compared against carbon-14 measurements. The first 12,410 years of the IntCal04 radiocarbon tree-ring dendrochronology curve and the coral curve are essentially identical.[49] From 12,410 to 26,000 calendar years BP, the majority of the IntCal04 curve is based on coral.

An additional correlation curve was developed by measuring the age of organic material in alternating layers of sediment in Lake Suigetsu, Japan. Small white blooms grow on the lake each year. At the end of their annual life cycle, the blooms settle to the bottom and form a white layer. During the remainder of the year, black sediment collects on the lake bottom. Alternating layers of black sediments and white blooms permit a visual count of annual layers. Scientists counted 29,100 alternating layers and built a floating radiocarbon correlation curve from 8,830 to 37,930 calendar years before present.[50] For the period of time they overlap, the Lake Suigetsu correlation curve and the IntCal04 calibration curve are essentially identical.

(2) Fractionation – A second important variable in radiocarbon dating is fractionation.[51] Earth's atmosphere contains a given ratio of carbon-12, carbon-13 and carbon-14. Fractionation occurs when living things retain carbon in a ratio different than the atmosphere. This happens because plants discriminate against slight chemical and physical differences among the three different atomic forms.

The amount that plants discriminate against carbon-13 has been measured. Plants classified as C4 plants such as maize, sorghum, amaranth, millet, buffalo grass, pigweed and sugar cane use the Hatch-Slack cycle during photosynthesis and retain an average of 13% less carbon-13 than found in the atmosphere. Plants classified as C3 plants such as rice, wheat, oats, soybeans and potatoes use the Calvin-Benson cycle during photosynthesis and retain an average of 26% less carbon-13 than found in the atmosphere.[52] Humans and animals which feed on these plants end up with a fractionation effect in their bones, teeth, fingernails, horns, claws and other tissues.

A third group of plants which show strong fractionation (normally not used for food) include warm evergreens, cool deciduous, hardwood and conifer trees. These trees retain approximately 27% less carbon-13 than found in the atmosphere.[53] Fractionation effects are retained in charcoal when wood is burned (carbonized) in short term experiments.[54]

The physical properties of carbon-14 cause it to have different fractionation values than carbon-13. The fractionation effect of carbon-14 in a sample is considered to be twice that of carbon-13.[55, 56] Fractionation makes things look older than they really are, and when dates are corrected in radiocarbon dating they are always younger (more recent).

(3) Matter transfer – A third variable in radiocarbon dating is the transfer of material in or out of a sample. In radiometric dating, when an igneous rock cools it becomes a sealed unit. Unless the rock is heated to its melting point or near its melting point, none of the radioactive isotopes can leave the rock. This ensures an accurate clock. This is not the case with organic material. Foreign matter can infiltrate organic material such as wood or bone to the extent it becomes completely fossilized. When wood and bone buried in soil begin to decay, atoms and molecules transfer in and out of their structures. Carbon, calcium, other minerals, proteins and amino acids are removed from bone as it slowly decays. Organic materials stored in dry, protected surroundings have potential for giving the best radiocarbon dates.

(4) Laboratory procedures – A fourth variable in radiocarbon dating is the facility performing the analysis. Readings among laboratories may vary based on calibration and accuracy of the equipment. The handling of samples during cleaning, preparation and analysis will vary among laboratories. Samples must typically be cleaned to remove contamination in the form of carbonates, soil humic, fluvic acids and other soluble organic matter absorbed from their surroundings.[57]

(5) Marine environment – A fifth variable relates specifically to marine animals. Because of special factors affecting ocean water and marine animals, a one-year old penguin can give a radiocarbon date hundreds of years old. This effect can cause a land animal, such as a polar bear, that has fed upon marine animals to falsely appear quite old when radiocarbon dated.

Knowledgeable individuals whose business is radiocarbon dating are aware of all five of the variables mentioned. Some of these variables can be accounted for, controlled and monitored, while others cannot. This dating method is similar to any other emerging technology, becoming more dependable as it matures.

In summary, radiocarbon dating supports the existence of life on Earth substantially before 4004 BC in contradiction to traditional young-Earth creation theology. The carbon-14 decay process has been directly checked against tree-ring chronologies and annual lake sediment accumulations. The resultant calibration and correlation curves are strong evidence of life on Earth beyond 30,000 years. *Exemplar* creation easily harmonizes with the evidence of carbon-14 dating and empirical science.

Conclusion

A number of physical indicators show the Earth and universe are very old. The measured distance to stars and the speed of light prove the heavens are millions of years old. Impact craters illustrate the implausibility of a 6,000 year old Earth and Moon. Ice cores from Greenland and Antarctica confirm the Earth is more than 100,000 years old. Coal, animal and plant fossils in Antarctica prove it had a warm climate before the temperature plunged and ice began accumulating more than 100 millennia ago. The oldest igneous rocks on Earth have been radiometric dated between 3.8 and 4.0 billion years old. Tree-ring chronologies calibrated against radiocarbon dating go back 12,410 calendar years, while the Lake Suigetsu sediment layers document 29,100 calendar years of time. This is compelling evidence that traditional young-Earth creation theology is terribly wrong in regard to time. The heavens, Earth and life existed substantially before 4004 BC.

Exemplar creation offers a solution to the time problem that respects the integrity of the Bible and easily harmonizes with empirical science. The Genesis creation story is an abbreviated, non-technical, historical account of origins that is truthful and trustworthy. Where the historical account of the Genesis creation story and empirical science intersect, they can be harmonized in regard to time. *Exemplar* creation permits the occurrence of events requiring extensive time in the undefined past prior to Genesis 1:3.

It is important to understand that extensive time in a building process is not equivalent to evolution. For example, the Great Pyramid of Khufu in Giza, Egypt was built during a twenty year period. An estimated 2.3 million blocks of cut stone were used in its construction.[58] No one denies the Great Pyramid is due to intelligent cause rather than natural cause. However, suppose the pyramid builders were immortal and laid one block every 100 years to take a total of 230 million years for its construction. In considering this amount of time, could we rightfully say the Great Pyramid had evolved? Of course not! Regardless of the time taken to build the Great Pyramid, it is undeniably a product of intelligent cause.

Given billions of years of time undirected phenomena cannot build the Great Pyramid. Uncontrolled natural mechanisms are incapable of cutting 2.3 million stone blocks to the required shape, transporting them to the job site and then assembling them. Immense time does not empower evolution with any significant constructive abilities.

For many Bible students and Christians, an old age for the universe is very disturbing. Although unsettling, we should reject the notion that if the Earth is billions of years old then evolution must be true. It is important to realize that more than enormous time is needed. The Big Bang, evolutionary star formation, the solar nebular hypothesis, chemical evolution and biological evolution require plausible natural constructive mechanisms, in addition to billions of years of time.

Empirical science based on observation, measurement and experimentation has determined that undirected phenomena are very limited as a constructive mechanism. Evolution is incapable of creating all of the information, complex functional systems, mathematically improbable structures and fine-tuned arrangements observed in nature. Billions of years of uncontrolled change cannot create the universe and all therein. Time is not a magic wand capable of curing the severe limitations of natural cause. Conversely, a powerful, intelligent creator identified in the Bible as God is fully capable of building the galaxies, stars, our solar system, Earth and all living things when given millions and billions of years of time.

Hope for the Bible believing community lies not in attempting to overturn the massive amount of evidence supporting a long age for living things, Earth and the heavens. Instead, hope rests in properly interpreting and translating the Genesis creation story. When key words and phrases in Genesis are proper understood, a biblically sound creation story emerges that easily harmonizes with extensive time.

These were more noble than those in Thessalonica, in that they received the word with all readiness of mind, and searched the scriptures daily, whether those things were so. (Acts 17:11) KJV

Chapter 5

The Words of Genesis

Interpreting the Hebrew of the Old Testament and translating it into English can be a contentious issue. Conflict occurs when a word has several legitimate translations and selection becomes a judgment call. Both English and Hebrew words have homonyms, homographs, heteronyms and multiple meanings. For example, the Hebrew word *erets* can mean "land" or "Earth" (as a planet). In the story of Noah's flood, does *erets* refer to land or the entire planet? The translation of *erets* dramatically affects our understanding of this event.

Biblical Hebrew is often quite limited in its choice of nouns, pronouns, adjectives, verbs and adverbs. The common language of Moses' day, reflected in the Bible, was sometimes inadequate to describe complex actions, physical phenomena and scientific concepts. *Strong's Exhaustive Concordance of the Bible* lists 8,674 Hebrew words in its Hebrew and Chaldee Dictionary, representing most of the words in the Old Testament. On the other hand, *Merriam-Webster's Collegiate Dictionary*, 11th edition, contains roughly 200,000 English words. With such a disparity, it is obvious that properly translating the Old Testament can be problematic. A single word in Hebrew may be capable of being translated into two, three or more English words having a specific meaning. In the King James Version, the Hebrew word *erets* is translated in descending frequency as "land, earth, country and ground," all of which have a fairly precise connotation in English.

When the context of a sentence forces a Hebrew word having several different legitimate meanings to be translated a specific way, proper selection is not difficult. Problems arise when the context is neutral, and the entire meaning of a sentence hinges on a single word. If a word can be legitimately translated two or three different ways, then selection becomes a judgment call. This is an issue in translating some key verses in the Genesis creation account.

The *Biblia Hebraica* (Hebrew Bible) presented in *The Lexham Hebrew-English Interlinear Bible* by Logos Research Systems of Bellingham, Washington is the basis for the *exemplar* translation of Genesis Chapters 1 and 2. The Lexham Bible was translated from Hebrew by the University of Stellenbosch in South Africa. The personal computer, electronic format of this Bible gives a clear presentation of the Hebrew text and allows for rapid searches using Boolean operators on Hebrew words of interest. Each Hebrew word is given in its original verse structure with a corresponding English translation below. The Hebrew text is presented in a right to left format (opposite the English style) identical to the original scripture.

A *Strong's Exhaustive Concordance of the Bible* (in addition to the eliyah.com Internet version) was used for all of the following word studies. Strong's concordance contains every English word in the King James Version of the Bible and assigns a numerical identifier to each corresponding Hebrew or Greek word. Strong's numbers provide a common method of identifying specific words in the original Hebrew text without resorting to the Hebrew alphabet and are used throughout Chapters 5 and 6.

Several Hebrew words in this chapter are transliterated (rather than translated) into English. The transliterations given in *Strong's Exhaustive Concordance of the Bible* are used since it is the most well-known biblical Hebrew reference book. The transliteration, pronunciation and English phonetic spelling of Hebrew words can change over time and will often vary among authorities. Individuals desiring a current reference should consult a Hebrew dictionary or Internet resources that give Hebrew to English transliterations.

Several key words in Genesis deserve an in-depth study. Their proper understanding is absolutely essential for a correct interpretation of the original text. Keep in mind that words used to describe the Genesis account have no special meaning applicable only to the creation story. They were words commonly used in the Hebrew language of the day and occur throughout the Old Testament.

The basis of study for the following key words in Genesis is not to explore every possible meaning, but to ensure the specific words used in the *exemplar* translation are legitimate and well supported elsewhere in the Old Testament. In this approach, Hebrew words and their frequency of translation into English are analyzed. For example, Hebrew words based on the root *bânâh* occur 376 times in the Old Testament and are translated 366 times as "build" in the KJV. Therefore, the traditional translation of *bânâh* as "made" in Genesis 2:22 is replaced with "built." Although the average person may not appreciate the small difference between *made* and *built*, subtleties such as this have caused the current errant translation of the Genesis creation account and its conflict with general revelation.

Bârâ – Create

The Hebrew word *bârâ* (Strong's 01254) occurs seven times in five verses in the first two chapters of Genesis and is consistently translated "create." The Brown-Driver-Briggs Hebrew Lexicon gives the English word "create" as an acceptable translation of the verb *bârâ.*[1] It is underlined and accompanied by its Strong's number in superscript in the following *exemplar* translation:

- Genesis 1:1 In the beginning, God created[01254] the heavens and the Earth.
- Genesis 1:21 And God *is the one who* created[01254] the great reptiles.
- Genesis 1:27 And God created[01254] man in His image, in God's image created[01254] him, male and female *God* created[01254] them.
- Genesis 2:3 That God did *when* creating[01254].
- Genesis 2:4 The *following is the* account of the heavens and the Earth when created[01254].

The root word *bârâ* (Strong's 01254) occurs 54 times in the Old Testament and is translated "created" 33 times, "create" 8 times and "creator" 3 times, in addition to other miscellaneous words in the KJV.[2] In the Pentateuch, *bârâ* is specifically reserved for actions performed by God and is translated "create" in the KJV except in two instances (Exodus 34:10 and Numbers 16:30). *Bârâ* can legitimately be translated create in the Genesis story.

Âsâh – Do, Did and Doing

The Hebrew word *âsâh* (Strong's 06213) occurs twelve times in the first two chapters of Genesis and can be translated a number of ways. The Brown-Driver-Briggs Hebrew Lexicon gives the English words "do" and "make" as primary translations of the verb *âsâh.*[3] An understanding of *âsâh* is essential for its proper translation in the Genesis creation story.

The word *âsâh* (Strong's 06213) occurs approximately 2,633 times in the Old Testament and is translated "do, did, done or doing" 1,333 times, "make or made" 653 times, "wrought" 52 times, "deal" 52 times, "commit" 49 times, "offer" 49 times, "execute" 48 times, "keep" 48 times, "shew" 43 times, "prepare" 37 times, "work" 29 times, "do so" 21 times, "perform" 18 times, "get" 14 times, "dress" 13 times, "maker" 13 times and "maintain" 7 times, in addition to other miscellaneous words in the KJV.[4]

In view of the different translations of *âsâh,* it is natural to question its meaning. The basic meaning of *âsâh* is equivalent to the English word "do," a general word that specifies the performance or execution of an action.

In the English language, the words "do, did, done or doing" can refer to a multitude of actions. For example, we might say, "She is doing her hair." In this sentence *doing* is a general word to indicate some type of action is being performed, which may include washing, rinsing, dyeing, drying, curling, combing, brushing, teasing or straightening hair. We could also say, "She is doing the dishes." In this case, *doing* may include picking up, scraping, rinsing, washing, drying or putting dishes away. We can say, "He is doing drugs," which might refer to sniffing, inhaling, smoking, swallowing or injecting drugs. The word "do" and its tenses in the English language can refer to an unlimited number of activities permissible within the context of a statement.

The general meaning of *âsâh* results in it being translated a number of different ways in the Old Testament based on verse context and the judgment of the translator. The problem occurs when the context of a verse is neutral and does not suggest a specific meaning. The translation of *âsâh* within a neutral context causes a small amount of confusion in understanding the Genesis creation story.

The Hebrew words *bârâ* and *âsâh* have different meanings. *Bârâ* means "create" and in the Pentateuch is specifically reserved to describe creative acts of God. *Âsâh* is a general word to denote some kind of action, which can include the action of creating. *Âsâh* is never translated "create" in the Old Testament in the KJV, but nevertheless is used in referring to creative acts of God as in Genesis 1:31 and 2:2.

Âsâh is often translated "made." *Âsâh* is used to describe the building of the tabernacle in Exodus 36:11-37 when the workman *made* loops, curtains, coverings, boards, bars, veils and pillars for the tabernacle. *Âsâh* is used to describe things God *made* in the Genesis creation story.

The word *âsâh* occurs twelve times in eleven verses in Genesis Chapters 1 and 2. *Âsâh* is underlined and accompanied by its Strong's number in superscript in the following *exemplar* translation:

- Genesis 1:7 And God *is the one who* made[06213] the expanse.
- Genesis 1:11 And the fruit tree making[06213] fruit *according* to *its* kind.
- Genesis 1:12 And the trees making[06213] fruit.
- Genesis 1:16 And God *is the one who* made[06213] two great luminaries.
- Genesis 1:25 And God *is the one who* made[06213] the animals of the Earth.
- Genesis 1:26 And God said, We *shall* make[06213] man in our image.
- Genesis 1:31 And God saw everything that *He* did[06213] and behold *it was* very good.
- Genesis 2:2 And on the seventh day God finished His work that *He* did[06213] and rested on the seventh day from all His work that *He* did[06213].
- Genesis 2:3 That God did[06213] *when* creating.
- Genesis 2:4 On the *sixth* day *when* Yahweh God did[06213] *things to the* Earth.
- Genesis 2:18 I *shall* make[06213] for him a counterpart as a helper.

The above verses represent a range of acceptable English translations for *âsâh*. Each word translation is based on the context of the verse and the explanatory framework of the chapter. In cases where there is legitimate doubt as to word selection, *âsâh* is translated "did" such as in Genesis 2:2-4. Translating *âsâh* in this manner permits it to be applicable to a wide range of possible actions.

In addition to the above, two other important verses contain *âsâh* in reference to the Genesis creation account.

• Exodus 20:11 For in six days the LORD made[06213] heaven and earth, the sea, and all that in them is, and rested the seventh day: wherefore the LORD blessed the sabbath day, and hallowed it. (KJV)
• Exodus 31:17 It is a sign between me and the children of Israel for ever: for in six days the LORD made[06213] heaven and earth, and on the seventh day he rested, and was refreshed. (KJV)

The above verses are used by believers in traditional young-Earth creation theology to support the idea that the heavens and Earth were created in six 24-hour days. Actually these verses are stating that God *did* (rather than made) the heavens, land, sea and all therein. The context of these two verses is a discussion of work and the Sabbath—not a discussion of how the universe was created. The context is the giving of the Ten Commandments. Keep in mind that in the KJV, *âsâh* is predominately translated "do, did, done or doing" in the Old Testament.

In the above two verses, the writer of Exodus discusses working six days and resting on the Sabbath. Physical things God *did* during the six-day Genesis chronicle include creating, forming, planting, growing and building. Non-physical things God *did* include inspecting, reviewing, describing, naming and blessing. The best translation for âsâh to describe the many things God accomplished in Genesis Chapter 1 is the general word "did."

In Exodus 20:10-11 quoted below, man is commanded to work six days and rest on the Sabbath, similar to God doing things for six days and resting on the seventh. Notice the shift in nomenclature from "work" in verse ten to "did" in verse eleven. In describing the creation account, *âsâh* is used to explain a broad range of activities equated with work. Also notice the expression "For in six days" in the KJV is changed to "For six days" in the *exemplar* translation of Exodus 20:11. The word "in" is clearly absent from the Hebrew text.

• Exodus 20:10 But the seventh day is the sabbath of the LORD thy God: in it thou shalt not do any work[04399], thou, nor thy son, nor thy daughter, thy manservant, nor thy maidservant, nor thy cattle, nor thy stranger that is within thy gates. (KJV)
• Exodus 20:11 For six days Yahweh did[06213] the sky[08064] and land[0776] the sea[03220] and all that is in them, and rested on the seventh day: wherefore Yahweh blessed the Sabbath day, and hallowed it. (ET)

In Exodus 31:15-17, the Hebrews were commanded to keep the Sabbath, and the creation account is again used as an example of resting from labor. Again notice the shift in nomenclature from "work" in verses fifteen and sixteen to "did" in verse seventeen. *Âsâh* is used to explain a broad range of activities equated with work. The expression "for in six days" in the KJV is changed to "for six days" in the *exemplar* translation of Exodus 31:17. The word "in" is clearly not a part of the Hebrew text.

• Exodus 31:15-16 Six days may work[04399] be done; but in the seventh is the sabbath of rest, holy to the LORD: whosoever doeth any work[04399] in the sabbath day, he shall surely be put to death. Wherefore the children of Israel shall keep the sabbath, to observe the sabbath throughout their generations, for a perpetual covenant. (KJV)

• Exodus 31:17 It is a sign between me and the children of Israel forever; for six days Yahweh did[06213] the heavens[08064] and Earth[0776], and on the seventh day He rested, and was refreshed. (ET)

Tôhûw and *Bôhûw* – Vacant and Empty

The two Hebrew words *tôhûw* (Strong's 08414) and *bôhûw* (Strong's 0922) are found in Genesis 1:2. Each is used once in the first two chapters of Genesis. These words are best translated "vacant" and "empty" to describe the complete absence of life on early Earth. Compare the KJV against the *exemplar* translation below.

• Genesis 1:2 And the earth was without form[08414] and void[0922] and darkness was upon the face of the deep. And the Spirit of God moved upon the face of the waters. (KJV)

• Genesis1:2 And the Earth was vacant[08414] and empty[0922] *of life*. And darkness *was* over the surface of the deep *waters*. And the Spirit of God hovered over the surface of the waters. (ET)

Of these two words, *bôhûw* is the least controversial. The Wilhelm Gesenius Lexicon gives "emptiness" and "voidness" as the only two translations of the noun *bôhûw*.[5] The Brown-Driver-Briggs Hebrew Lexicon gives "emptiness" as the only translation of the noun *bôhûw*.[6] The word *bôhûw* (Strong's 0922) occurs 3 times in the Old Testament in Genesis 1:2; Isaiah 34:11 and Jeremiah 4:23, and is translated "void" 2 times and "emptiness" 1 time in the KJV.[7] In Genesis 1:2, *bôhûw* is translated "empty" or "void" by virtually all English translations of the Bible.

Tôhûw is the more problematic of the two words. Most of the difficulty in understanding this word is due to the influence of theology. The Brown-Driver-Briggs Hebrew Lexicon gives the English word "emptiness" as an acceptable translation of the noun *tôhûw*.[8] Based on its usage in the Old Testament, the basic meaning of *tôhûw* is "empty" or "vacant."

The KJV and other Bibles translate *tôhûw* in an inconsistent manner. *Tôhûw* (Strong's 08414) occurs twenty times in nineteen verses in the Old Testament.[9] When referring to a place that is physically empty, *tôhûw* is often translated "wilderness, waste, empty place, barren, nothing, desert place, desolate or uninhabited." When *tôhûw* refers to moral or intellectual emptiness, it is often translated "vain, vanity, void, nothing or useless."

Different Bible versions are selected below to show that *tôhûw* has been properly translated in some instances. Notice how this word is rendered based on its basic meaning of *empty* or *vacant*. The expression "without form" will not fit into the majority of these verses.

- Deuteronomy 32:10 He found them in a desert land, in an empty[08414], howling wasteland. He surrounded them and watched over them; he guarded them as he would guard his own eyes. (NLT)
- 1 Samuel 12:21 And do not turn aside after empty[08414] things that cannot profit or deliver, for they are empty[08414]. (ESV)
- Job 6:18 Turn aside do the paths of their way, They ascend into emptiness[08414], and are lost. (Young's Literal Translation, 1898[10])
- Job 12:24 Turning aside the heart of the heads of the people of the land, And he causeth them to wander in vacancy[08414]—no way! (Young's Literal Translation)
- Job 26:7 He stretcheth out the north over the empty[08414] *place*, and hangeth the earth upon nothing. (KJV)
- Psalm 107:40 He is pouring contempt upon nobles, And causeth them to wander in vacancy[08414]—no way. (Young's Literal Translation) [The second half of this verse is identical to the second half of Job 12:24.]
- Isaiah 24:10 The ruined city lies desolate[08414]; the entrance to every house is barred. (NIV) [This is a perfect description of a defenseless town affected by war. Prior to the battle, the people lock their houses and flee the oncoming conflict. After the battle, the ruined city is empty of its inhabitants.]
- Isaiah 29:21 Who make a man an offender by a word, And lay a snare for him who reproves in the gate, And turn aside the just by empty[08414] *words*. (NKJV)
- Isaiah 34:11 But pelican and hedgehog will possess it, And owl and raven will dwell in it; And He will stretch over it the line of desolation[08414] And the plumb line of emptiness. (NASB) [This describes Edom as a destroyed, deserted, empty place, an object of God's judgment.]
- Isaiah 40:17 All the nations are as nothing before him, they are accounted by him as less than nothing and emptiness[08414]. (ESV)
- Isaiah 40:23 Who brings princes to nothing, and makes the rulers of the earth as emptiness[08414]. (ESV)
- Isaiah 41:29 Behold, all of them are false; Their works are worthless, Their molten images are wind and emptiness[08414]. (NASB)
- Isaiah 44:9 Framers of a graven image are all of them emptiness[08414], And their desirable things do not profit, And their own witnesses they are, They see not, nor know, that they may be ashamed. (Young's Literal Translation, 1898)
- Isaiah 45:18 For thus says the Lord, who created the heavens (he is God!), who formed the earth and made it (he established it; he did not create it empty[08414], he formed it to be inhabited!): "I am the Lord, and there is no

other." (ESV) [This verse points out that the Earth was made to be inhabited rather than remain empty.]

- Isaiah 45:19 I have not spoken in secret, in a dark place of the earth: I said not unto the seed of Jacob, Seek ye me in vain[08414]: I the LORD speak righteousness, I declare things that are right. (KJV)
- Isaiah 49:4 Then I said, I have laboured in vain[08414], I have spent my strength for nought, and in vain: *yet* surely my judgment *is* with the LORD, and my work with my God. (KJV)
- Isaiah 59:4 No one calls for justice, Nor does any plead for truth. They trust in empty[08414] words and speak lies; They conceive evil and bring forth iniquity. (NKJV)
- Jeremiah 4:23 *I* saw the land and *it* looked vacant[08414] and empty[0922] and to the heavens there was no light. (ET) [This is a prophecy from Jeremiah against the Hebrew people and Jerusalem because of their disobedience to God. A future time is described when the land would be empty owing to its people being carried off as prisoners of war. This verse has nothing to do with the creation or destruction of the Earth. All English Bibles mistranslate this verse due to the influence of traditional young-Earth creation theology since the expression *tôhûw* and *bôhûw* is the same as found in Genesis 1:2.]

When the Genesis creation story is reconnected with the remainder of the Old Testament, the meaning of *tôhûw* can be readily determined. From the above verses, *tôhûw* means "vacant or empty."

Tôhûw is translated "without form" two times in the KJV in Genesis 1:2 and Jeremiah 4:23, and is an incorrect translation in both instances. The word "form" is properly translated from the Hebrew root word *yâtsar* (Strong's 03335) which means to form or fashion as a potter forms clay. *Yâtsar* is used in Genesis 2:7, 2:8 and 2:19 to describe when God formed man and animals from the ground. There is no Hebrew equivalent of the English term "without form."

Use of the words *tôhûw* and *bôhûw* in Genesis 1:2 represent a style commonly found in the Old Testament. When Hebrew writers wanted to emphasize a point, they sometimes used two words with a similar meaning joined by a conjunction. Further emphasis was achieved through rhyming words as demonstrated by *tôhûw* and *bôhûw*. This style is often used in the English language. For example, a ghost town might be described as "desolate and deserted." A common expression found in Genesis 1:22 and elsewhere in the Old Testament is *râbâh* (Strong's 06509) and *pârâh* (Strong's 07235), translated be "fruitful and multiply." In 2 Chronicles 1:10, the words *chokmâh* (Strong's 02451) and *maddâe* (Strong's 04093) are used and translated "wisdom and knowledge." In 1 Chronicles 16:27, the words *hôwd* (Strong's 01935) and *hâdâr* (Strong's 01926) are used and translated "glory and honor." In Hebrew, the words in each of these expressions have a similar meaning and a poetic quality. When translated into English, a significant loss of impact occurs when the rhyming sound is destroyed.

In summary, *tôhûw* and *bôhûw* can be translated "vacant and empty" in Genesis 1:2 with full biblical support. The illegitimate translation of *tôhûw* as "without form" besmirches the integrity of biblical scholarship and is due to theological influence. This illicit translation is found in virtually all English Bibles. The Hebrew words *tôhûw* and *bôhûw* correctly describe the complete absence of life on Earth sometime after its formation.

Tehôwm – Deep

The Hebrew word *tehôwm* (Strong's 08415) occurs one time in the first two chapters of Genesis and is translated "deep." There is no need to complicate translation of this word based on theology. "Deep" followed by the explanatory "waters" is proposed as the best translation of *tehôwm* in Genesis 1:2. Compare the KJV against the *exemplar* translation below.

- Genesis 1:2 And the earth was without form and void, and darkness was upon the face of the deep[08415]. And the Spirit of God moved upon the face of the waters. (KJV)
- Genesis 1:2 And the Earth was vacant and empty *of life*. And darkness was over the surface of the deep[08415] *waters*. And the spirit of God hovered over the surface of the waters. (ET)

Brown-Driver-Briggs Hebrew Lexicon gives the noun *tehôwm* as meaning "deep of subterranean waters," "deep overwhelming sea," "abyss of sea," "deep of primeval ocean" and "deep depth of river."[11] The word *tehôwm* (Strong's 08415) occurs 36 times in the Old Testament and is translated "deep" 20 times, "depth" 15 times and "deep places" 1 time in the KJV.[12]

Several pertinent verses in the Bible contain *tehôwm* and show it is used in reference to deep water. English words translated from *tehôwm* are underlined and identified by their Strong's number in superscript as follows:

- Deuteronomy 8:7 For Jehovah thy God bringeth thee into a good land, a land of water-brooks, of springs, and of deep[08415] waters, that gush forth in the valleys and hills. (Darby, 1890[13])
- Job 38:30 The waters are hid as with a stone, and the face of the deep[08415] is frozen. (KJV) [The face of the deep can be frozen, which implies the deep is water.]
- Psalm 106:9 He rebuked the Red Sea, and it became dry, and he led them through the deep[08415] as through a desert. (ESV) [The bottom of the Red Sea is described as the deep.]

• Isaiah 51:10 Art thou not it which hath dried the sea, the waters of the great <u>deep</u>[08415]; that hath made the depths of the sea a way for the ransomed to pass over? (KJV) [Waters of the great deep were dried up by God so the Israelites could pass through the Red Sea.]
• Isaiah 63:13 That led them through the <u>deep</u>[08415], as an horse in the wilderness, that they should not stumble? (KJV) [This verse describes Moses when he divided the waters of the Red Sea and led the nation of Israel through the deep.]
• Ezekiel 26:19 For thus saith the Lord God; When I shall make thee a desolate city, like the cities that are not inhabited; when I shall bring up the <u>deep</u>[08415] upon thee and great waters shall cover thee. (KJV) [God is threatening to cover the city with deep water.]
• Jonah 2:5 The waters surrounded me, even to my soul; The <u>deep</u>[08415] closed around me; Weeds were wrapped around my head. (NKJV)

Use of *tehôwm* in the above verses refers to (1) a place where water came from, (2) a place covered by the Red Sea or (3) waters that covered Jonah. When the Genesis creation story is reconnected to the remainder of the Old Testament, the meaning of *tehôwm* becomes apparent.

By translating *tehôwm* as "deep" in Genesis 1:2, followed by the explanatory word "waters," clarity is added. Translating *tehôwm* only as "deep" leaves the reader wondering what is meant. Is Moses referring to the depths of a formless, dark, watery void? Is he describing a black, chaotic, fluidic mass? Such translations give theologians reason to wax eloquent to the total confusion of all who listen.

In ancient times, man did not have the same concept of the oceans that we have today. Humans were not capable of diving to the bottom of the deep seas or oceans to explore them. Deep waters were an unknown quantity, and a mystery described simply as the "deep." This is contrasted by shallow streams, rivers and ponds which could be investigated and held no mystery.

There are many mythological and pagan religious beliefs about the deep that have been found in ancient Babylonian and Ugaritic writings.[14] Although the word "deep" occurs in the Genesis creation story, it is not used in reference to these beliefs. Pagan mythology was kept out of the Bible and is not the basis for the creation account. A correct understanding of the word "deep" in Genesis 1:2 is based on its usage elsewhere in the Old Testament.

Shâmayim – Heaven and Sky

The Hebrew word *shâmayim* is translated either "heaven" or "air" by the KJV in the first two chapters of Genesis. In comparison, the New International Version translates *shâmayim* as "sky" six times in Genesis Chapters 1 and 2.

The word *shâmayim* (Strong's 08064) occurs 420 times in the Old Testament and is translated "heaven" or "heavens" 398 times, "air" 21 times and "astrologers" 1 time in the KJV.[15] The King James Version never translates *shâmayim* as sky. The English Standard Version translates *shâmayim* as sky two times while the New International Version translates it as sky approximately forty-five times.[16]

As used in the Old Testament, "heaven" has three meanings. Heaven can mean the sky where birds fly, from which rain falls and where clouds exist. Secondly, heaven can mean the cosmic heavens where the Sun, Moon, planets and stars are seen. Thirdly, heaven is God's abode.

The following two verses refer to heaven as a place where the birds fly and from which the rain falls.

- Jeremiah 15:3 And I will appoint over them four kinds, saith the Lord: the sword to slay, and the dogs to tear, and the fowls of the <u>heaven</u>[08064], and the beasts of the earth, to devour and destroy. (KJV)
- Deuteronomy 11:11 But the land, whither ye go to possess it, is a land of hills and valleys, and drinketh water of the rain of <u>heaven</u>[08064]. (KJV)

The following two verses refer to heaven as a place where the Sun, Moon and stars exist, visible from the surface of the Earth.

- Genesis 22:17 That in blessing I will bless thee, and in multiplying I will multiply thy seed as the stars of the <u>heaven</u>[08064], and as the sand which is upon the sea shore; and thy seed shall possess the gate of his enemies. (KJV)
- Deuteronomy 4:19 And lest thou lift up thine eyes unto <u>heaven</u>[08064], and when thou seest the sun, and the moon, and the stars, even all the host of <u>heaven</u>[08064], shouldest be driven to worship them. (KJV)

The following two verses refer to heaven as the abode of God. Heaven is a place above the Earth where God has His throne and looks down on man.

- Psalm 11:4 The Lord is in his holy temple, the Lord's throne is in <u>heaven</u>[08064] his eyes behold, his eyelids try, the children of men. (KJV)
- Psalm 53:2 God looked down from <u>heaven</u>[08064] upon the children of men, to see if there were any that did understand, that did seek God. (KJV)

The context of a verse determines whether *shâmayim* should be translated "heaven" or "sky." If the context does not imply a specific meaning, then translation becomes a judgment call.

When ancient man looked up at the sky, he saw a continuum from the surface of the Earth to distant galaxies—identical to what we see today. However, because he did not understand the scientific concept of atmosphere, barometric pressure, vacuum, gravity, the solar system, stars and galaxies, his perception was radically different from ours. Our current scientific knowledge can be used to more precisely translate *shâmayim* as either sky or heaven.

Erets – Land and Earth

The Hebrew noun *erets* is translated primarily as "land" and "earth" in the King James Version. *Erets* (Strong's 0776) occurs 2,504 times in the Old Testament and is translated "land" 1,543 times, "earth" 712 times, "country" 140 times, "ground" 98 times and "world" 4 times, in addition to other miscellaneous words in the KJV.[17]

Based on early man's knowledge of the Earth, *erets* should normally be translated "land." Translation of *erets* as "Earth" should rarely be done and only when referring to the entire planet from our current perspective. Ancient man (prior to 1000 BC) did not understand the concept of Earth as a planet. Adam did not know the Earth was one of eight planets in our solar system. Noah did not know the Earth was a giant sphere orbiting the Sun and in turn orbited by the Moon. Abraham did not know the Earth was a huge ball 12,700 km (7,900 miles) in diameter with 70 percent of its surface area covered by water. Jacob did not know the Earth had seven continents surrounded by four oceans. Moses did not know the Earth was a huge rotating globe with an equator and a spin axis passing through the north and south poles.

To ancient man, *erets* was simply the land upon which he lived and traveled. *Erets* consisted of the forests, fields, deserts and mountains where man journeyed. *Erets* was where the plants grew, a place animals roamed and over which the birds flew. *Erets* was all that man could see from the highest mountain. Man could walk on *erets* to the edge of the sea where it confined the waters of the deep.

A few verses in the Bible firmly show ancient man's concept of *erets* was equivalent to our modern concept of land as follows:

- Genesis 11:28 And Haran died before his father Terah in the land[0776] of his nativity, in Ur of the Chaldees. (KJV)
- Genesis 12:1 Now the LORD had said unto Abram, Get thee out of thy country[0776], and from thy kindred, and from thy father's house, unto a land[0776] that I will shew thee (KJV) [*Erets* is translated both "country" and "land" in this verse. Abram was moving from one land to another land.]
- Genesis 12:10 And there was a famine in the land[0776] and Abram went down into Egypt to sojourn there; for the famine was grievous in the land[0776]. (KJV) [There was a famine in the land of Canaan, not the whole Earth.]
- Genesis 13:12 Abram dwelled in the land[0776] of Canaan, and Lot dwelled in the cities of the plain, and pitched his tent toward Sodom. (KJV) [It is obvious *erets* means land rather than Earth.]
- Exodus 10:5 And they shall cover the face of the earth[0776], that one cannot be

able to see the earth[0776]. (KJV) [This verse is talking about locusts covering the land during the plagues in Egypt and is obviously not referring to the entire Earth.]

The Hebrew word *erets* should be translated "land" in the vast majority of occurrences in the Old Testament. When *erets* is referring to land, ground, territory, terrain or country, it should not be translated "earth" or "Earth." It should only be translated "Earth" when referring to the entire planet.

At the other extreme, to translate all occurrences of *erets* in the Old Testament as "land" would be absolutely wrong. Some verses of the Bible definitely apply to the entire planet. The actions of God should not be constrained because of inadequate Hebrew nomenclature and imperfect human knowledge during Moses' day. The translation of *erets* only as "land" in the first eleven chapters of Genesis seriously distorts the creation account.[18] We can use current knowledge to more specifically translate *erets* as either land or Earth depending on context.

Qâvâh – Wait

The Hebrew word *qâvâh* (Strong's 06960) is of significant importance in properly understanding the six-day Genesis chronicle. The Brown-Driver-Briggs Hebrew Lexicon gives the English word "wait" as the primary translation of the verb *qâvâh.*[19] The Wilhelm Gesenius Lexicon gives "to await" as an acceptable translation.[20]

In Genesis 1:9, *qâvâh* is translated "let be gathered together" in the KJV, but should be translated "wait." This word has been mistranslated for more than twenty-two hundred years due to the influence of traditional young-Earth creation theology. Compare the KJV against the *exemplar* translation below.

• Genesis 1:9 And God said, Let the waters under the heaven be gathered together[06960] unto one place, and let the dry land appear: and it was so. (KJV)
• Genesis 1:9 And God said, "The waters under the sky wait[06960] in one place, and the dry ground *is* seen." And *it* was so. (ET)

Jeremiah 3:17 is the second occurrence in the Bible where *qâvâh* is mistranslated "shall be gathered," rather than correctly translated "wait." This verse is one of many that teach waiting on Yahweh, and its mistranslation is without excuse. Compare the KJV against the *exemplar* translation below.

• Jeremiah 3:17 At that time they shall call Jerusalem the throne of the LORD; and all the nations shall be gathered[06960] unto it, to the name of the LORD, to Jerusalem: neither shall they walk any more after the imagination of their evil heart. (KJV)

• Jeremiah 3:17 At that time they shall call Jerusalem the throne of Yahweh. And all nations will <u>wait</u>[06960] to name Yahweh in Jerusalem and go no longer after the stubbornness of their evil heart. (ET)

Forty-nine words are identified by Strong's number 06960 and listed in Table 5.1. *Qâvâh* is translated "wait or waited" 31 times, "look, looked or looketh" 14 times, "gathered" 2 times, "patiently" 1 time and "tarrieth" 1 time in the KJV.[21] *Qâvâh* is consistently translated "wait, look, patiently and tarrieth" in 47 out of 49 occurrences or 96% of the time in the KJV. In only two verses, Genesis 1:9 and Jeremiah 3:17, *qâvâh* is translated "let be gathered together" and "shall be gathered" respectively. The translation in both of these verses is illegitimate as the following will show.

Table 5.1 – Occurrences of the Hebrew Verb *Qâvâh* in the Old Testament
Strong's 06960, based on the root (קוה)

Strong's No. with Suffix	Complete Hebrew Term	Basic Hebrew Word	KJV Translation	Verse Location
06960A	יקוו	קוו	let be gathered together	Genesis 1:9
06960A	יקוו	קוו	wait	Isaiah 51:5, 60:9
06960A	קוו	קוו	wait	Psalm 56:6, 119:95
06960A	לקוו	קוו	wait	Lamentations 3:25
06960A	ונקוו	קוו	shall be gathered	Jeremiah 3:17
06960A	קוו	קוו	waited	Job 6:19
06960B		קוה	wait, look, patiently, tarrieth	Job 7:2, 17:13; Psalm 27:14, 37:34, 40:1, 52:9, 69:20; Prov. 20:22; Isaiah 59:9, 59:11, 64:3; Jer. 8:15, 14:19, 14:22; Hosea 12:6; Micah 5:7
06960C		קוי	wait, look	Gen. 49:18; Job 30:26; Psalm 25:3, 25:5, 25:21, 37:9, 39:7, 40:1, 69:6, 130:5; Isaiah 5:4, 8:17, 25:9, 26:8, 33:2, 40:31, 49:23; Jer. 13:16; Lam. 2:16
06960D		קו	look, wait	Job 3:9; Ps. 130:5; Isaiah 5:2, 5:7

Note: The Hebrew root as identified by *The Lexham Hebrew-English Interlinear Bible, Strong's Exhaustive Concordance* and Brown-Driver-Briggs Hebrew Lexicon.
References:
(1) Eliyah.com (2006). Online Strong's Concordance.
(2) van der Merwe, Christo, ed. (2004) *The Lexham Hebrew-English Interlinear Bible.*

In the Old Testament, there are several Hebrew words translated "gather." The five most common words translated "gather," arranged in descending order are:

• *Âcaph* (Strong's 0622) – This Hebrew word occurs 200 times and is translated "gather" 86 times, "together" 51 times and "assemble" 15 times, in addition to other miscellaneous words in the KJV.[22]

• *Qâbats* (Strong's 06908) – This Hebrew word occurs 127 times and is translated "gather" 70 times, "gather together" 42 times, "assemble" 6 times and "gather up" 3 times, in addition to other miscellaneous words in the KJV.[23]

• *Lâqat* (Strong's 03950) – This Hebrew word occurs 37 times and is translated "gather" 23 times, "glean" 12 times and "gather up" 2 times in the KJV.[24]

• *Kânac* (Strong's 03664) – This Hebrew word occurs 11 times and is translated "gather" 5 times, "gather together" 4 times, "heap" 1 time and "wrap" 1 time in the KJV.[25]

• *Qâshash* (Strong's 07197) – This Hebrew word occurs 8 times and is translated "gather" 6 times and "gather together" 2 times in the KJV.[26]

Several scriptures in the Old Testament clearly describe a gathering of people in Israel and Jerusalem. The following are a collection of prophetic verses which describe a gathering. Notice these verses do not use *qâvâh*. The Hebrew words *âcaph* (Strong's 0622) and *qâbats* (Strong's 06908) are used exclusively.

• Isaiah 43:9 All the nations gather together[06908] and the peoples assemble[0622]. Who among them can declare this, and show us the former things? Let them bring their witnesses to prove them right, and let them hear and say, It is true. (ESV)

• Ezekiel 11:17 Therefore say, Thus saith the Lord GOD; I will even gather[06908] you from the people, and assemble[0622] you out of the countries where ye have been scattered, and I will give you the land of Israel. (KJV)

• Ezekiel 20:34 And I will bring you out from the people, and will gather[06908] you out of the countries wherein ye are scattered, with a mighty hand, and with a stretched out arm, and with fury poured out. (KJV)

• Ezekiel 20:41 I will accept you with your sweet savour, when I bring you out from the people, and gather[06908] you out of the countries wherein ye have been scattered; and I will be sanctified in you before the heathen. (KJV)

• Ezekiel 22:19 Therefore thus saith the Lord GOD; Because ye are all become dross, behold, therefore I will gather[06908] you into the midst of Jerusalem. (KJV)

• Ezekiel 36:24 For I will take you from among the heathen, and gather[06908] you out of all countries, and will bring you into your own land. (KJV)

• Micah 4:11 Now also many nations are gathered[0622] against thee, that say, Let her be defiled, and let our eye look upon Zion. (KJV)

• Micah 2:12 I will surely[0622] gather[0622] all of you, O Jacob; I will surely bring together the remnant of Israel. I will bring them together like sheep in a pen, like a flock in its pasture; the place will throng with people. (NIV) [The Hebrew text repeats the word "gather" for emphasis and actually says, "I will gather gather all of you."]

• Zechariah 12:3 On that day I will make Jerusalem a heavy stone for all the peoples. All who lift it will surely hurt themselves. And all the nations of the earth will gather[0622] against it. (ESV)
• Zechariah 14:2 For I will gather[0622] all nations against Jerusalem to battle; and the city shall be taken, and the houses rifled, and the women ravished; and half of the city shall go forth into captivity, and the residue of the people shall not be cut off from the city. (KJV)
• Zechariah 14:14 And Judah also shall fight at Jerusalem; and the wealth of all the heathen round about shall be gathered together[0622], gold, and silver, and apparel, in great abundance. (KJV)

Just as the idea of a *gathering* is taught in the Old Testament, the idea of *waiting* is also taught. Seventeen scriptures strongly promote the idea of waiting on Yahweh (commonly translated LORD or Jehovah). The following, which includes Jeremiah 3:17, are a sample of verses which teach this concept. The Hebrew word *qâvâh* (Strong's 06960) is used exclusively to indicate waiting.

• Psalm 25:5 Lead me in thy truth, and teach me: for thou art the God of my salvation; on thee do I wait[06960] all the day. (KJV)
• Psalm 27:14 Wait[06960] on the LORD, be of good courage, and he shall strengthen thine heart, wait[06960] I say, on the LORD. (KJV)
• Psalm 52:9 I will praise thee for ever, because thou hast done it: and I will wait[06960] on thy name; for it is good before thy saints. (KJV)
• Psalm 69:6 Let not them that wait[06960] on thee, O Lord GOD of hosts, be ashamed for my sake; let not those that seek thee be confounded for my sake, O God of Israel. (KJV)
• Proverbs 20:22 Say not thou, I will recompense evil; but wait[06960] on the LORD, and he shall save thee. (KJV)
• Isaiah 8:17 And I will wait[06960] upon the LORD, that hideth his face from the house of Jacob, and I will look for him. (KJV)
• Isaiah 25:9 And it shall be said in that day, Lo, this is our God; we have waited[06960] for him, and he will save us: this is the LORD; we have waited[06960] for him, we will be glad and rejoice in his salvation. (KJV)
• Isaiah 33:2 O LORD, be gracious unto us; we have waited[06960] for thee: be thou their arm every morning, our salvation also in the time of trouble. (KJV)
• Isaiah 40:31 But they that wait[06960] upon the LORD shall renew their strength; they shall mount up with wings as eagles; they shall run, and not be weary; and they shall walk, and not faint. (KJV)
• Jeremiah 3:17 At that time they shall call Jerusalem the throne of Yahweh. And all nations will wait[06960] to name Yahweh in Jerusalem and go no longer after the stubbornness of their evil heart. (ET) [All English Bibles incorrectly translate "wait" as "gather" in this verse.]

• Lamentations 3:25 The LORD is good unto them that wait[06960] for him, to the soul that seeketh him. (KJV)
• Hosea 12:6 Therefore turn thou to thy God: keep mercy and judgment and wait[06960] on thy God continually. (KJV)

Some Hebrew verbs clearly describe inactive behavior while others describe physical action. *Qâvâh* is translated "wait, look, patient and tarry" in the KJV and denotes inactive behavior. In contrast, Hebrew verbs such as *âcaph* and *qâbats,* translated "gather" in the KJV, denote physical action to collect things and describe gathering food (Gen. 41:48), gathering fruit (Exodus 23:10), gathering quails (Num. 11:32), gathering ashes (Num. 19:9), gathering corn (Deut. 11:14), gathering the spoil of battle (Deut. 13:16) and gathering grain (Ruth 2:7).

An examination of the Hebrew text gives additional proof for translating *qâvâh* in Genesis 1:9 as "wait." Forty-nine occurrences in 45 verses identified by Strong's number 06960 are translated from the Hebrew root (קוה). Word variations and their spelling based on this root are identified in Table 5.1.

Eight verses, including Genesis 1:9 and Jeremiah 3:17, contain identical spelling for *qâvâh* and are identified as Strong's 06960A in Table 5.1. Six of these eight verses translate *qâvâh* as "wait." These six verses give compelling support for translating Genesis 1:9 and Jeremiah 3:17 as "wait" rather than "gather" as shown below.

• Job 6:19 The troops of Tema looked, the companies of Sheba waited[06960A] for them. (KJV)
• Isaiah 51:5 My righteousness is near; my salvation is gone forth, and mine arms shall judge the people; the isles shall wait[06960A] upon me, and on mine arm shall they trust. (KJV)
• Isaiah 60:9 Surely the isles shall wait[06960A] for me, and the ships of Tarshish first, to bring thy sons from far, their silver and their gold with them, unto the name of the LORD thy God, and to the Holy One of Israel, because he hath glorified thee. (KJV) [The isles are an inanimate object and can only wait. They cannot expect, look, trust or hope.]
• Psalm 56:6 They gather themselves together, they hide themselves, they mark my steps, when they wait[06960A] for my soul. (KJV)
• Psalm 119:95 The wicked have waited[06960A] for me to destroy me: but I will consider thy testimonies. (KJV)
• Lamentations 3:25 The LORD is good unto them that wait[06960A] for him, to the soul that seeketh him. (KJV)

In respect to the above eight verses, the influence of the Septuagint can be clearly seen. In Isaiah 51:5, 60:9; Psalm 56:6, 119:95 and Lamentations 3:25, *qâvâh* is translated in the Septuagint[27] as "wait" (identical to the KJV). In Genesis 1:9 and Jeremiah 3:17, the Septuagint translates *qâvâh* as "gathered" (identical to the KJV).

In Job 6:19, *qâvâh* is not translated by the Septuagint (it was left out) while translated "wait" in the KJV.

In summary, several verses in the Old Testament teach gathering and waiting. In reference to a gathering in Israel, Jerusalem or elsewhere, all verses use the Hebrew words *âcaph* (Strong's 0622) or *qâbats* (Strong's 06908). Many verses in the Old Testament also teach the idea of waiting on Yahweh and use the Hebrew word *qâvâh* (Strong's 06960). *Qâvâh* should be translated "wait" in Jeremiah 3:17.

The incorrect translation of *qâvâh* represents the pervasive influence of traditional young-Earth creation theology. Mistranslations of "gather" in Genesis 1:9 and Jeremiah 3:17 represent a fabrication without biblical support. This illegitimate translation is found in all English Bibles. Regardless of the good intentions of those involved, this rendering is a stain on the integrity of biblical scholarship.

The correct translation of *qâvâh* as "wait" in Genesis 1:9 is overwhelmingly supported by its usage elsewhere in scripture. When the Genesis creation story is reconnected to the remainder of the Old Testament, the true meaning of *qâvâh* becomes apparent. On the third day, God observed the waters under the sky waiting in one place. He was describing a passive scene. There was no "gathering together" of the waters on that day.

Miqveh **– Collection**

The best translation for the Hebrew word *miqveh* (Strong's 04723) in Genesis 1:10 is "collection." Unfortunately, *miqveh* is translated "gathering together" in the KJV to support the belief that God gathered the waters together on the third day. Compare the KJV against four other translations where *miqveh* is underlined.

- Genesis 1:10 And God called the dry land Earth; and the gathering together[04723] *of the* waters called he Seas: and God saw that it was good. (KJV)
- Genesis 1:10 And God calleth to the dry land "Earth," and to the collection[04723] of the waters He hath called "Seas;" and God seeth that it is good. (Young's Literal Translation, 1898)
- Genesis 1:10 And God called the dry land Earth, and the collection[04723] of waters he called Seas: and God saw that it was good. (Webster's Translation, 1833[28])
- Genesis 1:10 And the Lord called the dry land Earth, and the place of the collection of waters He called Sea. And the Lord saw that it was good. (Targum of Onkelos, English translation by J. W. Etheridge, 1862)
- Genesis 1:10 And God called the dry ground "land" and called the collection[04723] of waters "seas." And God saw that *it was* good. (ET)

The Brown-Driver-Briggs Hebrew Lexicon gives the English word "collection" as the only translation of the noun *miqveh.*[29] The word *miqveh* (Strong's 04723 and 04724) occurs 13 times in the Old Testament and is translated "linen yarn" 4 times, "hope" 4 times, "gathering together" 1 time, "pool" 1 time, "ditch" 1 time, "plenty" 1 time and "abiding" 1 time in the KJV.[30] The inconsistent translation of *miqveh* in the KJV and other translations obscure its meaning.

Table 5.2 – Occurrences of the Hebrew Verb *Miqveh* in the Old Testament
(Strong's 04723 and 04724)

Strong's Number	Complete Hebrew Term	Basic Translation	Basic Hebrew Word	No.	Verse Location
04723	ולמקוה	and to collection	מקוה	1	Genesis 1:10
04723	מקוה	collection	מקוה	1	Exodus 7:19
04723	מקוה	collection	מקוה	1	Leviticus 11:36
04723	ומקוה	and collection	מקוה	2	1 Kings 10:28
04723	מקוה:	collection	מקוה	1	1 Chronicles 29:15
04723	ומקוא	and collection	מקוא	2	2 Chronicles 1:16
04723	מקוה	collection	מקוה	1	Ezra 10:2
04724	מקוה	collection	מקוה	1	Isaiah 22:11
04723	מקוה	collection	מקוה	3	Jeremiah 14:8, 17:13, 50:7

References:
(1) Eliyah.com (2006). Online Strong's Concordance.
(2) van der Merwe, Christo, ed. (2004) *The Lexham Hebrew-English Interlinear Bible.*

In most verses where *miqveh* occurs, it can be translated "collection" for improved clarity and consistency. Keep in mind that *miqveh* can refer to a collection of water (cistern, spring, pool or reservoir), a collection of people (the nation Israel), a collection of belief (consensus of opinion), a collection of goods (herd of horses) and a collection of money (an offering).

Table 5.2 and the following ten verses show how *miqveh* can be translated "collection" in the Old Testament. *Miqveh* is identified by the underlined words and their accompanying Strong's number in superscript.

• Exodus 7:19 Yahweh said to Moses, "Say to Aaron, Take your staff and stretch out your hand over the waters of Egypt, over their streams, over their rivers, over their ponds and over all <u>collections</u>[04723] of their water. And be blood and be blood in all the land and in all the wood *vessels* and all the stone *vessels.*" (ET) [The expression "and be blood" is repeated for emphasis in the original Hebrew. All collections of water in Egypt would turn to blood.]

• Leviticus 11:36 Only—a fountain or pit, a collection[04723] of water, is clean, but that which is coming against their carcass is unclean. (Young's Literal Translation, 1898) [A fountain (spring) and pit (cistern) represent a collection of water.]
• 1 Kings 10:28 And Solomon brought out horses from Egypt. And from a collection[04723] *of horses,* the king's traders took from the collection[04723] at a price. (ET) [As described in this verse, a collection of horses were brought out of Egypt and Solomon's traders took horses from that collection for the king's use. His traders probably received first pick from any herd of horses imported into Israel. Solomon paid for the horses that he took at an agreed upon price. Today in businesses where the quality of a product naturally varies, such as gemstones, vegetable produce, livestock or lumber, preferred customers are often given first pick of new merchandise.]
• 1 Chronicles 29:15, 16 For we are strangers before thee, and sojourners, as were all our fathers: our days on the land are as a shadow and without a collection[04723] *for you.* Yahweh our God, as for all this abundance that we have provided for building you a temple for your Holy Name, it comes from your hand, and all of it belongs to you. (ET) [Verses 29:7-10 in this chapter discuss the collection of money and materials given by David and the people of Israel for the building of the temple by Solomon, while 29:11-19 is a prayer about this offering.]
• 2 Chronicles 1:16 And Solomon brought out horses from Egypt. And from a collection[04723] *of horses,* the king's traders took from the collection[04723] at a price. (ET) [This verse is essentially the same as 1 Kings 10:28.]
• Ezra 10:2 And Shechaniah the son of Jehiel, one of the sons of Elam, answered and said unto Ezra, We have trespassed against our God, and dwell with strange wives of the people of the land. And now there is a collection[04723] *of belief* in Israel concerning this thing. (ET) [Collection is used in the sense of a *consensus of belief* about the strange wives they had married and the course of action to take. The strange wives were subsequently sent away.]
• Isaiah 22:11 You made a collection[04724] between the two walls for the water of the old pool. But you did not look to him who did it, or see him who planned it long ago. (ET) [Collection is translated "ditch" in the KJV and "reservoir" in the NKJV, NIV, ESV, RSV, ASV and NASB in this verse.]
• Jeremiah 14:8 O collection[04723] of Israel, He saves in time of trouble. Why be as a stranger in the land and like a wanderer spread out to spend night? (ET) [The expression "congregation of Israel" or "congregation of the children of Israel" appears more than 39 times in the Old Testament. In this instance, Jeremiah uses the expression "collection of Israel."]
• Jeremiah 17:13 O collection[04723] of Israel, all who forsake Yahweh are ashamed, disloyal and departed. May it be written in the land, they have forsaken Yahweh the fountain of living water. (ET) [The expression "collection of Israel" is used instead of "congregation of Israel."]

• Jeremiah 50:7 All *who* found them devoured them. And their enemies said, "We are not guilty *for destroying them*, for they have sinned against Yahweh. Righteous habitation and <u>collector</u>[04723] is their father Yahweh." (ET) [This is a poetical verse. God is the collector of the Hebrew people, providing a righteous abode for them. They would otherwise be as scattered sheep.]

In summary, all uses of the word *miqveh* relating to water refer to a "collection of water" rather than a "gathering together of water." Reconnecting the Genesis creation story to the rest of the Old Testament helps clarify this word. On the third day, God called the collection of waters "seas."

Hâyâh **– Be**

Hebrew verbs based on the root *hâyâh* (היה) occur several times in Genesis Chapters 1 and 2. The Wilhelm Gesenius Lexicon gives "to be" and "to exist" as the first meanings of the root *hâyâh.*[31] The Brown-Driver-Briggs Hebrew Lexicon gives "be" and "exist" as two acceptable translations of *hâyâh.*[32]

Verbs in the English language can be categorized as either dynamic or stative. *The Oxford Dictionary of English Grammar* points out that a dynamic verb implies action or change.[33] Stative verbs express a state or condition.[34] According to the *American Heritage College Dictionary,* stative English verbs include the words "be, is, was, were, been, am and being."[35] In addition, "are" is a tense of be.[36] The word "exist" is defined as, "To have actual being, be real,"[37] and is a stative verb. The root *hâyâh* (היה) and its derivatives are translated into English primarily as stative verbs rather than dynamic verbs.

Hâyâh (Strong's 01961) and its many derivatives occur approximately 3,482 times in the Old Testament.[38] Table B.1 in Appendix B lists forty-eight derivative verbs based on the Hebrew root (היה) and the number of times each occurs in the Bible. In the King James Version, *hâyâh* and its derivatives are translated roughly 70% of the time as "be, is, was or were." In the remaining occurrences they are translated "came" such as in "came to pass" 501 times, "come" such as in "come to pass" 125 times, "been" 75 times, "had" 68 times, "become" 66 times, "became" 63 times, "have" 62 times and "continue" 10 times.[39]

The derivatives (יהי), (ויהי), (והיו), (והיה), (יהיה), (היות) and (היתה) from the Hebrew root *hâyâh* (היה) occur in several verses in the first two chapters of Genesis. (To simplify discussion of these derivatives, they will be referred to as *hâyâh* followed by their Hebrew spelling.)

A word search using a Strong's online concordance shows that *hâyâh* (Strong's 01961) occurs only four times in the first two chapters of Genesis. In contrast, a search of the Hebrew text shows *hâyâh* occurs 33 times in 23 separate verses.[40] Some of the more significant verses in the first two chapters of Genesis containing *hâyâh* are given below. Words in the *exemplar* translation have been replaced with "be" to illustrate the nature of this stative Hebrew verb.

- Genesis 1:2 And the Earth be[01961] vacant and empty of life.
- Genesis 1:3 And God said, "be[01961] light and light be[01961]."
- Genesis 1:5 And evening be[01961] and morning be[01961]—day one.
- Genesis 1:6 And God said, "Be[01961] an expanse in the middle of the water, and it be[01961] to separate between water *above* from water *below*."
- Genesis 1:8 And evening be[01961] and morning be[01961]—second day.
- Genesis 1:14 And God said, "Be[01961] luminaries in the expanse of the heavens to distinguish between the day and between the night, and *they* be[01961] a sign *for* appointed times and days and years."
- Genesis 1:15 And luminaries be[01961] in the expanse of the heavens.
- Genesis 1:29 As food *they* be[01961] for you.
- Genesis 1:31 And evening be[01961] and morning be[01961]—the sixth day.
- Genesis 2:5 And each bush of the field be[01961] not yet on the land.
- Genesis 2:7 And the man be[01961] as a living creature.
- Genesis 2:18 God said, "*It is* not good that the man be[01961] alone."
- Genesis 2:24 And they *shall* be[01961] as one flesh.
- Genesis 2:25 And the two be[01961] naked, the man and his wife.

What is the most important single word in the Genesis creation story? Some people will identify the words "create, make, or beginning" as the most important. Others mention "day" or perhaps "evening and morning." In truth, the Hebrew verb *hâyâh* is probably the most significant word in the creation account and is one of the keys to properly understanding this historical event. To grasp the importance of *hâyâh*, we need to know some of is history and usage.

Gap theology proposes that between Genesis 1:1 and 1:2 there was an enormous period of time. Extensive time is allowed for geological change and the fossil record to occur. As mentioned in Scofield's notes on Genesis, at the end of this time the face of the Earth underwent a cataclysmic change as the result of divine judgment.[41] God destroyed the surface of the Earth and then began reforming it in Genesis 1:3. Gap theology came into prominence in the early 1800s when it was promoted by Thomas Chalmers in an attempt to harmonize the Bible and a long age for the Earth.[42]

To make scripture support gap theology, a different rendering for Genesis 1:2 was proposed. The new translation said, "But the earth *had become* a ruin and a desolation,"[43] in contrast to the KJV which says, "And the earth *was* without form and void."

Arthur C. Custance wrote *Without Form and Void* in support of gap theology and the new translation required. His entire book focuses on promoting the translation of *hâyâh* (היתה) in Genesis 1:2 as "had become" rather than "was." In his

English translation, Custance replaced a stative verb with one that appeared to show action. The acceptance of gap theology is based on a desperate attempt to harmonize scripture with science, rather than sound biblical hermeneutics and exegesis. Most people knowledgeable of both Hebrew and science reject this translation.

Hâyâh (הָיְתָה), translated "was" in Genesis 1:2, is shown in Table 5.3. This specific Hebrew word occurs 116 times in the Old Testament and its occurrences are listed in Table B.2 of Appendix B. *Hâyâh* (הָיְתָה) is translated in the KJV using the stative verbs "was, had, is, been, be and are" in 90 out of 116 times it occurs. *Hâyâh* (הָיְתָה) is translated "became" in only four instances.

Table 5.3 – Verbs Based on the Hebrew Root *Hâyâh* in the Creation Story
(Strong's 01961)

Scripture	Complete Hebrew Term	Basic Translation	*Exemplar* Translation (the word "be" is underlined)
Genesis 1:2	הָיְתָה	be	was vacant and empty
Genesis 1:3	יְהִי	be	There is light
Genesis 1:3	וַיְהִי־אוֹר	and be light	and light exists
Genesis 1:5	וַיְהִי־עֶרֶב	and be evening	and there was evening
Genesis 1:5	וַיְהִי־בֹקֶר	and be morning	and there was morning
Genesis 1:6	יְהִי	be	There is an expanse in the middle
Genesis 1:6	וִיהִי	and be	And it exists to separate between
Genesis 1:7	וַיְהִי־כֵן	and be so	and *it* was so
Genesis 1:14	יְהִי	be	There are luminaries
Genesis 1:14	וְהָיוּ	and be	And *they* are a sign for seasons
Genesis 2:7	וַיְהִי	and be	And the man existed as a living

Reference: van der Merwe, Christo, ed. (2004) *The Lexham Hebrew-English Interlinear Bible.*

Derivative of *hâyâh* occur in key places in Genesis 1:3, 6 and 14 as shown in Table 5.3. The KJV and other versions translate these verses in a misleading manner to indicate action is occurring when it is not (similar to what gap theology proposes for Genesis 1:2). These three verses are traditionally explained as commands—God speaks and things are instantly created from nothing. The word "let" is added to indicate a command in the expression, "Let there be."

Hâyâh identified as (יְהִי) and (וִיהִי) occur in five key places in Genesis 1:3, 6 and 14 as follows:

- Genesis 1:3 And God said, Let there be light and there was light. (KJV)

• Genesis 1:6 And God said, Let there be a firmament in the midst of the waters and let it divide the waters from the waters. (KJV)
• Genesis 1:14 And God said, Let there be lights in the firmament of the heaven to divide the day from the night. (KJV)

Compare the KJV above with the *exemplar* translation below where the derivatives of *hâyâh* are properly translated as non-command, stative verbs. Notice these are descriptive statements and no action is occurring or indicated.

• Genesis 1:3 And God said, "There is light, and light exists." (ET)
• Genesis 1:6 And God said, "There is an expanse in the middle of the water, and *it* exists to separate between water *above* from water *below*." (ET)
• Genesis 1:14 And God said, "There are luminaries in the expanse of the heavens to distinguish between the day and between the night." (ET)

Genesis 1:3 describes the existence of light—nothing more. The literal Hebrew text without any added words is, "And say God, be light and be light." Nothing is being created, made or formed in this verse. Bible translators have incorrectly taken this simple statement and turned it into a command. Traditional translations treat the first *hâyâh* (יְהִי) as a command, while the second *hâyâh* (וַיְהִי) is stative. The word "let" is added to indicate a command.

Genesis 1:6 describes the existence of an expanse (firmament). In the Hebrew, the first occurrence of *hâyâh* (יְהִי) is literally translated, "And say God, be expanse in [the] middle [of] the water." This simple descriptive statement has been incorrectly turned into a command. Worse yet, the second occurrence of *hâyâh* (וִיהִי), translated "and let," implies an additional command. Traditional translations incorrectly treat these two stative verbs as command verbs where action is occurring.

Genesis 1:14 describes the existence of two luminaries in the heavens. A literal translation of the Hebrew text says, "And say God, be luminaries in [the] expanse [of] the heavens." God is only describing the existence of two luminaries—the Sun and Moon—not their creation. Translators have incorrectly translated *hâyâh* (יְהִי) as "let there be" and have turned this verse into a command.

Hâyâh (יְהִי) in the above three verses occurs 50 times in the Old Testament and its occurrences are listed in Table B.3 of Appendix B. This specific verb is translated as the stative verb "be" 18 times, in addition to "let be" 15 times, "let there be" 5 times and "let" 2 times. All of these occurrences can easily be translated as stative verbs.

The English words "exist, be, is, was and were" are always stative verbs. They never switch back and forth between stative and dynamic. If they are the only verbs in a sentence, the sentence will be stative. For example, in the English language it is difficult if not impossible to construct a sentence which uses "exist" in a dynamic sense. In like manner, it is difficult to believe *hâyâh* in Genesis 1:3, 1:6 and 1:14 is other than a stative verb.

In summary, *hâyâh* in Genesis 1:2 can be translated as the stative verb "was" in rejection of gap theology. Likewise, in Genesis 1:3, 1:6 and 1:14 *hâyâh* can be translated as a stative verb in rejection of traditional young-Earth creation theology. Reconnecting the Genesis account with the rest of the Old Testament helps in understanding and properly translating this verb.

It is interesting to note that all Hebrew lexicons show "exist" as a legitimate translation of *hâyâh*. However, in approximately 3,482 occurrences of *hâyâh* and its derivatives, it is not translated "exist" in a single instance in the King James Version and other popular English translations. The *exemplar* translation has several verses where *hâyâh* is appropriately translated "exist."

Hebrew Verbs

Hebrew verbs comprise one of the more complicated verbal systems in existence. They consist of a root verb plus seven stems. Each *verb stem* is developed from the same root. The seven verb stems for Hebrew strong verbs are qal, niphal, piel, pual, hiphil, hophal and hithpael.[44, 45] The spelling for the qal stem is identical to the root, from which the other six stems are derived.[46]

In addition to the above, each verb has several *verb states* (sometimes called tenses) which include perfect, imperfect, imperative, cohortative, jussive, infinitive and participle.[47] Also included in verb states are the masculine, feminine, singular, plural, first person, second person and third person. This permits building a verb table based on a matrix of verb stems and verb states. A single root can have more than 170 *verb forms* when all stems and states are considered.[48]

Most verb forms are constructed by adding a prefix or suffix to the root, though in some instances no change in spelling occurs. For example, standard verb tables for strong verbs show that an imperfect 3ms verb (third person masculine singular) and a jussive 3ms verb have the same spelling for all stems except the hiphil.[49] For verbs forms that are spelled the same, determining a specific verb form is based on sentence context and can be a judgment call.

The astute Bible student will notice that verb forms with the same spelling in a verb table have different vowel points and accent marks. For example, qal jussive, niphal jussive and piel jussive strong verbs are spelled the same, but have different vowel points and accent marks[50] (small dots, dashes and other symbols added above, below or within the letters). The primary purpose of these marks is to guide in the pronunciation of words when the Hebrew Bible is read aloud. As previously mentioned, vocalization marks were not a part of the original Hebrew text and were added by Masoretic scribes starting around 500 AD.

Hebrew verbs can be totally intimidating; however, help is available to aid to their study. The *Old Testament Parsing Guide* by Beall, Banks and Smith lists every single verb in the Old Testament and is a good starting point in the study of verb stems, verb states and verb forms. Several Hebrew grammar books provide additional help in learning about verbs. Understanding verb forms is absolutely essential for a correct interpretation of the Genesis creation story.

And God said

Ten key verbs in Genesis 1:3, 6, 9, 11, 14, 20 and 24 control interpretation of the Genesis creation story. These verbs are traditionally classified as jussive (meaning they are command verbs) and transform the seven verses in which they occur into commands. Each verse begins with the expression, "And God said." God speaks and allegedly commands things into existence. According to traditional interpretations, during six consecutive days God instantaneously created the heavens, Earth and all life therein. These ten "jussive/command" verbs drive the incorrect translation of the six-day Genesis chronicle.

As mentioned in Chapter 3, the Septuagint (a translation of the Hebrew Old Testament into Greek) is the basis for traditional thinking about the Genesis creation account. The first five books (the Pentateuch) were translated around 275 BC. Jewish scribes were familiar with the Hebrew and Greek languages of the day and ideally suited for this task; however, their poor scholarship and theological beliefs resulted in a less than accurate translation. They used this opportunity to embed their theology into the Old Testament for all time by translating ten key verbs and seven verses as commands. These translators helped establish an interpretive paradigm which has existed for more than twenty-two hundred years. The Septuagint translation of the Pentateuch is the basis for current incorrect views about the Genesis creation story.

The *exemplar* translation explains Genesis 1:3, 6, 9, 11, 14, 20 and 24 as descriptive statements rather than commands. Below are seven brief reasons why these verses are not commands.

(1) Weak scriptural support – Traditional theology that God spoke the heavens and Earth into existence is not directly supported by scripture. The best verses in support of this belief are Psalm 33:6 and Hebrews 11:3 (discussed in Chapter 4). No scripture in the Bible specifically mentions that God created the heavens and Earth *in six days*. No scripture specifically states that God *instantly* created the heavens and Earth.

(2) Incorrect verb state – The *Old Testament Parsing Guide* by Beall, Banks and Smith lists every Hebrew verb in the Old Testament. As shown in Table 5.4, the seven verses of Genesis 1:3, 6, 9, 11, 14, 20 and 24 contain ten key verbs that are assigned a jussive or jusm verb state.[51] Jussive verbs are considered "command verbs" and force these seven verses to be explained as commands. Appendix B discusses these ten Hebrew verbs in detail and gives justification for treating them as imperfect verbs (with no requirement for a command) based on their usage elsewhere in the Old Testament.

(3) No jussive verb spelling – Based on spelling, four of the verbs listed in Table 5.4 are not jussive verbs. The four verbs in Genesis 1:9 and 1:20 are listed as "jusm" rather than "jussive" by the *Old Testament Parsing Guide*. Jusm is explained to mean, "jussive in meaning, but no unique form to indicate jussive."[53] In other words, the spelling of these words (their unique form) does not show they are jussive, yet they have been assigned a jussive meaning.

Table 5.4 – Ten Key Verbs in Seven Verses of Genesis Chapter One
(Traditionally considered jussive/command verbs)

Verse	Hebrew Term[1]	Verb Root[2]	KJV Translation	Traditional Assigned Verb Form[2]	*Exemplar* Translation
Gen. 1:3	יְהִי	הָיָה	let there be	Qal jussive	there is
Gen. 1:6	יְהִי	הָיָה	let there be	Qal jussive	there is
Gen. 1:6	וִיהִי	הָיָה	and let	Qal jussive	and exists
Gen. 1:9	יִקָּווּ	קָוָה	let be gathered together	Niphal jusm	wait
Gen. 1:9	ותראה	ראה	and let appear	Niphal jusm	and is seen
Gen. 1:11	תדשא	דשא	let bring forth	Hiphil jussive	sprouting
Gen. 1:14	יְהִי	הָיָה	let there be	Qal jussive	there are
Gen. 1:20	ישרצו	שרץ	let bring forth	Qal jusm	swarm
Gen. 1:20	יעופף	עוף	that may fly	Polel jusm	flying
Gen. 1:24	תוצא	יצא	let bring forth	Hiphil jussive	brings forth

Note: According to the *Old Testament Parsing Guide*, when the verb form is given as "jusm" rather than "jussive," the spelling of the word does not indicate jussive. The four verbs of Genesis 1:9 and 1:20 are not jussive based on spelling.
References:
(1) van der Merwe, Christo, ed. (2004). *The Lexham Hebrew-English Interlinear Bible.*
(2) Beall, Todd S. and William A. Banks, Colin Smith (2000). *Old Testament Parsing Guide: Revised and Updated Edition.*

Treating these verbs as jussive/command verbs has forced the mistranslation of other Hebrew words in the text as previously discussed. In Genesis 1:9, God allegedly commands the waters to separate from dry ground, forcing the mistranslation of *qâvâh* as "let be gathered together" rather than "wait."

(4) No command required – The seven key verses of Genesis 1:3, 6, 9, 11, 14, 20 and 24 begin with the expression "And God said." Outside of Genesis Chapter 1, the same expression where God is speaking occurs sixteen times.[52] A command is issued in eight verses in Genesis 6:13, 17:9, 17:15, 17:19, 21:12, 35:1; Exodus 3:14 and Numbers 22:12. A statement is issued in the remaining eight verses of Genesis 9:8, 9:12, 9:17, 46:2; 1 Kings 3:5, 3:11; 2 Chronicles 1:11 and Jonah 4:9. These verses clearly show that a command does not necessarily follow the expression "And God said." It cannot be dogmatically required that Genesis 1:3, 6, 9, 11, 14, 20 and 24 be interpreted as commands.

(5) Use of stative verbs – In Genesis 1:3, 6 and 14, the use of *stative* verbs in the Hebrew text rather than *dynamic* verbs discredits the idea God is creating anything. Stative verbs are used to describe the way things "are, is, was, be or exist" rather than to show action or change. The stative verb *hâyâh* in these three verses can mean "be or exist" and is primarily translated "be, is, was or were" in the Old Testament.

When God commands, the appropriate dynamic verbs (rather than stative verbs) are used. For example, dynamic verbs are used in eight verses outside of Genesis Chapter 1 where the Hebrew expression, "And God said" is followed by a command. Dynamic verbs are used in commanding to "make thee an ark" (Gen. 6:13-14), "shall keep my covenant" (Gen. 17:9), "not call her name" (Gen. 17:15), "shall bear thee a son" (Gen. 17:19), "hearken to her voice" (Gen. 21:12), "arise and go" (Gen. 35:1), "thus shalt thou say" (Exodus 3:14) and "thou shalt not go" (Num. 22:12).

(6) Missing dynamic verbs – In the seven key verses of Genesis 1:3, 6, 9, 11, 14, 20 and 24, the expression, "And God said" is not used with the appropriate dynamic verbs to show God is causing action to take place. With the exception of Genesis 1:11, the Hebrew verbs translated "create, make, form, plant, grow and build" are not used anywhere in these seven verses—important dynamic verbs are clearly missing.

When the seven alleged "command" verses are compared with other action verses in the creation account, the difference is striking. Notice the pattern of Genesis 1:1 when God *created* the heavens and Earth, of 1:26 when God proposed to *make* man, of 1:27 when God *created* man, of 2:7 when God *formed* man, of 2:8 when God *planted* the garden, of 2:9 when God *grew* the trees of the garden, of 2:19 when God *formed* the animals of the garden and of 2:22 when God *built* Eve from Adam's rib. In these verses, the appropriate dynamic Hebrew verbs translated "create, make, form, plant, grow and build" are used to indicate past and present action. Also notice the expression, "And God said" is absent in all of these verses.

(7) Conflict with general revelation – The belief that Genesis 1:3, 6, 9, 11, 14, 20 and 24 were commands that resulted in creation of the universe during six days in 4004 BC is in irreconcilable conflict with physical evidence in regard to time. An incorrect explanation of the six-day Genesis chronicle causes God's special revelation to conflict with general revelation. The time frame forced upon Genesis by traditional young-Earth creation theology strongly conflicts with empirical science.

Conclusion

The eighteen key words (6 nouns and 12 verbs) discussed in this chapter and their use in the *exemplar* translation is well supported. Reconnecting these key words with the rest of the Bible significantly aids in their proper understanding. In the *exemplar* translation, there are no illegitimate word translations or metaphorical interpretations. There are no special definitions that apply only to the creation account. There are no forced interpretations unsupported by the remainder of the Old Testament.

Key words in the Genesis creation account control its meaning. After God created the Earth, it was vacant and empty of life rather than "without form." The "deep" is a clear reference to the depths of the seas and oceans. On day one, God did not create light but described its presence through use of the stative Hebrew verb *hâyâh.* On the second day, God did not create an expanse but described its existence by using the stative verb *hâyâh.* On the third day, the waters calmly waited in one place, and there was no "gathering together." On the fourth day, God did not create the Sun and Moon but described their existence through use of the stative *hâyâh.* The expression, "And God said" in Genesis 1:3, 6, 9, 11, 14, 20 and 24 is followed by descriptive statements of what God saw. The acceptance and application of this information are the basis of the *exemplar* translation.

Although not necessarily an exciting topic, a proper understanding of key Hebrew words is essential for a correct explanation of the Genesis creation story. A misinterpretation of key words and expressions has caused the current problem that has existed for more than twenty-two hundred years. When key words are placed in their scriptural context, their significance becomes apparent. (As key words are reintroduced in Chapter 6, the reader should refer to this material)

At this point, Hebrew scholars and serious Bible students should read Appendix B for additional discussion of the ten key verbs identified above. Due to the specialized academic nature of this material, it has been separated from the main text. Appendix B discusses in detail the ten "jussive/command" verbs identified in Table 5.4 which control interpretation of the six-day Genesis chronicle.

Lead me in your truth and teach me, for you are the God of my salvation; for you I wait all the day long. (Psalm 25:5) ESV

Chapter 6

The Genesis Creation Story

The first two chapters of Genesis are a fascinating study, as attested by innumerable books and commentaries. The Genesis creation story is a source of controversy, not only in the religious world but also the secular. Theologians, philosophers and scientists alike seek its meaning, yet a proper understanding seems elusive. Any interpretation acceptable to Bible believing theologians, philosophers and layman must treat the Old Testament as truthful, trustworthy and inspired of God. On the other hand, Bible believing scientists expect special revelation and general revelation to harmonize. *Exemplar* creation satisfies these fundamental requirements.

The development of the *exemplar* translation is explained in this chapter. The *Biblia Hebraica* (Hebrew Bible) presented in *The Lexham Hebrew-English Interlinear Bible* is the starting point. The Lexham translation is checked for accuracy on a word-for-word basis against the original Hebrew text. Any word translated from Hebrew into English must be justifiable and supported elsewhere in the Old Testament. In a few instances, words are changed when the Lexham translation is not correct or an alternate rendering is preferred. The King James Version (KJV) of the Bible is used as the standard reference for the English translation of words. In some instances, the English Standard Version (ESV), New King James Version (NKJV), New American Standard Bible (NASB), New International Version (NIV) and other versions are referenced.

The following format is used to show development from the Hebrew text to the *exemplar* translation. The A-form of the verse is directly copied from *The Lexham Hebrew-English Interlinear Bible*, except presented in a left-to-right format rather than the reverse. The word order of the Hebrew text is retained, and slash marks separate each Hebrew word or term. The B-form of the verse reflects word changes from the Lexham version, and all substitutions are underlined. Some simple verses may not have a B-form. The C-form of the verse rearranges the text into the English style and adds words enclosed in brackets to smooth out language flow. The D-form shows the completed version of the verse. Brackets are dropped, final minor changes are made and italics are added.

Progression from the A-form to the D-form allows the Bible student and Christian to follow verse development. Underlined words indicate changes from the previous verse form. A Strong's number is often attached in superscript to identify specific English words with their corresponding Hebrew word. In the final D-form, significant words not in the Hebrew text are identified by italics similar to the KJV.

Some Hebrew words have legitimate multiple meanings, and choosing the appropriate one is a judgment call. The *exemplar* translation is not immune from this issue. Word selection from available valid choices is controlled by a strict appeal to the original Hebrew text, verse context and a strong explanatory framework guided by general revelation. As practiced in Chapter 5, Hebrew words and their frequency of translation into English are occasionally analyzed.

Hebrew verbs do not have a past, present or future tense similar to verbs in the English language. Hebrew verbs can be divided into the *perfect* and *imperfect* state. The perfect expresses completed action or state of being while the imperfect expresses incomplete action.[1] Neither the perfect nor imperfect have a tense (time of action) apart from context and issues of syntax.[2] When the context is unclear, determining past or present tense is a judgment call. For example, in the *exemplar* translation of Genesis 1:7, 16, 21 and 25, "made" and "create" refer to actions that occurred in the past, but are mentioned in the present.

Keep in mind that Hebrew was written from right to left—opposite the English style of writing. A sentence in Hebrew typically follows the order of verb-subject-object. In contrast, English sentence structure is typically subject-verb-object. When Hebrew is translated into English, the verb and subject order are normally reversed.

The Genesis creation story cannot be interpreted in a historical and scientific vacuum. For example, it is impossible to know what a camel looks like from only reading scripture. Personal knowledge of the world helps flesh out this animal. In like manner, to understand the creation account, information from general revelation must be referenced. *Exemplar* creation uses inspired scripture, in conjunction with lasting evidence from the past, to build a strong interpretative framework for the creation story. Romans 1:20 says, "For since the creation of the world God's invisible qualities—his eternal power and divine nature—have been clearly seen, being understood from what has been made, so that men are without excuse." (NIV) Therefore, man can understand God's past actions to an extent by looking at what He has made. Knowledge of general revelation is invaluable in understanding the biblical creation account.

The *exemplar* translation was developed with the goal of being a highly literal, yet understandable rendering of the original Hebrew text. Although many Bible translations allege to be literal, those claims are debatable. For example, verses in the A-form copied from *The Lexham Hebrew-English Interlinear Bible* are literally translated, but quite difficult to read. Few languages translate into another in an easy flowing style. Words must be rearranged or added to improve readability, sentence structure and clarity (rearrangements and word additions are sometimes debatable).

In addition, words capable of being legitimately translated several different ways must be narrowed to a single term (also debatable in some instances). The following is the most credible word-for-word translation of Genesis Chapters 1 and 2 yet developed in the English speaking world.

Genesis Chapter One

1:1A. In beginning/create/God/the heaven/and/the earth. (LHEIB[3])
1:1C. In [the] beginning[07225], God create[01254] the heaven[08064] and the Earth[0776].
1:1D. In the beginning, God created the heavens and the Earth.

This verse is traditionally translated:

- In the beginning God created the heaven and the earth. (KJV)
- In the beginning God made the heaven and the earth. (Septuagint[4])

The Septuagint poorly translates this simple verse. The Hebrew word *bârâ* is translated "made" rather than "create." The Septuagint also mistranslates *bârâ* in Genesis 1:21, 27 and 2:4 in the same manner. In Genesis 2:3 it does not translate *bârâ* at all. This verse clearly illustrates the inaccuracy of the Septuagint.

Dr. Bruce Waltke in *Genesis: A Commentary* says some theology considers Genesis 1:1 as a summary of the entire six-day creation account that is to follow.[5] Accordingly, placement of this verse at the beginning has no chronological significance. Some verses in Genesis Chapter 1 are explained as non-chronological and must be shuffled around to achieve a logical flow of events.

Traditional young-Earth creation theology, proposed by Dr. Henry Morris, explains "heavens" in Genesis 1:1 as the equivalent of "space."[6] Initially space was empty and remained so until the fourth day. He says, "In like manner the term 'earth' refers to the component of matter in the universe. At the time of the initial creation, there were no other planets, stars, or other material bodies in the universe; nor did they come into being until the fourth day."[7] "So this verse [Genesis 1:1] must speak essentially of the creation of the basic elements of matter, which thereafter were to be organized into the structured earth and later into other material bodies."[8]

Dr. John Sailhamer believes the heavens and Earth in Genesis 1:1 refer to the entire universe. He points out that "beginning" is translated from the Hebrew *rê'shîyth* and refers not to an instant in time but an undefined period of time when God created.[9] *Rê'shîyth* can refer to a beginning period of time ranging from one day to several years. For example, in Genesis 10:10 the beginning of Nimrod's kingdom was Babel, Erech, Accad, and Calneh. These cities were built or acquired over a period of time. In Job 8:7, the beginning period of Job's life is mentioned as being small, but the latter end was to be greater. In addition, empirical science supports the idea that the Earth, stars and galaxies are of different ages and did not originate at the same moment in time.

Exemplar creation does not require a metaphorical interpretation of the words "heaven" and "earth" in Genesis 1:1. The heavens refer to everything man can see when looking upward from the surface of the Earth including galaxies, globular star clusters, nebulae, stars, solar systems, planets, moons and other objects. The Hebrew word *erets* (Strong's 0776), here translated "Earth," refers to all the land from our current knowledge perspective. From the beginning, our planet's size, mass and spherical shape have remained essentially constant.

Genesis 1:1 attributes God as creator of the heavens and Earth, without defining the time of their creation. Undefined time could be one day or billions of years ago. The exact time, sequence and method God used to create the heavens and Earth are not revealed in scripture, and man is left to his own resources to fill in the missing information.

- This is the first occurrence in the Bible of the Hebrew word *shâmayim* (Strong's 08064), here translated "heavens." It occurs 420 times in the Old Testament and is translated "heaven" 398 times, "air" 21 times and "astrologers" 1 time in the KJV.[10] See Chapter 5 for additional discussion of this word.
- This is the first occurrence of *bârâ* (Strong's 01254), here translated "create." *Bârâ* occurs eleven times in the first book of the Bible in Genesis 1:1, 1:21, 1:27, 2:3, 2:4, 5:1, 5:2 and 6:7 and is always translated create. See Chapter 5 for additional discussion of this word.
- This is the first occurrence of *rê'shîyth* (Strong's 07225), here translated "beginning." It occurs 51 times in the Old Testament and is translated "beginning" 18 times, "firstfruits" 11 times, "first" 9 times and "chief" 8 times, in addition to other miscellaneous words in the KJV.[11] The Brown-Diver-Briggs Hebrew lexicon gives "beginning" as an acceptable translations of the noun *rê'shîyth.*[12]
- This is the first occurrence of *erets* (Strong's 0776), here translated "Earth," which occurs 2,504 times in the Old Testament.[13] See Chapter 5 for additional discussion of this word.

1:2A. And the earth/be/formless/and empty/and darkness/over surface/the deep/and spirit/God/hover/over surface/the waters. (LHEIB)
1:2B. And the earth/be/<u>vacant</u>/and empty/and darkness/over surface/the deep/and spirit/God/hover/over surface/the waters.
1:2C. And the Earth[0776] be[01961] vacant[08414] and empty[0922] [of life]. And darkness[02822] [was] over[05921] [the] surface[06440] [of] the deep[08415] [waters]. And [the] spirit [of] God hover[07363] over[05921] [the] surface[06440] [of] the waters.
1:2D. And the Earth <u>was</u> vacant and empty *of life*. And darkness *was* over the surface of the deep *waters*. And the Spirit of God hovered over the surface of the waters.

This verse is traditionally translated:

• And the earth was without form and void; and darkness was upon the face of the deep. And the Spirit of God moved upon the face of the waters. (KJV)

• Yet the earth was invisible and unformed and darkness was over the abyss, and a divine wind was being carried along over the water. (Septuagint[14])

• Earth was a soup of nothingness, a bottomless emptiness, an inky blackness. God's Spirit brooded like a bird above the watery abyss. (*The Message*) [This version totally misrepresents the Hebrew text]

• The erth was voyde and emptie, ad darckneffe was vpon the depe, and the fpirite of god moved vpon the water. (William Tyndale, 1530[15])

• And ye earth was voyde and emptie, and darcknes was vpon the depe, & ye sprete of God moued vpo the water. (Myles Coverdale, 1535[16])

• And the earth was void and empty, and darkness was upon the face of the deep; and the spirit of God moved over the waters. (Douay-Rheims, 1899)

According to Waltke, some theology explains Genesis 1:1 as a summary of the entire six-day creation account that is to follow. Genesis 1:2 is a description of the chaotic state of the Earth immediately after its creation. Earth is described as an unproductive and uninhabited place.[17] The problem with this explanation is that Genesis 1:1 contains the only mention of God creating the Earth. If Genesis 1:1 is considered to be a summary of the entire creation account, then theologians admit that, "There is no word of God creating the planet earth or darkness or the watery chaos" as described in Genesis 1:2.[18]

In a similar vein, Sailhamer discusses Genesis 1:1 as a title for the creation account. If Genesis 1:1 is a summary of the first chapter, then the actual creation process starts with Genesis 1:2.[19] It begins with an Earth that is without form and void. He says, "We are left with the uncomfortable discovery that the passage does not tell us when, or who, created the earth."[20] "Since the creation story starts with an unformed Earth, the traditional concept of "creation out of nothing" is not invalidated but is not taught in Genesis 1."[21]

Traditional young-Earth creation theology proposed by Morris explains the "heavens and earth" in Genesis 1:1 as equivalent to "space and raw matter." In regard to Genesis 1:2 Morris says, "The matter so created was at first unformed and uninhabited."[22] His explanation of heaven and earth in Genesis 1:1-2 is metaphorical and not a literal or straightforward reading of the text.

The above explanations illustrate the confusion over interpretation of these two verses. The KJV, *The Message* and Septuagint represent traditional translations, all of which are incorrect. Theological influence has obfuscated the understanding of Genesis 1:1-2.

Exemplar creation explains Genesis 1:1-2 as occurring in the correct chronological order. Genesis 1:1 describes the initial creation of the heavens and Earth in the past, without defining the time frame. Genesis 1:2 is a description of the Earth after some time has elapsed, which could be one day or billions of years. At this point, the Earth is (1) vacant and empty of life, (2) has deep water covered by darkness and (3) the Spirit of God is hovering over the waters. No special or metaphorical meanings need be applied to any of the words.

What is meant by the statement, "And the Earth was vacant and empty?" The logical answer is that it was empty of life since this is the primary topic of the six-day Genesis chronicle. The similar meaning of the two rhyming Hebrew words *tôhûw* (Strong's 08414) and *bôhûw* (Strong's 0922) strongly emphasize this point. In addition to the biblical text, scientific evidence indicates the Earth was at once a molten ball of material. The resultant high temperatures (above 800° F) ensured a sterilized planet completely devoid of all life. See Chapter 5 for additional discussion of *tôhûw* and *bôhûw*.

The translation of *tôhûw* as "without form" in the KJV is illegitimate and a result of theological influence. A careful study of all verses in the Bible containing *tôhûw* show it cannot be correctly translated in this manner. William Tyndale's Pentateuch of 1530, the Myles Coverdale Bible of 1535 and the Douay-Rheims version (all quoted above) translate *tôhûw* and *bôhûw* as "void and empty." Theological influence changed the translation to "without form and void" in the Geneva Bible of 1560, and most subsequent English Bibles adopted the same words.

- This is the first occurrence of a Hebrew verb based on the root *hâyâh* (Strong's 01961), here translated "was." When used as a standalone word or part of a Hebrew compound word, derivatives of the root *hâyâh* occur 33 times in 23 separate verses in Genesis 1:2, 3, 5, 6, 7, 8, 9, 11,13,14, 15, 19, 23, 24, 29, 30, 31, 2:5, 2:7, 2:10, 2:18, 2:24 and 2:25.
- Arthur Custance in his book *Without Form and Void* quotes an English translation of Genesis 1:2 from Martin Luther which says, "But the earth was empty and void, and darkness was over the face of the abyss."[23]

1:3A. And say/God/be/light/and be light. (LHEIB)
1:3C. And God say, "Be^{01961} light^{0216} and light^{0216} be^{01961}."
1:3D. And God said, "<u>There is</u> light, and light <u>exists</u>."

This verse is traditionally translated:

- And God said, Let there be light: and there was light. (KJV)
- And God said, Let there be light, and there was light. (Septuagint[24])
- God spoke: "Light!" And light appeared. (*The Message*)

Genesis 1:3 is traditionally explained as when God created light to shine on the first day. This interpretation is driven by more than twenty-two hundred years of errant young-Earth creation theology embedded in the Septuagint. In this verse, the Hebrew verb *hâyâh* (Strong's 01961) occurs two times. The first occurrence of *hâyâh* (יְהִי) is translated “let there be” in the KJV and traditionally assigned a jussive verb state.[25] When treated as jussive, the word “let” becomes a part of the translation. Jussive verbs are traditionally considered “command verbs” and turn this simple declarative statement into a command. The second occurrence of *hâyâh* (וַיְהִי) is translated “and there was” in the KJV and assigned an imperfect state (not a command). *The Message* represents conventional errant thinking about this text.

Traditional explanations of Genesis 1:3 have God creating some type of intermediate light on the first day, before the Sun is created on the fourth day; however, the details are rather vague. Was this light coming from the presence of God? Was it from a distant point light source similar to the Sun? Some have proposed that events on the first day are out of sequence and should be placed within the fourth day when God created the Sun and Moon. As Waltke says, “Certain ‘difficulties’ in the order of the days seem clearly to represent a dischronologization.”[26] An incorrect translation of Genesis 1:3, accompanied by a faulty explanatory framework, leads one down a path that only gets worse with subsequent verses.

Exemplar creation explains Genesis 1:3 as a statement by God about one of the most important things in the solar system—light. That light is coming from the Sun which has been shining for many years. Nothing is created in this verse. There is a complete absence of dynamic Hebrew verbs in Genesis 1:3-5 to indicate God is creating, making, forming, planting, growing or building anything.

Exemplar creation views Genesis 1:3 as a statement rather than a command. This is the first of seven verses in Genesis 1:3, 6, 9, 11, 14, 20 and 24 that start with the expression, “And God said.” According to traditional interpretations, God commanded light to come forth. Yet if God created light in this verse, it would follow the pattern of Genesis 1:1 when God *created* the heavens and Earth, of 1:26 when God proposed to *make* man, of 1:27 when God *created* man, of 2:7 when God *formed* man, of 2:8 when God *planted* the garden, of 2:9 when God *grew* the trees of the garden, of 2:19 when God *formed* the animals of the garden and of 2:22 when God *built* Eve from Adam’s rib. All of these verses use dynamic Hebrew verbs to describe action. In Genesis 1:3, these action words are not used and God is not commanding anything to happen.

The identical Hebrew expression translated, “And God said” is not restricted to the seven verses of the creation story, and is used elsewhere in the Old Testament. Outside of Genesis Chapter 1, when God is speaking, this expression occurs sixteen times.[27] A command is issued in eight verses in Genesis 6:13, 17:9, 17:15, 17:19, 21:12, 35:1; Exodus 3:14 and Numbers 22:12. A statement is issued in the remaining eight verses of Genesis 9:8, 9:12, 9:17, 46:2; 1 Kings 3:5, 3:11; 2 Chronicles 1:11 and Jonah 4:9. In eight of these sixteen occurrences, no command follows the expression—only a quotation of what God says is given.

In addition, the English expression, “The LORD said” occurs 219 times in the KJV of the Old Testament.[28] In the vast majority of occurrences, God is making a statement rather than issuing a command. Therefore, when God speaks, a command does not necessarily follow.

We can reject dogmatic claims that Genesis 1:3 is a command based on (1) an analysis of verses containing the Hebrew expression, "And God said," (2) verses containing the English phrase, "The LORD said," (3) knowledge that *hâyâh* in Genesis 1:3 is primarily if not totally a stative verb and (4) information given in Chapter 5 and Appendix B.

Exemplar creation explains Genesis 1:1-3 as occurring in a chronological sequence, but in terms of undefined time. In Genesis 1:1, God created the heavens and Earth. In Genesis 1:2, the Earth is described as being a lifeless planet with a dark surface covered by deep water sometime after its formation. Between Genesis 1:2 and 1:3, God created marine life, raised the continents and formed the ocean basins. He created plants, land animals and flying creatures. The transition from undefined time to literal days occurs at Genesis 1:3 as shown in Table 1.1. Verse three begins the first of six chronological 24-hour days leading up to the creation of man on the sixth day. The appearance of man on Earth, discussed in Chapters 10 and 11, dates the occurrence of the six-day Genesis chronicle.

As mentioned, the events prior to Genesis 1:3 are undefined in regard to time, and one day or billions of years may have transpired. The Bible is not specific on this issue. However, general revelation reveals that significant time passed prior to Genesis 1:3. During this time God created the stars, galaxies, our solar system and Earth. Extensive time is allowed for stellar aging, distant starlight travel, Moon impact crater formation, the removal of impact craters on Earth by weathering, continental drift, coal formation, fossil formation, Antarctic ice accumulation, radiometric isotope decay and carbon-14 decay.

Starting with Genesis 1:3, God describes the Earth and heavens immediately before the creation of man on the sixth day. The sequence God used to *create things* prior to Genesis 1:3 is not necessarily the sequence used to *describe them* in Genesis 1:3-25. The exact time, sequence and methods God used when creating are not revealed in scripture, and man is left to his own resources to fill in the missing information.

In the *exemplar* translation of Genesis 1:3, the first occurrence of *hâyâh* (יהי) is translated "there is" and the second *hâyâh* (ויהי) is translated "and exists." Both are stative verbs and denote the existence of light and nothing more. See Chapter 5 for additional discussion of the Hebrew verb *hâyâh*.

- This is the first occurrence of the noun *ôwr* (Strong's 0216), here translated "light." In the first two chapters, it occurs six times in Genesis in 1:3, 4, 5 and 18. *Ôwr* should not be confused with *mâ'ôwr* (Strong's 03974), translated in Genesis 1:14, 15 and 16 as "luminaries" or light giving bodies. The KJV translates both *ôwr* and *mâ'ôwr* as light in the first two chapters.
- In the Hebrew text, the pattern of Genesis 1:3 is reflected in Genesis 1:6. Notice the similar underlined structure, "And God said, There is an expanse. . . and

it exists to separate between water *above* from water *below*."

• In the Hebrew text, the pattern of Genesis 1:3 is also reflected in Genesis 1:14-15. Notice the underlined structure, "And God said, There are luminaries. . . And luminaries exist in the expanse of the heavens. . ."

1:4A. And see/God/the light/that good/and cause to separate/God/between/the light/and between/the darkness. (LHEIB)

1:4B. And saw/God/the light/that good/and distinguish/God/between/the light/and between/the darkness.

1:4C. And God saw that the light[0216] [was] good. And God distinguish[0914] between[0996] the light[0216] and between[0996] the darkness[02822].

1:4D. And God saw that the light *was* good. And God distinguished between the light and between the darkness.

This verse is traditionally translated:

• And God saw the light, that it was good: and God divided the light from the darkness. (KJV)

• And God saw that the light was good. And God separated the light from the darkness. (ESV)

• And the Lord saw the light that it was good. And the Lord distinguished between the light and between the darkness. (Targum of Onkelos, English translation by J. W. Etheridge, 1862)

Traditional theology explains this verse as God physically dividing or separating light from darkness, rather than God distinguishing light from darkness.

In English, "separating" and "distinguishing" are to closely related words with a slight difference. Contrasting physical objects can be readily separated from each other. Goats can be separated from sheep. Wheat can be separated from chaff. Men can be separated from women. However, non-physical phenomena are normally distinguished. We distinguish between darkness and light, cold and heat, silence and sound. Darkness, cold and silence are not substances. Darkness is the absence of light, cold is the absence of heat and silence is the absence of sound.

Newer translations of the Bible such as the ESV, NKJV and NIV recognize the subtle difference between separating and distinguishing. In Leviticus 10:10 and Ezekiel 22:26, the holy is distinguished from the common. In Leviticus 11:47 and 20:25, the clean is distinguished from the unclean.

The Hebrew word *bâdal* (Strong's 0914) can be legitimately translated several different ways, but in Genesis 1:4 is best translated "distinguish" rather than "divide." *Bâdal* can mean to physically separate as in Numbers 16:21 when God told Moses and Aaron to *separate* from the people so He could destroy them. In Exodus 26:33, the veil of the tabernacle *divided* between the holy and the most holy. However, *bâdal* can be legitimately translated "distinguish" in the sense of distinguishing between holy and common, clean and unclean, or light and darkness.

In Genesis 1:4, God is doing nothing more than distinguishing light from darkness in preparation for naming day and night. In the *exemplar* translation, *bâdal* is translated "distinguish" in Genesis 1:4, 14 and 18, and "separate" in 1:6 and 7.

- The word *bâdal* (Strong's 0914) occurs 42 times in the Old Testament and is translated "separate" 25 times, "divide" 8 times, "difference" 4 times, "sever" 3 times and "asunder" 2 times, in addition to other miscellaneous words and in the KJV.[29] In the first two chapters of Genesis, *bâdal* occurs in Genesis 1:4, 1:6, 1:7, 1:14 and 1:18.
- Gesenius' Hebrew Lexicon defines *bâdal* as primarily "to separate" or "to distinguish,"[30] Strong's Concordance defines *bâdal* as "to divide" and can mean literally or figuratively to separate or distinguish.[31]
- The English Standard Version (ESV) translates *bâdal* as "distinguish" in Leviticus 10:10, 11:47 and Ezekiel 22:26.
- The New King James Version (NKJV) translates *bâdal* as "distinguish" in Leviticus 10:10, 11:47, 20:25 and Ezekiel 22:26
- The New International Version (NIV) translates *bâdal* as "distinguish" in Leviticus 10:10, 11:47, 20:25 and Ezekiel 22:26.
- This is first occurrence of the Hebrew word *bêyn* (Strong's 0996), here translated "between." In the first two chapters, *bêyn* occurs nine times in Genesis 1:4, 1:6, 1:7, 1:14 and 1:18.

1:5A. And call/God/to the light/day/and to the darkness/call/night/and be evening/and be morning/day/one. (LHEIB)
1:5C. And God call[07121] to the light[0216] day[03117] and call[07121] to the darkness[02822] night. And evening be[01961] and morning be[01961] day[03117] one[0259].
1:5D. And God called the light "day" and called the darkness "night." And evening was, and morning was—day one.

This verse is traditionally translated:

- And God called the light Day, and the darkness he called Night. And the evening and the morning were the first day. (KJV)
- And he called the light Day, and the darkness Night; and there was evening and morning one day. (Douay-Rheims, 1899)

In Genesis 1:5, God names day and night and defines the first day. One complete cycle of evening and morning defines one day. One cycle marks 24 hours of time, signifies one revolution of the Earth and corresponds to three hundred sixty degrees of planetary rotation.

Traditional young-Earth creation theology, as explained by Morris, has a major problem at this point. The first day is now complete. The Earth has been created and is rotating on its axis. The planet is being illuminated by some kind of temporary light source corresponding to the Sun. Light and darkness are occurring to produce day and night.[32] Everything appears logical except there is no mention of God creating the Earth. As explained by Morris, the word "earth" in Genesis 1:1-2 is the basic matter from which the planet Earth and universe are constructed.[33] According to his interpretative framework, Genesis fails to record the creation of Earth as a planet.

The Hebrew expression translated into English in verse 1:5C above as, "And evening be and morning be—day one," is identical in the Hebrew text for all six days except for the numbers and "the" used on the sixth day.

In Genesis 1:5, the Hebrew word *echâd* (Strong's 0259), meaning "one," is used to number the day. *Echâd* is a cardinal number employed for counting (one, two, three, etc.) in the Hebrew language.[34] On the remaining six days, ordinal numbers are used (second, third, fourth, etc.), indicating position in Hebrew.[35] This is *day one* of the six-day Genesis chronicle, rather than the *first day* of the universe. The Douay-Rheims Bible correctly translates this cardinal number as "one."

- This is the first occurrence of the Hebrew word *qârâ* (Strong's 07121), here translated "call." It occurs 735 times in the Old Testament and is translated "called" 528 times, "cried" 98 times, "read" 38 times and "proclaim" 36 times, in addition to other miscellaneous words in the KJV.[36]
- In the first two chapters, "call" occurs eight times in Genesis 1:5, 1:8, 1:10, 2:19, 2:20 and 2:23. In every instance when "call" is used, something is being named.
- The Hebrew verb *hâyâh* (ויהִי) occurs twice and is translated "and be" in 1:5C in the phrase, "And evening *be* and morning *be*—day one." "Be" is a stative verb rather than a dynamic verb, and no action is implied. Compare the use of "be" in this phrase with that of Genesis 1:3C.
- This is the first occurrence of *echâd* (Strong's 0259), here translated "one." It occurs 952 times in the Old Testament and is translated "one" 687 times, "first" 36 times, "another" 35 times, "other" 30 times and "any" 18 times, in addition to other miscellaneous words in the KJV.[37]

1:6A. And say/God/be/expanse/in middle/the water/and be/cause to separate/between/water/to water. (LHEIB)

1:6B. And say/God/be/expanse/in middle/the water/and be/<u>to separate</u>/between/water/to water.

1:6C. And God said, "Be[01961] [an] expanse[07549] in [the] middle[08432] [of] the water, and [it] be[01961] to separate[0914] between[0996] water [above from] to water [below]."

1:6D. And God said, "<u>There is</u> an expanse in the middle of the water, and it <u>exists</u> to separate between water *above* from water *below*."

This verse is traditionally translated:

• And God said, Let there be a firmament in the midst of the waters, and let it divide the waters from the waters. (KJV)

• And God said, Let there be a firmament in the midst of the water, and let it be a division between water and water, and it was so. (Septuagint[38])

Genesis 1:6 is traditionally explained as when God created an expanse and separated the waters on the second day. This interpretation is driven by more than twenty-two hundred years of errant young-Earth creation theology embedded within the Septuagint. In this verse, the Hebrew verb *hâyâh* (Strong's 01961) occurs two times. The first *hâyâh* (יְהִי) is translated "let there be" and the second (וַיְהִי) is translated "and let" in the KJV. Both occurrences are traditionally assigned a jussive verb state.[39] When *hâyâh* is treated as a jussive/command verb, the word "let" becomes a part of the translation. The treatment of *hâyâh* as a command verb is misleading and turns a simple descriptive statement of God into a command.

Exemplar creation explains Genesis 1:6 as a statement by God about Earth's atmosphere. The sky is a clear expanse that has existed for many years. Nothing is being moved around, separated or reorganized. There is a complete absence of dynamic Hebrew verbs in this verse to indicate God is creating, making, forming, planting, growing or building anything.

Exemplar creation views Genesis 1:6 as a statement rather than a command. This is the second of seven verses in Genesis 1:3, 6, 9, 11, 14, 20 and 24 that start with the expression, "And God said." As pointed out in the discussion of Genesis 1:3, when God speaks a command does not necessarily follow. Based on analysis given in Chapter 5 and Appendix B, we can reject claims that Genesis 1:6 must be a command.

In the *exemplar* translation, the stative terms "there is" and "and exists" are translated from the Hebrew verb *hâyâh* (Strong's 01961), meaning primarily to "be" or "exist." The structure of this verse and its use of *hâyâh* are similar to Genesis 1:3. Genesis 1:6 denotes the existence of an expanse and nothing more. See Chapter 5 for additional discussion of the verb *hâyâh*.

• The word *râqîya* (Strong's 07549) occurs 17 times in the Old Testament and is consistently translated "firmament" in the KJV.[40] The word "expanse" in the ESV, NASB and NIV is more descriptive and used in the *exemplar* translation.

1:7A. And make/God/the expanse/and separate/between/the waters/which/from under/to the expanse/and between/the waters/which/from over/to the expanse/and be so. (LHEIB)

1:7C. And God [is the one who] make[06213] the expanse[07549] and separate[0914] between[0996] the waters which [were] from under to the expanse[07549] and between[0996] the waters which [were] from over[05921] to the expanse[07549]. And [it] be[01961] so[03651].

1:7D. And God *is the one who* made the expanse and separated between the waters which *were* under the expanse and between the waters which *were* over the expanse. And *it* was so.

This verse is traditionally translated: "And God made the firmament, and divided the waters which were under the firmament from the waters which were above the firmament: and it was so." (KJV)

In traditional translations, God repeats himself in four separate instances in Genesis 1:6-7, 1:14-16, 1:20-21 and 1:24-25. These supposed repetitive creative events show the fallacy of traditional interpretations. In Genesis 1:6, God allegedly spoke the expanse into existence and then needlessly repeated the act in Genesis 1:7.

To believe Earth's atmosphere was created on the second day strongly conflicts with general revelation in regard to time. There is overwhelming evidence that our atmosphere has existed for more than 100,000 years. An incorrect explanation of the time issue makes the Genesis creation story equivalent to pagan mythology. Physical evidence discussed in Chapter 4 supports an old age for the Earth and indicates its existence extends significantly beyond 4004 BC.

Exemplar creation explains these verses as a chronological sequence of events. God made the expanse prior to Genesis 1:3. On the second day in Genesis 1:6, God describes the expanse as separating water from water. In Genesis 1:7, God is firmly identified as the one who made the expanse. Since the expanse is described as existing in Genesis 1:6, there is no need for God to make it in Genesis 1:7.

This is the first occurrence of the Hebrew word *âsâh* (Strong's 06213), here translated "make." *Âsâh* can indicate many different types of action. The basic meaning if *âsâh* is "do;" however, "made" is a legitimate translation in this verse. See Chapter 5 for additional discussion of this word.

- In the first two chapters, the Hebrew word *âsâh* (Strong's 06213) occurs twelve times in Genesis 1:7, 1:11, 1:12, 1:16, 1:25, 1:26, 1:31, 2:2, 2:3, 2:4 and 2:18 and is translated "made, make, making and did" in the *exemplar* translation.
- This is the first occurrence of the term translated "and *it* was so." This term occurs in Genesis 1:7, 1:9, 1:11, 1:15, 1:24, 1:30; Judges 6:38; 2 Kings 15:2 and Amos 5:14.[41]

1:8A. And call/God/to the expanse/heaven/and be evening/and be morning/day/second. (LHEIB)

1:8C. And God call[07121] to the expanse[07549] heaven[08064]. And evening be[01961] and morning be[01961] second day[03117].

1:8D. And God called the expanse "sky." And evening was, and morning was—second day.

This verse is traditionally translated: "And God called the firmament Heaven.

And the evening and the morning were the second day." (KJV)

In the *exemplar* translation, the Hebrew word *shâmayim* (Strong's 08064) is translated "sky." This verse is obviously referring to the Earth's atmosphere. The expanse has water above and water below as described in Genesis 1:7. The word *shâmayim* as used by the Hebrews in the Old Testament could refer to (1) the sky where birds flew and from which rain fell, (2) the cosmic heavens where the Sun, Moon and stars were or (3) to heaven as the abode of God. See Chapter 5 for additional discussion of this word.

1:9A. And say/God/be gathered/the waters/from under/the heaven/to place/one/and be seen/the dry ground/and be so. (LHEIB)
1:9B. And say/God/wait/the waters/from under/the sky/to place/one/and is seen/the dry ground/and be so.
1:9C. And God said, "The waters from under the sky[08064] wait[06960] [in] to one place[04725] and the dry ground[03004] is seen[07200]". And [it] be[01961] so[03651].
1:9D. And God said, "The waters under the sky wait in one place and the dry ground is seen." And *it* was so.

This verse is traditionally translated:

- And God said, "Let the waters under the heaven be gathered together unto one place, and let the dry land appear: and it was so." (KJV)
- And God said, "Let the water that is under the sky be gathered into one gathering, and let the dry land appear." And it became so. And the water that was under the sky was gathered into their gatherings, and the dry land appeared. (Septuagint[42])

Genesis 1:9 is traditionally explained as when God commanded the waters to separate from dry land on the third day to form the oceans and seas. This interpretation is driven by more than twenty-two hundred years of errant young-Earth creation theology embedded within the Septuagint. The Hebrew word *qâvâh* (Strong's 06960) is translated "let be gathered together" and *râ'âh* (Strong's 07200) is translated "let appear" in the KJV. Both are traditionally considered command verbs and assigned a jussive verb state.[43] When *qâvâh* and *râ'âh* are treated as jussive/command verbs, the word "let" becomes a part of the translation. This interpretation is misleading and turns a simple descriptive statement of God into a command.

In Genesis 1:9, the traditional translation of *qâvâh* (Strong's 06960) as "let be gathered together" is illegitimate. The original text of the Old Testament strongly supports the correct translation of *qâvâh* as "wait." See Chapter 5 for a discussion of this word.

In the Septuagint, this verse is poorly translated from the Hebrew into Greek, and it is easy to see the original language has been expanded. A new last sentence has been added to this verse with an emphasis on the "gathering" of the waters and illustrates the strong influence of errant theology on the Septuagint translation.

The Hebrew word *râ'âh* (Strong's 07200) can be legitimately translated several different ways, but in Genesis 1:9 is best translated "seen" rather than "appear." *Râ'âh* is translated "see, saw, seen, seest and seeing" 67% of the time in the KJV and as "appear" only 5% of the time. Use of the word "appear" in this verse is misleading since it implies ground is being thrust upward out of the water. The word "seen" more accurately describes this passive scene.

Exemplar creation explains Genesis 1:9 as a statement by God about the Earth's water and land. The waters are tranquilly waiting in one place and dry ground is visible. Earth's land masses and oceans have been in their respective positions for many years. Land is not being uplifted nor are the waters being gathered together on the third day. There is a complete absence of Hebrew verbs in Genesis 1:9-13 to indicate God is creating, making, forming, planting, growing or building anything.

Exemplar creation views Genesis 1:9 as a statement rather than a command. This is the third of seven verses in Genesis 1:3, 6, 9, 11, 14, 20 and 24 that start with the expression, "And God said." As pointed out in the discussion of Genesis 1:3, when God speaks a command does not necessarily follow. Based on analysis given in Chapter 5 and Appendix B, we can reject claims that Genesis 1:9 must be a command.

- The word *râ'âh* (Strong's 07200), here translated "see," is found nine times in the first two chapters of Genesis in 1:4, 1:9, 1:10, 1:12, 1:18, 1:21, 1:25, 1:31 and 2:19. Hebrew words of various spellings based on the root (ראה) occur approximately 1,313 times in the Old Testament and are translated "see, saw, seen, seest, seeing" 879 times, "look" 104 times, "behold" 83 times, "shew" 68 times and "appear" 66 times, in addition to other miscellaneous words in the KJV.[44]

1:10A. And call/God/to the dry ground/land/and to collection/the waters/call/seas/ and see/God/that good. (LHEIB)

1:10C. And God call[07121] to the dry ground[03004] land[0776] and call[07121] to the collection[04723] [of] waters seas. And God see that [it was] good.

1:10D. And God called the dry ground "land" and called the collection of waters "seas." And God saw that *it was* good.

This verse is traditionally translated: "And God called the dry land Earth; and the gathering together of the waters called he seas: and God saw that it was good." (KJV)

Traditional young-Earth creation theology uses this verse to support the belief that God "gathered the waters" in Genesis 1:9. However, the Hebrew word *miqveh* (Strong's 04723) is best translated "collection" rather than "gathering together." Chapter 5 discusses the translation of this word in detail. The original text of the Bible and general revelation support translating *miqveh* as “collection” for a superior word choice. Based on scientific knowledge, the continents and oceans have been in their current positions for millions of years.

In Genesis 1:10, God continues to name and define terms. The Hebrew word *erets* (Strong’s 0776) is translated “land” in the *exemplar* translation rather than “earth” as in the KJV. God specifically defines all the dry areas not covered by water as “land” and the collected waters as “seas.”

As God reviews His work, He repeatedly declares it good throughout the six-day Genesis chronicle. God declares His work “good” in Genesis 1:4, 1:10, 1:12, 1:18, 1:21 and 1:25, and in 1:31 summarizes all He sees as “very good.”

1:11A. And say/God/produce/the earth/green/plant/give seed/seed/tree/fruit/make/fruit/according to kind/which/seed in/on the earth/and be so. (LHEIB)
1:11B. And say/God/sprouting/the land/sprouts/plant/seeding/seed/tree/fruit/make/fruit/to kind/which/seed in/over the land/and be so.
1:11C. And God say, “The land[0776] [is] sprouting[01876] sprouts[01877]: plants[06212] seeding[02232] seed[02233] [and the] fruit[06529] tree make[06213] fruit[06529]—to which kind [is] in [the] seed[02233]—over[05921] the land[0776].” And [it] be[01961] so[03651].
1:11D. And God said, “The land is sprouting sprouts: plants seeding seeds and the fruit trees making fruit—to which kind is in the seed—over the land.” And *it* was so.

This verse is traditionally translated:

- And God said, Let the earth bring forth grass, the herb yielding seed, and the fruit tree yielding fruit after his kind, whose seed is in itself, upon the earth: and it was so. (KJV)
- And God said, “Let the earth sprout vegetation, plants yielding seed, and fruit trees bearing fruit in which is their seed, each according to its kind, on the earth.” And it was so. (ESV)

Genesis 1:11 is traditionally explained as when God created all plants on the third day by speaking them into existence. This interpretation is driven by more than twenty-two hundred years of errant young-Earth theology embedded within the Septuagint. The Hebrew word *dâshâ* (Strong’s 01876) is traditionally assigned a jussive verb state[45] and translated “let bring forth” in the KJV. When *dâshâ* is treated as a jussive/command verb, the word “let” becomes a part of the translation. This interpretation is misleading and turns a simple descriptive statement of God into a command.

Exemplar creation explains Genesis 1:11 as a statement by God about Earth’s vegetation. Plants and trees cover the land, and their seeds have been sprouting for many years. Nothing is being created on the third day. There is a complete absence of dynamic Hebrew verbs in Genesis 1:9-13 indicating God created, made, formed, planted, grew or built anything.

Exemplar creation views Genesis 1:11 as a statement rather than a command. This is the fourth of seven verses in Genesis 1:3, 6, 9, 11, 14, 20 and 24 that start with the expression, "And God said." As pointed out in the discussion of Genesis 1:3, when God speaks a command does not necessarily follow. Based on analysis given in Chapter 5 and Appendix B, we can reject claims that Genesis 1:11 must be a command.

Genesis 1:11-12 appear to be almost identical verses, with the only significant difference being the word *dâshâ* (Strong's 01876) in Genesis 1:11 and *yâtsâ* (Strong's 03318) in Genesis 1:12. In the *exemplar* translation, *dâshâ* and *yâtsâ* are respectively translated "sprouting" and "brings forth." In these two verses, God is describing the sprouting of sprouts and the growth of sprouts—two completely different mechanisms involved in plant maturation.

This is the first occurrence of *cognate accusatives* in Genesis. In English, some basic words can be modified and used as either a noun or verb. For example, "run" can refer to running (a verb) or to a runner (a noun). We could say, "The running runner is moving very fast." The Hebrew text often uses verb and noun combinations in this manner, called cognate accusatives. In Genesis 1:11 the land is "sprouting sprouts" and the plants are "seeding seeds." Other cognate accusatives are "swarming swarmers" in Genesis 1:20, "flying fliers" in Genesis 1:20 and "moving movers" in Genesis 1:26. These are literal translations of the Hebrew text.

- The Hebrew word *dâshâ,* here translated "sprouting" and identified as Strong's 01876, is based on the Hebrew root (דשא). *Dâshâ* is used as a verb 2 times in Genesis 1:11 and Joel 2:22 in the Old Testament.[46]
- The word *deshe,* here translated "sprouts" and identified as Strong's 01877, is also based on the Hebrew root (דשא). *Deshe* is used as a noun 14 times and is translated "grass" 7 times, "herb" 6 times and "green" 1 time in the KJV. *Deshe* occurs in Gen. 1:11, 1:12; Deut. 32:2; 2 Sam. 23:4; 2 Kings 19:26; Job 6:5, 38:27; Psalm 23:2, 37:2; Prov. 27:25; Isaiah 15:6, 37:27, 66:14 and Jer. 14:5 in the Old Testament.[47] *Deshe* refers to tender vegetation, green growth or sprouts that have just emerged from the ground.
- The word *zâra,* here translated "seeding" and identified as Strong's 02232, is based on the Hebrew root (זרע). *Zâra* is used as a verb 56 times in the Old Testament and is translated "sow" 47 times and "yielding" 3 times, in addition to other miscellaneous words in the KJV.[48]
- The word *zera,* here translated "seed" and identified as Strong's 02233, is also based on the root (זרע). *Zera* is used as a noun 229 times in the Old Testament and is translated "seed" 221 times in addition to other miscellaneous words in the KJV.[49]
- The words *dâshâ* and *deshe* (sprouting sprouts) occur only in Genesis 1:11 as cognate accusatives. The words *zâra* and *zera* (seeding seeds) are used as cog-

nate accusatives in Genesis 1:11, 12 and 29.

• This is the first occurrence of the word *eseb* (Strong's 06212), here translated "plant." It occurs 33 times in the Old Testament and is translated "herb" 17 times and "grass" 16 times in the KJV.[50] *Eseb* is found in Genesis 1:11, 1:12, 1:29, 1:30 and 2:5 and is translated "plant" in the first two chapters of Genesis in the *exemplar* translation. The word "plant" is superior to "grass" or "herb" since it is more inclusive of Earth's growing things. Elsewhere, the term "plant of the field" is used to indicate plants used for food. In other instances, the terms "plants of the mountains" and "plants of the earth" refer to wild plants.

1:12A. And cause to go forth/the earth/green/plant/give seed/seed/according to kind/ and tree/make fruit/which/seed in/according to kind/and saw/God/that good. (LHEIB)
1:12B. And brings forth/the land/sprouts/plant/seeding/seed/to kind/and tree/make fruit/which/seed in/to kind/and saw/God/that good.
1:12C. "And the land[0776] brings forth[03318] sprouts[01877]: plant[06212] seeding[02232] seed[02233] [according] to [their] kind, and [the] tree make[06213] fruit[06529]—in which seed[02233] [is] to kind." And God saw that [it was] good.
1:12D. "And the land brings forth sprouts: plants seeding seeds *according* to *their* kind, and the trees making fruit—in which seed is to kind." And God saw that *it was* good.

This verse is traditionally translated: "And the earth brought forth grass, and herb yielding seed after his kind, and the tree yielding fruit, whose seed was in itself, after his kind: and God saw that it was good." (KJV)

Traditionally, Genesis 1:11-12 are explained as two repetitious verses. God commands the earth to bring forth grass, herbs and fruit trees in Genesis 1:11 and then reiterates the process in Genesis 1:12. The writer of Genesis tends to repeat himself.

Exemplar creation explains Genesis 1:11 and 1:12 as two different aspects of plant maturation that involve seed sprouting and sprout growth. Both processes are made possible by the information and complex structure within seeds.

To believe all plants were created on the third day strongly conflicts with general revelation in regard to time. Ice core data indicate plant life has existed on Earth for more than 100,000 years. An incorrect explanation of the time issue tends to make the Genesis creation story appear as myth to the scientific community. Evidence discussed in Chapter 4 supports an old age for the Earth and its plant life.

• The word *yâtsâ* (Strong's 03318), here translated "brought forth," occurs 1,069 times in the Old Testament and is translated ". . . out" in the sense of going out 518 times, ". . . forth" in the sense of going forth 411 times, "bring" 24 times, "come" 24 times and other miscellaneous words in the KJV.[51]

1:13A. And be evening/and be morning/day/third. (LHEIB)
1:13C. And evening be[01961] and morning be[01961] third day[03117].
1:13D. And evening was, and morning was—third day.

1:14A. And say/God/be/light source/in dome/the heaven/cause to separate/between/the day/and between/the night/and be/as sign/and as appointed time/and as day/and year. (LHEIB)
1:14B. And say/God/be/luminaries/in expanse/the heaven/to distinguish/between/the day/and between/the night/and be/as sign/and as appointed time/and as day/and year.
1:14C. And God say, "Be[01961] luminaries[03974] in [the] expanse[07549] [of] the heaven[08064] to distinguish[0914] between[0996] the day[03117] and between[0996] the night, and [they] be[01961] as [a] sign[0226] and as [for] appointed times[04150] and as days[03117] and years[08141]."
1:14D. And God said, "There are luminaries in the expanse of the heavens to distinguish between the day and between the night, and *they* are a sign *for* appointed times and days and years."

1:15A. And be/as light source/in dome/the heaven/give light/on the earth/and be so. (LHEIB)
1:15B. And be/as luminaries/in expanse/the heaven/to give light/over the earth/and be so.
1:15C. "And as luminaries[03974] be[01961] in [the] expanse[07549] [of] the heaven[08064] to give light[0215] over[05921] the earth[0776]." And [it] be[01961] so[03651].
1:15D. "And luminaries exist in the expanse of the heavens to give light over the Earth." And *it* was so.

Genesis 1:14-15 are traditionally translated:
- 1:14 And God said, Let there be lights in the firmament of the heaven to divide the day from the night; and let them be for signs, and for seasons, and for days, and years: (KJV)
- 1:15 And let them be for lights in the firmament of the heaven to give light upon the earth: and it was so. (KJV)
- God spoke: "Lights! Come out! Shine in Heaven's sky! Separate Day from Night. Mark seasons and days and years, Lights in Heaven's sky to give light to Earth." And there it was. (*The Message*) [This version severely distorts the original text]

Genesis 1:14-15 are traditionally explained as when God created the Sun, Moon and stars on the fourth day by speaking them into existence. This interpretation is driven by more than twenty-two hundred years of errant young-Earth creation theology embedded within the Septuagint. Why create a "special light" on the first day and then replace it with the Sun on the fourth day? Creating the Sun on the first day is more logical, shows better planning, less waste and superior use of available resources. There is no mention in scripture of the alleged special light being removed by God or replaced by the Sun.

In Genesis 1:14, the Hebrew verb *hâyâh* (Strong's 01961) occurs two times. The first occurrence of *hâyâh* (יהי) is translated "Let there be" in the KJV and the second (והיו) is translated "and let be." The Septuagint, KJV and most other translations treat both occurrences as command verbs. However, the *Old Testament Parsing Guide* assigns a jussive verb state to the first occurrence of *hâyâh*, but not the second.[52] When *hâyâh* is treated as a jussive/command verb, the word "let" becomes a part of the translation. This interpretation is misleading and turns a simple descriptive statement of God into a command.

In Genesis 1:15, *hâyâh* (והיו) is translated "And let be," but not identified as a jussive verb by the *Old Testament Parsing Guide*.[53] However, the Septuagint, KJV and other translations treat it as a command verb through use of the word "let." According to most Bible translations, God commanded the lights to be created in Genesis 1:14, and for good measure commanded them again in Genesis 1:15. Since *hâyâh* (והיו) is not identified as a jussive verb by the *Old Testament Parsing Guide* in either Genesis 1:14 or 15, it is not included in Table 5.4 as a key verb.

Exemplar creation explains Genesis 1:14-15 as a statement by God about heavenly objects as viewed from the surface of the Earth. The Sun, Moon and stars have existed for many years. Nothing is being created in Genesis 1:14-15. There is a complete absence of dynamic Hebrew verbs in these two verses to indicate God is creating, making, forming, planting, growing or building anything.

Exemplar creation views Genesis 1:14 as a statement rather than a command. This is the fifth of seven verses in Genesis 1:3, 6, 9, 11, 14, 20 and 24 that start with the expression, "And God said." As pointed out in the discussion of Genesis 1:3, when God speaks a command does not necessarily follow. Based on analysis given in Chapter 5 and Appendix B, we can reject claims that Genesis 1:14-15 must be a command.

In the *exemplar* translation, the stative English words "there are," "are," "exist" and "was" are translated from the Hebrew verb *hâyâh* (Strong's 01961), meaning primarily to "be" or "exist." In Genesis 1:14-15, this verb denotes the existence of luminaries and nothing more. No dynamic action is occurring. See Chapter 5 for additional discussion of the verb *hâyâh*.

1:16A. And make/God/two/the light source/the great/the light source/the great/as authority/the day/and the light source/the small/as authority/the night/and/the stars. (LHEIB)
1:16B. And do/God/two/the luminaries/the great/the luminary/the great/to rule/the day/and the luminary/the small/to rule/the night/and/the stars.
1:16C. And God [is the one who] do[06213] the two great[01419] luminaries[03974]: the great[01419] luminary[03974] to rule[04475] the day[03117] and the small[06996] luminary[03974] to rule[04475] the night and the stars.

1:16D. And God *is the one who* <u>made</u> two great luminaries: the greater luminary to rule the day and the smaller luminary to rule the night and the stars.

This verse has been translated a number of ways:

• God made two big lights, the larger to take charge of Day, The smaller to be in charge of Night; and he made the stars. (*The Message*) [This is one of the poorest versions available.]

• And God made two great lights; the greater light to rule the day, and the lesser light to rule the night: *he made* the stars also. (KJV) [The translators put the words "he made" in italics to show they are added to the original Hebrew text. The Hebrew letter vav (ו), normally translated "and," is translated "also" in this instance. The word order is reversed from "and the stars" to "the stars also," making an unusual translation.]

• God made the two great lights, the greater light to govern the day, and the lesser light to govern the night; {He made} the stars also. (NASB) [The words "He made" are in brackets to clearly identify them as additions to the original text.]

• And the Lord made the two great luminaries: the greater luminary to rule in the day; and the smaller luminary to rule in the night, and the stars. (English translation of the Targum of Onkelos by J. W. Etheridge, 1862)

• And God made two great lights: a greater light to rule the day; and a lesser light to rule the night: and the stars. (Douay-Rheims, 1899)

• And God made the two great luminaries, the great luminary for rulership of the day and the lesser luminary for rulership of the night, and the stars. (Septuagint[54])

• God made the two great lights: the greater light to govern the day, the smaller light to govern the night, and the stars. (New Jerusalem Bible, 1985)

Genesis 1:16 is traditionally explained as when God made the Sun, Moon and stars on the fourth day. A careful reading of this verse shows that only two lights are mentioned as being made on the fourth day—the greater light and the lesser light. All translations of this verse specifically state, "And God made two great lights." Stars are mentioned as existing on the fourth day, but not created.

This verse shows the error of treating both Genesis 1:14 and 1:16 as creative acts. God allegedly commanded the luminaries into existence in Genesis 1:14 and then needlessly repeated the act in Genesis 1:16. In traditional translations, God repeats himself in four separate instances in Genesis 1:6-7, 1:14-16, 1:20-21 and 1:24-25.

Virtually all traditional versions of the Bible struggle with Genesis 1:16 because of their incorrect explanatory framework. Of the above seven translations, the first three incorrectly attempt to show how God created the stars on the fourth day, while the last four correctly translate the Hebrew text. The Etheridge English version of the

Targum of Onkelos (a Hebrew to Aramaic translation probably before 200 AD), the Douay-Rheims Bible (based on Jerome's translation from Hebrew to Latin completed in 405 AD), the Pietersma English translation of the Septuagint and the New Jerusalem Bible are the most accurate. Probably sometime after Jerome, theology emerged that required creation of the stars on the fourth day, forcing the mistranslation of this verse.

Genesis 1:16 clearly states the function of the two identified luminaries—to rule. The larger luminary rules the day and the smaller luminary rules over the night and the stars. The Sun totally dominates the day with its brightness. The Moon, as the brightest object at night, dominates the dark night and the faint stars.

Exemplar creation explains Genesis 1:14-16 as a sequence of events. God made the Sun, Moon and stars prior to Genesis 1:3. On the fourth day in Genesis 1:14-15, God describes the Sun and Moon as existing to give light on the Earth. In Genesis 1:16, God is firmly identified as maker of the Sun and Moon. Since the Sun and Moon are described as existing in Genesis 1:14-15, there is no need for God to make them in Genesis 1:16.

To believe the Sun, Moon and stars were created on the fourth day strongly conflicts with general revelation in regard to time. Starlight from distant stars and galaxies is millions of years old. An incorrect explanation of the time issue tends to make the Genesis creation story appear as mythological to virtually all scientists. Overwhelming evidence from empirical science discussed in Chapter 4 supports an old age for the universe and indicates the Sun, Moon and stars are significantly older than 4004 BC.

1:17A. And place/they/God/in dome/the heaven/give light/upon the earth. (LHEIB)
1:17B. And gave/they/God/in expanse/the heaven/to give light/over the earth.
1:17C. And God gave[05414] they in [the] expanse[07549] [of] the heaven[08064] to give light[0215] over[05921] the Earth[0776].

1:17D. And God gave them in the expanse of the heavens to give light over the Earth.

This verse is traditionally translated: "And God set them in the firmament of the heaven to give light upon the earth." (KJV)

The KJV translation of this verse shows the influence of traditional young-Earth creation theology. The reader is led to believe God set or placed the Sun and Moon in the heavens on the fourth day. While "set" is a legitimate translation, "gave" is a superior word choice in this instance.

Exemplar creation explains this verse as a companion to the previous verse. In Genesis 1:16 and 17, God is identified respectively as the *maker* and *giver* of these luminaries.

The Hebrew word *nâthan* (Strong's 05414) can be legitimately translated several different ways. *Nâthan* is translated "gave" in the KJV 54% of the time and "set" only 5% of the time. *Nâthan* also occurs in Genesis 1:29 when God *gave* plants to man for food and in Genesis 3:6 when Eve *gave* the fruit from the tree of knowledge of good and evil to Adam. In the *exemplar* translation, the more common “gave” rather than “set” in used.

• This is the first occurrence of the Hebrew word *nâthan* (Strong’s 05414), here translated “gave.” It occurs 2,008 times in the Old Testament and is translated “give” 1078 times, “put” 191, “deliver” 174 times, “made” 107 times, “set” 99 times and “up” 26 times, in addition to other miscellaneous words in the KJV.[55] In the first two chapters, *nâthan* occurs in Genesis 1:17 and 1:29.
• This is the second occurrence of the Hebrew verb *ôwr* (Strong’s 0215), here translated “give light.” It occurs as a verb in Genesis 1:15 and 17. *Ôwr* occurs as a noun (Strong’s 0216) in Genesis 1:3, 4, 5 and 18 where it is translated “light.”[56]

1:18A. And to rule/over the day/and over the night/and cause to separate/between/the light/and between/the darkness/and see/God/that good. (LHEIB)
1:18B. And to rule/in the day/and in the night/and to distinguish/between/the light/ and between/the darkness/and see/God/that good.
1:18C. And to rule[04910] in the day[03117] and in the night and to distinguish[0914] between[0996] the light[0216] and between[0996] the darkness[02822]. And God saw that [it was] good.
1:18D. And to rule in the day and in the night and to distinguish between the light and between the darkness. And God saw that *it was* good.

This verse is traditionally translated:
• And to rule over the day and over the night, and to divide the light from the darkness: and God saw that it was good. (KJV)
• And to rule in the day and in the night, and to distinguish between light and darkness. And the Lord saw that it was good. (Targum of Onkelos, English translation by J. W. Etheridge, 1862)

The Hebrew word *bâdal* (Strong’s 0914) can be legitimately translated several different ways, but in Genesis 1:18 is best translated “distinguish” rather than “divide.” The Sun and Moon do not divide light and darkness—they are incapable of dividing anything. The Sun distinguishes the day (light) and the Moon distinguishes the night (dark). In the *exemplar* translation, *bâdal* is translated “distinguish” in Genesis 1:4, 14 and 18 and “separate” in Genesis 1:6 and 7. The English translation of the Targum of Onkelos by J. W. Etheridge translates *bâdal* in Genesis 1:18 as “distinguish.”

1:19A. And be evening/and be morning/day/fourth. (LHEIB)

1:19C. And evening be[01961] and morning be[01961] fourth day[03117].
1:19D. And evening was, and morning was—fourth day.

1:20A. And say/God/swarm/the waters/swarm/creature/living/and bird/fly/above the earth/across surface/dome/the sky. (LHEIB)
1:20B. And say/God/swarm/the waters/swarm/creature/living/and flier/fly/over the land/over surface/expanse/the sky.
1:20C. And God said, "The waters swarm[08317] [with] swarm[08318] [of] living[02416] creature[05315] and flier[05775] fly[05774] over[05921] the land[0776] [and] over[05921] [the] surface[06440] [of the] expanse[07549] [of] the sky[08064]."
1:20D. And God said, "The waters swarm *with* swarms of living creatures, and fliers flying over the land and over the surface of the expanse of the sky."

This verse is traditionally translated:
- And God said, Let the waters bring forth abundantly the moving creature that hath life, and fowl that may fly above the earth in the open firmament of heaven. (KJV)
- And God said, "Let the waters bring forth creeping things among living creatures and birds flying on the earth against the firmament of the sky." (Septuagint[57])
- And God said, "Let the waters swarm with swarms of living creatures, and let birds fly above the earth across the expanse of the heavens." (ESV)
- And God said, Let the waters swarm with swarms of living creatures, and let birds fly above the earth in the open firmament of heaven. (ASV)
- God said, "Let the waters swarm with swarms of living creatures, and let birds fly above the earth in the open expanse of sky." (World English Bible)

Genesis 1:20 is traditionally explained as when God created marine life and flying creatures on the fifth day by speaking them into existence. This interpretation is driven by more than twenty-two hundred years of errant young-Earth creation theology embedded in the Septuagint. In all traditional translations the Hebrew word *shârats* (Strong's 08317) and *ûwph* (Strong's 05774) are assigned a jussive verb state.[58] When *shârats* and *ûwph* are treated as jussive/command verbs, the word "let" becomes a part of the translation. This interpretation is misleading and turns a simple descriptive statement of God into a command.

The expression "let the waters bring forth" in the Septuagint and KJV is a gross mistranslation and makes God's method of creation appear to be an evolutionary process. In fact, some theistic evolutionists interpret this verse to say God used evolution over millions of years as His method of creating marine life. From their viewpoint, the waters (primordial soup) brought forth (evolved) living creatures. The English Standard Version, American Standard Version, World English Bible and *exemplar* translation more accurately translate this expression as "the waters swarm with swarms of living creatures."

Exemplar creation explains Genesis 1:20 as a statement by God about Earth's abundant marine life and flying creatures. The pure water and air support a multitude of living things. Nothing is being created in Genesis 1:20. There is a complete absence of Hebrew verbs in this verse to indicate God is creating, making, forming, planting, growing or building anything.

Exemplar creation views Genesis 1:20 as a statement rather than a command. This is the sixth of seven verses in Genesis 1:3, 6, 9, 11, 14, 20 and 24 that start with the expression, "And God said." As pointed out in the discussion of Genesis 1:3, when God speaks a command does not necessarily follow. Based on analysis given in Chapter 5 and Appendix B, we can reject claims that Genesis 1:20 must be a command.

This is the first occurrence of the Hebrew word *ôwph* (Strong's 05775), here translated "fliers." This same word also occurs in Genesis 1:21, 22, 26, 28, 30, 2:19 and 2:20, and is translated fliers in all instances in the *exemplar* translation. The word *ôwph* properly applies to all flying or winged creatures, including bats, birds and insects. Leviticus 11:20, 21 and 23 mention an *ôwph* with four legs. Other Hebrew words used to identify flying creatures are *tsippôwr* (Strong's 06833) translated "bird," *nesher* (Strong's 05404) translated "eagle" and *ôrêb* (Strong's 06158) translated "raven."

- The word *shârats,* here translated "swarm" and identified as Strong's 08317, is based on the Hebrew root (שרץ). *Shârats* is used as a verb 14 times in the Old Testament and is translated "creep" 6 times, "bring forth abundantly" 5 times, "move" 1 time, "breed abundantly" 1 time and "increase abundantly" 1 time in the KJV.[59]
- The word *sherets,* here translated "swarms" and identified as Strong's 08318, is also based on the Hebrew root (שרץ). *Sherets* is used as a noun 15 times in the Old Testament and is translated "creeping thing" 11 times, "creep" 2 times, "creature" 1 time and "move" 1 time in the KJV.[60] Swarming things can live on the land, air or water. In Leviticus 11:29, swarming things that live on *land* are mentioned. In Leviticus 11:20-21, swarming things can fly in the *air*. In Leviticus 11:10, swarming things inhabit the *waters*.
- The word *ûwph,* here translated "flying" and identified as Strong's 05774, is based on the Hebrew root (עוף). *Ûwph* is used as a verb approximately 30 times in the Old Testament and is translated some form of "fly" 24 times in addition to other miscellaneous translations in the KJV.[61]
- The word *ôwph,* here translated "fliers" and identified as Strong's 05775, is also based on the root (עוף). *Ôwph* is used as a noun approximately 71 times in the Old Testament and is translated "fowl" 59 times and "bird" 9 times, in addition to other miscellaneous translations in the KJV.[62]
- The words *shârats* and *sherets* (swarming swarmers) are used together in

Genesis 1:20, 7:21; Leviticus 21:29, 11:41, 11:42 and 11:43 as cognate accusatives. The words *ûwph* and *ôwph* (flying fliers) are used together in Genesis 1:20 and Hosea 9:11.[63] (See Genesis 1:11 in this chapter for a discussion of cognate accusatives.)

1:21A. And create/God/the sea monsters/the great/and/every kind of creature/the living/the move/which/swarm/the waters/according to kind/and/every kind of bird/wing/according to kind/and see/God/that good. (LHEIB)
1:21B. And create/God/the reptiles/the great/and/every creature/the living/the move/which/swarm/the waters/to kind/and/every flier/wing/to kind/and see/God/that good.
1:21C. And God [is the one who] create[01254] the great[01419] reptiles[08577] and every[03605] move[07430] living[02416] creature[05315], [with] which the waters swarm[08317], [according] to [their] kind, and every[03605] wing[03671] flier[05775] [according] to [its] kind. And God saw that [it was] good.
1:21D. And God *is the one who* created the great reptiles and every moving living creature with which the waters swarm, *according* to *their* kinds, and every winged flier *according* to *its* kind. And God saw that *it was* good.

This verse is traditionally translated:

- And God created great whales, and every living creature that moveth, which the waters brought forth abundantly, after their kind, and every winged fowl after his kind: and God saw that it was good. (KJV)
- So God created the great sea creatures and every living creature that moves, with which the waters swarm, according to their kinds, and every winged bird according to its kind. And God saw that it was good. (ESV)

Traditional explanations have Genesis 1:21 as a virtual repeat of Genesis 1:20. God created all marine creatures and flying fliers in Genesis 1:20 by instantly speaking them into existence and then created them again for good measure in Genesis 1:21. In both instances, God created all animals *ex nihilo* (out of nothing). In traditional translations, God repeats himself in four separate instances in Genesis 1:6-7, 1:14-16, 1:20-21 and 1:24-25.

To believe the great reptiles, marine creatures and winged fliers were created on the fifth day strongly conflicts with general revelation in regard to time. Evidence supports the existence of marine life on Earth for more than 100,000 years. An incorrect explanation of the time issue makes the Genesis creation story an embarrassment to scientifically minded Bible believers. Evidence discussed in Chapter 4 supports an old age for the Earth and indicates animal life has existed on Earth much longer than 6,000 years.

Exemplar creation explains these verses as a chronological sequence of events. God created the great reptiles, marine creatures and winged fliers during an undefined period of time between Genesis 1:2 and 1:3. On the fifth day in Genesis 1:20,

God describes swarming swarms of marine creatures and flying fliers (but no great reptiles). In Genesis 1:21, God is identified as creator of the great reptiles, swarms of marine creatures and winged fliers. Since marine life and flying creatures are described as existing in Genesis 1:20, there is no need for God to create them in Genesis 1:21.

The Hebrew word *bârâ* (Strong's 01254), here translated "created," occurs first in Genesis 1:1 and not again until Genesis 1:21. "Create" is not used in any verse between Genesis 1:1 and 1:21. If God created things on the first, second, third and fourth days, then why is the verb "create" not used in describing events on those days? Its absence strongly discredits traditional interpretations of these verses.

In Genesis 1:20, God describes the existence of marine life and flying creatures on Earth, but not great reptiles. From this verse, we can infer that no great reptiles are mentioned because they were extinct at this time. Yet God clearly created them in the distant past as proclaimed in Genesis 1:21. Based on current evidence, the great reptiles were extinct by the time man appeared on Earth. When Genesis 1:20 and 1:21 are carefully examined in the proper explanatory framework, they fully harmonize with the evidence of empirical science.

The *exemplar* translation renders the Hebrew word *tannîyn* (Strong's 08577) as "reptiles" rather than "whales" found in the KJV. *Tannîyn* appears to be a general word for reptile while *nâchâsh* (Strong's 05175) is a specific reptile called a snake. The key to understanding these two words is the description given to Moses' rod. When Moses threw down his rod, it sometimes became a *tannîyn* as described in Exodus 7:9, 10 and 12, while at other times it became a *nâchâsh,* as in Exodus 4:3 and 7:15. Both words refer to the same animal. When Moses' rod miraculously changed, it was called both a *tannîyn* (reptile as a general name) and a *nâchâsh* (snake as a specific reptile).

The existence of great reptiles mentioned in Genesis 1:21 was unknown to scientists until relatively recently, and their discovery is confirmation of scripture. The great *tannîyn* mentioned in Genesis probably includes dinosaurs and reptiles. Under the current taxonomic system, dinosaurs are classified separate from reptiles. Through Isaiah 27:1, the *tannîyn* can be connected to the leviathan mentioned and described by God in Job 41:4.

• This is the first occurrence of the word *tannîyn* (Strong's 08577), here translated "reptiles." It occurs 28 times in the Old Testament and is translated "dragon" 21 times, "serpent" 3 times, "whale" 3 times and "sea monster" 1 time in the KJV.[64]

• In Exodus 7:9, 10 and 12 when Moses cast his rod down before Pharaoh it became a *tannîyn.* Isaiah 27:1 says the *tannîyn* living in the sea will be slain. Psalm 91:13 predicts how a *tannîyn* will be trampled under the feet. Deuteronomy 32:33 references a *tannîyn* with poison. Often *tannîyn* are mentioned in describing a desolate place such as in Job 30:29; Isaiah 34:13; Isaiah 43:20 and

Jeremiah 10:22. In general, reptiles such as lizards and snakes live in desolate places rather than towns and villages. *Tannîyn* is the general name given to reptiles.

• The word *nâchâsh* (Strong's 05175) occurs 31 times in the Old Testament and is translated "serpent" every time it occurs in the KJV.[65] *Nâchâsh* is a snake, which in turn is a reptile.

• It was a *nâchâsh* (snake) that tempted Eve to sin and was cursed by God to crawl on its belly in Genesis 3:1-14. In Exodus 4:3 and 7:15, Moses' rod turned into a *nâchâsh* when it was thrown down. In Numbers 21:6-7, God sent *nâchâsh* among the people to bite them. Moses made a *nâchâsh* of brass in Numbers 21:9 and put it on a pole for the people to look at and be saved. In 2 Kings 18:4, Hezekiah destroyed the brass *nâchâsh* the people had been worshiping.

1:22A. And bless/they/God/say/be fruitful/and multiply/and fill/the waters/in the sea/ and the bird/multiply/on the earth. (LHEIB)
1:22C. And God bless they say, "Be fruitful[06509] and multiply[07235] and fill[04390] the waters in the sea and multiply[07235] the flier[05775] on the land[0776]."
1:22D. And God blessed them saying, "Be fruitful and multiply and fill the waters in the seas and multiply the fliers on the land."

God blesses the marine animals and flying creatures that fill the land. He commands them to be fruitful and increase from their present state. This concludes the activities of day five.

• The Hebrew words *râbâh* (Strong's 07235) and *pârâh* (Strong's 06509) are translated "be fruitful and multiply." These two words rhyme in Hebrew, have a very similar meaning and are used to emphasize the idea of reproducing. A similar phrase occurs in Genesis 1:22, 1:28, 8:17, 9:1, 9:7, 47:27; Jeremiah 23:3 and Ezekiel 36:11.[66]

• This is the first occurrence of the word *pârâh* (Strong's 06509), here translated "be fruitful." It occurs 29 times in the Old Testament and is translated "fruitful" 20 times and "increased" 3 times, in addition to other miscellaneous words in the KJV.[67]

• This is the first occurrence of the word *râbâh* (Strong's 07235), here translated "multiply." It occurs 226 times in the Old Testament and is translated "multiply" 74 times, "increase" 40 times, "much" 29 times, "many" 28 times and "more" 12 times, in addition to other miscellaneous words in the KJV.[68]

1:23A. And be evening/and be morning/day/fifth. (LHEIB)
1:23C. And evening be[01961] and morning be[01961] fifth day[03117].
1:23D. And evening was, and morning was—fifth day.

1:24A. And say/God/bring forth/the earth/creature/living/according to kind/domestic animal/and creeping/and wild animal earth/according to kind/and be so. (LHEIB)
1:24B. And say/God/bring forth/the land/creature/living/to kind/beasts/and moving/ and animal land/to kind/and be so.
1:24C. And God said, "The land[0776] brings forth[03318] living[02416] creature[05315] [according] to [their] kind: beasts[0929] and moving[07431] [things] and animal [of the] land[0776] [according] to [their] kind." And [it] be[01961] so[03651].
1:24D. And God said, "The land brings forth living creatures *according* to *their* kind: beasts and moving *things* and animals of the land, *according* to *their* kind." And *it* was so.

This verse is traditionally translated:

• And God said, Let the earth bring forth the living creature after his kind, cattle, and creeping thing, and beast of the earth after his kind: and it was so. (KJV)

• And God said, "Let the earth bring forth the living creature according to kind: quadrupeds and creeping things and wild animals of the earth according to kind." And it became so. (Septuagint[69])

Genesis 1:24 is traditionally explained as when God created the land animals on the sixth day by speaking them into existence. This interpretation is driven by more than twenty-two hundred years of errant young-Earth creation theology embedded within the Septuagint. The Hebrew word *yâtsâ* (Strong's 03318) is traditionally assigned a jussive verb state[70] and translated "let bring forth" in the KJV. When *yâtsâ* is treated as a jussive/command verb, the word "let" becomes a part of the translation. This interpretation is misleading and turns a simple descriptive statement into a command.

Exemplar creation explains Genesis 1:24 as a statement by God about Earth's land creatures. The bounty of the land supports a multitude of beasts, moving things and animals. Nothing is being created in Genesis 1:24. There is a complete absence of Hebrew verbs in this verse to indicate God is creating, making, forming, planting, growing or building anything.

Exemplar creation views Genesis 1:24 as a statement rather than a command. This is the last of seven verses in Genesis 1:3, 6, 9, 11, 14, 20 and 24 that start with the expression, "And God said." As pointed out in the discussion of Genesis 1:3, when God speaks a command does not necessarily follow. Based on analysis given in Chapter 5 and Appendix B, we can reject claims that Genesis 1:24 must be a command.

• This is the first occurrence of the Hebrew word (חית) here translated "animal." This word occurs 55 times where it is translated "animal" by *The Lexham Hebrew-English Interlinear Bible.*[71] The KJV uses the term "beast" rather than animal.

1:25A. And make/God/wild animal/the earth/according to kind/and the domestic animal/according to kind/and/all creeping thing/the ground/according to kind/and see/God/that good. (LHEIB)
1:25B. And make/God/animal/the Earth/to kind/and the beasts/to kind/and/every moving/the ground/to kind/and see/God/that good.
1:25C. And God [is the one who] make[06213] [the] animal [of] the Earth[0776] [according] to [their] kind: and the beasts[0929] [according] to [their] kind and every[03605] moving[07431] [thing upon] the ground[0127] [according] to [its] kind. And God saw that [it was] good.
1:25D. And God *is the one who* made the animals of the Earth *according* to *their* kind: the beasts *according* to *their* kind and every moving *thing* upon the ground *according* to *its* kind. And God saw that *it was* good.

This verse is traditionally translated: "And God made the beast of the earth after his kind, and cattle after their kind, and every thing that creepeth upon the earth after his kind: and God saw that it was good." (KJV)

Traditionally, Genesis 1:24-25 are explained as two repetitious verses. God spoke and instantly created all land animals *ex nihilo* in Genesis 1:24 and then proceeded to make them again in Genesis 1:25. In traditional translations, God needlessly repeats himself in four separate instances in Genesis 1:6-7, 1:14-16, 1:20-21 and 1:24-25.

To believe all land animals were created on the sixth day strongly conflicts with general revelation in regard to time. Fossil and ice core evidence indicate land animals have been on the Earth for more than 100,000 years. Traditional translations make the Genesis creation story unbelievable to scientifically literate people. Physical evidence discussed in Chapter 4 supports the appearance of land animals significantly before 4004 BC.

Exemplar creation explains these verses as a chronological sequence of events. God created all land animals between Genesis 1:2 and 1:3. On the fifth day in Genesis 1:24, He describes the presence of abundant animal life. In Genesis 1:25, God is firmly identified as their creator. Since the land animals are described as existing in Genesis 1:24, there is no need for God to create them in Genesis 1:25.

In the *exemplar* translation, words used to describe land animals in this verse are translated in the broadest terms possible since all land dwelling creatures are being described. Terms such as domestic animals, cattle, livestock or wild animals are too narrow and not used.

1:26A. And say/God/make/humankind/in image + we/according to likeness + us/and rule over/over fish/the sea/and over bird/the heaven/and over domestic animal/and over all the earth/and over all the creeping thing/the move/on the earth. (LHEIB)
1:26B. And say/God/make/man/in image + our/after likeness + our/and have dominion/on fish/the sea/and on flier/the sky/and on beasts/and on all the Earth/and on all

the moving/the move/over the land.
1:26C. And God say, "[We shall] make[06213] man in our image[06754], after our likeness[01823]. And [he shall] have dominion[07287] on fish [of] the sea and on fliers[05775] [in] the sky[08064], and on beasts[0929] and on all[03605] the Earth[0776] and on all[03605] the moving[07431] [things that] move[07430] over[05921] the land[0776]."
1:26D. And God said, "*We shall* make man in our image, after our likeness. And *he shall* have dominion on fish of the sea and on fliers in the sky and on beasts and on all the Earth and on all the moving *things* that move over the land."

This verse is traditionally translated: "And God said, Let us make man in our image, after our likeness: and let them have dominion over the fish of the sea, and over the fowl of the air, and over the cattle, and over all the earth, and over every creeping thing that creepeth upon the earth." (KJV)

In the rendering of this verse, there is no significant difference between the KJV and the *exemplar* translation. In the KJV, the word "let" in the phrase, "Let us make man," is not in the Hebrew text, but is implied based on context. "Let" is accompanied by the dynamic verb "make" and is legitimate. In the *exemplar* translation, the insipid "let" is replaced by the stronger "shall" to show God's intent to create man.

• The Hebrew word *râmas,* here translated "move" and identified as Strong's 07430, is based on the Hebrew root (רמש). *Râmas* is used as a verb 16 times in the Old Testament and translated "creep" 11 times and "move" 5 times in the KJV.[72, 73]

• The word *remes,* here translated "moving things" and identified as Strong's 07431, is also based on the Hebrew root (רמש). *Remes* is used as a noun 18 times in the Old Testament and translated primarily as "creeping thing" in the KJV.[74, 75]

• The words *râmas* and *remes* (moving movers) are used together in Gen. 1:26, 7:14, 8:17, 8:19 and Ezekiel 38:20 as cognate accusatives. The most literal translation of Gen. 1:26 is, "And on all the Earth and on all the moving movers over the land." (See Genesis 1:11 in this chapter for a discussion of cognate accusatives.)

• The Hebrew verb *âsâh* (Strong's 06213), here translated "make," is assigned a qal cohortative 1cp verb form by the *Old Testament Parsing Guide.*[76] When the Hebrew letter nun (נ) is added as a prefix to the root (עשה) it becomes (נעשה), the verb found in this verse. A cohortative verb expresses desire or determination. God expresses His intent to make man.

1:27A. And create/God/the humankind/in image + he/in image/God/create/he/male/ and female/create/they. (LHEIB)
1:27C. And God create[01254] the humankind in He image, in God image create[01254] he, male and female [God] create[01254] they.

1:27D. And God created man in His image, in God's image created him, male and female *God* created them.

This verse is translated in the Septuagint as: "And God made man, according to the image of God he made him, male and female he made them." (Septuagint[77])

The Septuagint poorly translates this verse by rendering *bârâ* as "made" rather than "create" in three different places. Even though there is a corresponding Greek word for "create," the text was not translated in this manner, clearly illustrating the inaccuracy of the Septuagint.

Genesis 1:27 strongly attributes God as the creator of man. This verse is reinforced by the similar words of Genesis 5:1-2 which say, "This is the book of the generations of Adam. When God *created* man, he made him in the likeness of God. Male and female he *created* them, and he blessed them and named them Man when they were *created*." (ESV, italics added for emphasis)

1:28A. And bless/they/God/and say/to + they/God/be fruitful/and multiply/and fill/the earth/and subdue/and rule/over fish/the sea/and over bird/the heaven/and over all animal/the move/on the earth. (LHEIB)
1:28B. And bless/they/God/and say/to + they/God/be fruitful/and multiply/and fill/the earth/and subdue/and have dominion/on fish/the sea/and on flier/the sky/and on all living/the move/over the land.
1:28C. And God bless they, and God said to they, "Be fruitful[06509] and multiply[07235] and fill[04390] the Earth[0776] and subdue [it] and have dominion[07287] on [the] fish [of] the sea and on [the] flier[05775] [of] the sky[08064] and on all[03605] living[02416] [things that] the move[07430] over[05921] the land[0776]."
1:28D. And God blessed them, and God said to them, "Be fruitful and multiply and fill the Earth and subdue it and have dominion on the fish of the sea and on the fliers of the sky and on all living *things* that move over the land."

This verse is often considered a cultural mandate. God commands man to populate the Earth and by implication create civilization, governments and social structures. Man is to dominate the land, sky and marine animals, and by implication use them for food, clothing and beasts of burden.

- The Hebrew word *chay* (Strong's 02416), here translated "living," occurs 501 times in the Old Testament and is translated "live" 197 times, "life" 144 times, "beast" 76 times, "alive" 31 times and "creature" 15 times, in addition to other miscellaneous words in the KJV.[78]

1:29A. And say/God/look/give/to + you/every plant/give seed/seed/which/on surface/all the earth/and every kind of the tree/which in/fruit tree/give seed/seed/for + you/be/as food. (LHEIB)

1:29B. And say/God/behold/give/to + you/every plant/seeding/seed/which/over surface/all the land/and every the tree/which in/fruit tree/seeding/seed/for + you/be/as food.

1:29C. And God said, "Behold[02009], [I] give[05414] to you every[03605] plant[06212] seeding[02232] seed[02233], which [is] over[05921] [the] surface[06440] [of] all[03605] the land[0776] and every[03605] tree in which tree fruit[06529] [is] seeding[02232] seed[02233]. As food [they] be[01961] for you."

1:29D. And God said, "Behold, *I* give to you every plant seeding seed which is over the surface of all the land, and every tree in which fruit is seeding seed. As food *they* are for you."

1:30A. And to every kind of animal/the earth/and to every kind of bird/the heaven/and to every kind of/moving/on the earth/which in/breath/life/every kind of green/plant/as food/and be so. (LHEIB)

1:30B. And to every animal/the land/and to every flier/the sky/and to all/moving/over the land/which in/creature/living/every green/plant/as food/and be so.

1:30C. "And to every[03605] animal [of] the land[0776] and to every[03605] flier[05775] [in] the sky[08064] and to all[03605] [that] moving[07430] over[05921] the land[0776]—in which [is the] living[02416] creature[05315]— every[03605] green[03418] plant[06212] as food." And [it] be[01961] so[03651].

1:30D. "And to every animal of the land and to every flier in the sky and to all *that* moves over the land—in which is the living creature—every green plant is food." And *it* was so.

In Genesis 1:29-30, God plainly gives green plants to Adam and the animals to eat. These verses identify plants as the base of the food chain. We can deduce from this verse that humans and animals within the Garden of Eden were primarily vegetarian. However, to say the human race was vegetarian until God specifically allowed the killing of animals after the flood in Genesis 9:3 is an unsupported assumption. To require all of Earth's animals to be herbivorous is unwarranted.

We can infer that no significant carnivorous or predatory animals existed within the garden. If there were predator-prey relationships, they existed on a small scale and were not a threat to man. God wanted the garden to be a place of peace, safety and security from the world outside.

1:31A. And see/God/everything that/make/and look good/very/and be evening/and be morning/day/the sixth. (LHEIB)

1:31C. And God see everything[03605] that [He] make[06213] and behold[02009] [it was] very good. And evening be[01961] and morning be[01961] the sixth day[03117].

1:31D. And God saw everything that *He* did, and behold *it was* very good. And evening was, and morning was—the sixth day.

In this verse, the Hebrew word *âsâh* (Strong's 06213) is translated "did" in reference to various things God *did* during the six-day Genesis chronicle. Some of God's accomplishments are not fully detailed until Genesis Chapter 2.

Throughout Genesis Chapter 1, God follows a pattern of describing and than pronouncing things as "good." On day one, light is proclaimed as *good* in Genesis 1:4. On the third day, the waters and dry land are described as *good* in Genesis 1:10 and vegetation is called *good* in Genesis 1:12. On the fourth day, the Sun and Moon are recognized as *good* in Genesis 1:18. On the fifth day, marine life and flying creatures are labeled as *good* in Genesis 1:21. On the sixth day, land animals are identified as *good* in Genesis 1:25. Finally in Genesis 1:31, God summarizes everything as "very good." *Exemplar* creation views this pronouncement to include what He created in the undefined past prior to Genesis 1:3 and what He created on the sixth day.

Proponents of traditional young-Earth creation theology say God could not have called His creation "very good" in Genesis 1:31 if there had been millions of years of disease, pain, suffering, death and extinction.[79] Adam's sin is commonly known as the "fall of man." From their viewpoint, the expression "and it looked very good" indicates no animal death occurred on Earth before the fall.[80] According to their belief, no animals died or killed each other for food.[81] Earth was perfect before the fall.[82] Supporters of this theology say that before the fall there were no thorns or thistles, no struggle for survival, no carnivorous animals, no predator-prey relationships and no death on the entire Earth.

The "no death before Adam's sin" theology has been refined over the years. According to this belief, the curse (due to the fall of man) affected the physical elements, the animal kingdom and mankind.[83] Vertebrate animals (including fish) did not die before the fall; however, invertebrates and insects are permitted to be killed and eaten.[84] No vertebrates were diseased, died of natural causes or were killed by predators before Adam sinned.[85] Plants are allowed to die and be eaten. This theology is founded primarily on Romans 5:12; 1 Corinthians 15:20-22; Rom. 8:19-22 and Genesis 1:31.

Romans 5:12 says, "Therefore, just as sin came into the world through one man, and death through sin, and so death spread to all men because all sinned." (ESV) This verse clearly teaches that sin entered the world by one man and death came to all men because of Adam's sin. Animals are not mentioned in this verse. There is no statement that Adam's sin caused animals to die. There is no implication that animals sin or die because they sin.

1 Corinthians 15:21-22 says, "For as by a man came death, by a man has come also the resurrection of the dead. For as in Adam all die, so also in Christ shall all be made alive." (ESV) This verse states that death came to all mankind through Adam. Animals are not a topic in this verse. There is no mention that Adam's sin caused animal death or that animals can be made alive through Jesus Christ.

Romans 8:18-25 says, "For I consider that the sufferings of this present time are not worth comparing with the glory that is to be revealed to us. For the creation waits with eager longing for the revealing of the sons of God. For the creation was subjected to futility, not willingly, but because of him who subjected it, in hope that the creation itself will be set free from its bondage to corruption and obtain the freedom of the glory of the children of God. For we know that the whole creation has been groaning together in the pains of childbirth until now. And not only the creation, but we ourselves, who have the firstfruits of the Spirit, groan inwardly as we wait eagerly for adoption as sons, the redemption of our bodies. For in this hope we were saved. Now hope that is seen is not hope. For who hopes for what he sees? But if we hope for what we do not see, we wait for it with patience." (ESV)

In the above scripture, several key items are not mentioned. There is no declaration that bondage to corruption started in the Garden of Eden. There is no reference to Adam's sin and punishment. There is no specific statement that Adam's sin caused the bondage to corruption the creation is experiencing. As Dr. C. John Collins points out, the terms of Genesis 3:16-19 are not used in these verses.[86]

A combination of scripture and general revelation give the best explanation of Romans 8:18-25 about the decay of God's creation. These verses say "creation was subjected to futility" and creation is in "bondage to corruption" in reference to the initial creation of the heavens and Earth as described in Genesis 1:1. Since the very beginning, the universe has been deteriorating and running down. The Second Law of Thermodynamics and the other laws of chemistry and physics have always applied. God subjected the physical universe to decay at its inception.

These verses compare the results of human sin and freedom from the curse of sin to the decay of creation and its eventual release from decay. Romans 8:18-25 compares the current state of the "sons of God" (Rom. 8:14) and the "children of God" (Rom. 8:16) with God's creation. Eventually, creation will be born again as in "childbirth" and "set free from its bondage to corruption." The release of creation from bondage to decay will be similar to the freedom the children of God have in glory. Someday Christians will have new bodies and there will be a new heaven and new Earth (Rev. 21:1).

In Genesis 2:16-17, Adam is commanded not to eat of the tree of knowledge of good and evil, and warned that if he eats of the tree he will die. It is doubtful this command applied to the animals, since they are incapable of discerning good from evil.

Exemplar creation proposes that animals existed on Earth for an undefined period of time previous to the creation of Adam and Eve. Animals were created by God between Genesis 1:2 and 1:3. Animal struggle, disease, pain, suffering, death and extinction had occurred since inception.

If Earth had been a perfect place, the Garden of Eden would have been unnecessary. Obviously, there was death and danger in the world because God made a protected enclosure where man could develop. He created a place where food was plentiful, without thorns and thistles. Only animals harmless to man lived in the garden. There were no dangerous bears, lions, tigers, hyenas, wolves, poisonous spiders and venomous snakes. We can infer from scripture that no significant predator-prey relationships existed in the garden.

After Adam and Eve ate of the tree of knowledge of good and evil, they were punished along with the serpent in Genesis 3:14-19. As part of Adam and Eve's punishment, they were cast out of the garden into a harsh world of sweat, toil, thorns and thistles described as "cursed ground." The serpent was punished and condemned to crawl on its belly. There is no specific mention of other animals inside or outside the garden being cursed. It is a mistake to assume that all animals on the Earth were cursed along with the serpent.

God's relationship in the Bible has primarily been with humans—not animals. Man communed with God in the garden, not animals. Man sinned against God, not animals. Man was cast out of the garden, not animals. Man was given the Ten Commandments, not animals. Man was commanded to offer sacrifices for sins in the Old Testament, not animals. Jesus Christ came to die for the sins of humankind, not animals. There is no hint in scripture that animals follow a moral code, sin when killing each other or need redemption from sin.

Many animals are obviously predators, and if traditional young-Earth creationists are correct, substantial redesign would have been required after Adam's sin. Lions have powerful jaws, large fangs, sharp claws and powerful leg muscles to make them good predators. Venomous snakes have sharp fangs, poison sacs and deadly venom to kill prey. Spiders weave webs and paralyze prey with their poisonous bite. If these animals were not meant to be predators, why were they created this way? "No death before Adam's sin" theology cannot explain the physical features of these animals. This theology requires predatory animals to have acquired their features after Adam's fall.[87] The most plausible biblical explanation is that God created these animals with all their predatory features at inception, long before Adam sinned.

Without predatory animals, the biology of Earth would quickly become unbalanced. Rabbits, rats, mice and other fast multiplying animals would quickly overrun the land. Ultimately, starvation would be the primary means for restricting population growth. We can view predator-prey relationships as intentionally designed by God as the chief method of population control.

When God described the Earth in Genesis 1:31, it looked "very good." The Earth and its fine-tuned biosphere were working as planned. Plants were sprouting and growing in abundance as designed. Marine life and birds moved about in swarming swarms. Animals filled the land and reproduced faithfully after their kind. A multitude of carefully balanced dependent, symbiotic and synergistic relationships were working as expected. Predator-prey relationships existed as part of a food chain ultimately based upon plant life. Millions of complex living things were living, growing and reproducing as intended. Although not perfect, God was pleased with what He saw and called it "very good."

In summary, "no death before Adam's sin" theology is a prime example of thinking gone awry. This theology is little more than religious philosophy and is not directly supported by either the Bible or general revelation. The absence of animal violence, disease and death on the entire Earth before the fall negates the reason for building the Garden of Eden as a protective enclosure for Adam and Eve. This theology supposes that animal death and predator-prey relationships are evil and sinful, and misapplies God's pronouncement of "very good" upon the Earth. (*Peril in Paradise* by Mark Whorton gives a more complete discussion of this issue.)

The view supported by both scripture and general revelation is that the universe has been deteriorating since the beginning. At inception God subjected His creation to physical decay. On Earth, God created animals that experienced disease, pain, suffering, death and extinction. The Garden of Eden was a protected place from the violence, danger, thorns, thistles and cursed ground outside. Humankind was given a means of escaping decay through access to the tree of life. If Adam and Eve had never sinned, from a distance they could have looked at the world outside and not been subject to its harmful effects. Having fallen, they were cast out of the garden into a world of predatory animals and dangerous things.[88] The great evil was not the events taking place outside the garden, but Adam and Eve's personal sin which caused them to suffer the judgment of God in being cast out of paradise.

Genesis Chapter Two

2:1A. And be finished/the heaven/and the earth/and all array + they. (LHEIB)
2:1C. And the heaven[08064] and the Earth[0776] [were] finished[03615], and all[03605] their array[06635] [of miscellaneous things]
2:1D. And the heavens and Earth *were* finished, and all their multitude *of miscellaneous things.*

This verse is traditionally translated: "Thus the heavens and the earth were finished, and all the host of them." (KJV)

The expression "host of heaven" and "LORD of hosts" is used in several places in the Old Testament. "Host" is also used in reference to armies or multitudes of fighting men. In this verse, host refers to a multitude of details not mentioned. Genesis Chapter 1 is an abbreviated, non-technical, historical account of God's activities and many minutiae are not discussed.

- This is the first occurrence of the Hebrew word *tsâbâ* (Strong's 06635), here translated "host." It occurs 485 times in the Old Testament and is translated "host" 393 times, "war" 41 times and "army" 29 times, in addition to other miscellaneous words in the KJV.[89]

2:2A. And finish/God/on day/the seventh/work + he/that/do/and rest/on day/the seventh/from all work + he/that/do. (LHEIB)
2:2C. And on the seventh day God finish[03615] he work[04399] that [he] do[06213], and

rest[07673] on the seventh day[03117] from all[03605] he work[04399] that [he] do[06213].
2:2D. And on the seventh day God finished His work that *He* did, and rested on the seventh day from all His work that *He* did.

Genesis 2:2 discusses the seventh day, work and resting. God's work consisted of things He did. The Hebrew word *âsâh* (Strong's 06213) is translated "did" in this verse to indicate the broadest possible range of work. Physical work God *did* includes creating, forming, planting, growing, building, placing and bringing things. Non-physical work God *did* includes inspecting, observing, describing, distinguishing, naming, pronouncing, uniting and blessing. The following is a list of the work God did during the six-day Genesis chronicle:

- Day one: God *observes* light and *pronounces* it good. He *distinguishes* between light and darkness and *names* "day" and "night."
- Second day: God *observes* and *describes* the expanse. He *names* the expanse "sky."
- Third day: God *observes* the dry ground and the seas, and *pronounces* them good. He *names* the dry ground "land" and the collected waters "seas." He *observes* and *describes* Earth's abundant vegetation. God probably *inspected* the vegetation before *pronouncing* it good.
- Fourth day: God *observes* and *describes* the Sun and Moon. He *describes* their function in regard to signs, seasons, days and years. He probably *inspected* the Sun and Moon before *pronouncing* them good.
- Fifth day: God *observes*, *describes* and *blesses* marine and bird life. He probably *inspected* the marine creatures and flying fliers before *pronouncing* them good.
- Sixth day: God *observes*, *describes* and *pronounces* the land animals and as good. He *creates* (Gen. 1:27) and *forms* (Gen. 2:7) man. He *plants* a garden for man to live in (Gen. 2:8) and *grows* (Gen. 2:9) trees in the garden. He *places* man in the garden (Gen. 2:8, 15). He *forms* animals (Gen. 2:19) from the ground specifically for the garden. He *brings* the animals (Gen. 2:19) of the garden to Adam for naming. He *creates* (Gen. 1:27) and *builds* (Gen. 2:22) a woman from Adam's rib. He *unites* Adam and Eve as a permanent couple in marriage (Gen. 2:22-24). God *blesses* man and woman (Gen. 1:28) to be fruitful and multiply.

2:3A. And bless/God/day/the seventh/and hallow/he/because/on + he/rest/from all work + he/that create/God/do. (LHEIB)
2:3C. And God bless the seventh day[03117] and hallow he, because on he [he] rest from all[03605] his work[04399]—that God do[06213] [when] create[01254].
2:3D. And God blessed the seventh day and sanctified it, because on it *He* rested from all His work—that God did *when* creating.

2:4A. These/account/the heaven/and the earth/at create/on day/make/Yahweh/God/ earth/and heaven. (LHEIB)
2:4B. These/account/the heaven/and the earth/when create/on day/did/Yahweh/God/ earth/and heaven.
2:4C. These [following is the] account[08435] [of] the heaven and the Earth[0776] when create[01254]; on [the sixth] day[03117] [when] Yahweh God did[06213] [things to the] Earth[0776] and heaven[08064].
2:4D. The *following is the* account of the heavens and the Earth when created ; on the *sixth* day *when* Yahweh God did *things to the* Earth and heavens.

This verse is traditionally translated:

- These are the generations of the heavens and of the earth when they were created, in the day that the LORD God made the earth and the heavens. (KJV)
- This is the account of the heavens and the earth when they were created. When the LORD God made the earth and the heavens. (NIV)
- This is the book of the origin of heaven and earth, when it originated, on the day that God made the heaven and the earth. (Septuagint[90])

The Hebrew word *tôwledâh* (Strong's 08435) is normally translated "generations" or "account." When *tôwledâh* is used, a genealogy normally follows such as in Genesis 10:1, 11:10, 11:27, 25:12, 25:19, 36:1 and 36:9. In some instances, a historical account is given such as the story of Noah in Genesis 6:9 or the story of Joseph in Genesis 37:2. In a similar manner, Genesis 2:4 begins a historical account that ends in Genesis 4:26. The details covered in Genesis 2:4-25 are best explained in reference to events within or related to the Garden of Eden on the sixth day.

Traditional young-Earth creation theology emphasizes that the word "day" used throughout Genesis Chapter 1 means a literal 24-hour day. *Exemplar* creation agrees with this position. However, according to young-Earth reasoning, the creation account suddenly becomes metaphorical in Genesis 2:4 when "day" refers to all six days of creation rather than a single day. This explanation is inconsistent and misleading. In Genesis 2:4, the Hebrew noun is the singular day (יום) in contrast to Genesis 6:4, 7:4, 7:10, 21:4, 21:34, 40:12 and other verses where (ימים) occurs and is commonly translated as the plural "days."

The NIV leaves out "day" in its translation of Genesis 2:4, yet (ביום) translated "on day," in Genesis 2:4A above plainly occurs in the text and is a combination of the noun day (יום) and the letter (ב). The Hebrew term (ביום) occurs approximately 580 times in the Old Testament and is translated primarily "on day" in *The Lexham Hebrew-English Interlinear Bible.*[91] In Genesis 2:4, the NIV deliberately deletes what the Bible clearly states. Supporters of traditional young-Earth creation theology embrace the NIV translation because it agrees with their interpretation.[92]

Exemplar creation proposes that during the six-day chronicle described in Genesis 1:3-31, God only created on the last day. The sixth day of Genesis 1:24-31 corresponds with the day of Genesis 2:4. When properly understood, the creation story flows in a logical order. No convoluted hermeneutics, twisted exegesis and incorrect word translations are required.

The Septuagint translation given above highlights its many inaccuracies. The word "account" is mistranslated as "book of the origin." "Created" is mistranslated as "originated." The term "Yahweh God" is shortened to "God." The phrase "earth and heavens" is incorrectly reversed to "heaven and the earth."

The word order in the phrase "heavens and Earth" occurs 27 times in the KJV translation.[93] The reverse expression, "Earth and heavens" occurs in the latter part of Genesis 2:4 and in Psalm 148:13.[94] Normally, heavens are mentioned first because they are more significant. With an estimated 125 billion galaxies, and each galaxy containing an estimated 400 billion stars, the heavens are considerably greater than the Earth. Reversal of word order to "Earth and heavens" at the end of Genesis 2:4 indicates a reversal in significance and suggests God primarily "did things" to the Earth on the sixth day, and the heavens were of less importance.

- This is the first occurrence of the Hebrew word *tôwledâh* (Strong's 08435), here translated "account." It occurs 39 times in the Old Testament and is translated "generation" 38 times in the KJV.[95] The NIV, NASB, NLT and NAS use the word "account" rather than "generations" in Genesis 2:4, which is more precise. The NKJV, Darby and Hebrew Names Version use the word "history."
- This is the first occurrence of the Hebrew word (יהוה) translated as Yahweh (Strong's 03068). C. John Collins says this special name is given to God, the covenant-making and covenant-keeping God who made promises to the patriarchs of Israel.[96]

2:5A. And all/plant/the field/not yet/be/in the earth/and all plants/the field/not yet/spring up/because/not/cause to fall rain/Yahweh/God/on the earth/and humankind/was not/cultivate/the ground. (LHEIB)

2:5B. And each/bush/the field/not yet/be/on the land/and each plant/the field/not yet/grow/because/not/cause rain/Yahweh/God/over the land/and man/was not/cultivate/the ground.

2:5C. And each[03605] bush[07880] [of] the field[07704] be[01961] not yet[02962] on the land[0776], and each[03605] plant[06212] [of] the field[07704] [did] not yet[02962] grow[06779], because Yahweh God [had] not cause rain[04305] over[05921] the land[0776] and [there] was not man [to] cultivate[05647] the ground[0127] [of the garden].

2:5D. And each bush of the field was not yet on the land, and each plant of the field *did* not yet grow, because Yahweh God had not caused rain over the land and there was no man to cultivate the ground *of the garden*.

This verse is traditionally translated:

• And every plant of the field before it was in the earth, and every herb of the field before it grew: for the LORD God had not caused it to rain upon the earth, and there was not a man to till the ground. (KJV)

• When no bush of the field was yet in the land and no small plant of the field had yet sprung up—for the Lord God had not caused it to rain on the land, and there was no man to work the ground. (ESV)

This verse is traditionally interpreted as applying to the entire Earth, although such an explanation does not logically fit anywhere into Genesis Chapter 1. According to this view, the entire planet is barren of plants and bushes, no rain has fallen anywhere and by implication the ground is so dry plants will not grow. This interpretation strongly conflicts with traditional explanations for the first three day of creation. Allegedly the Earth was a watery mass for the first two days. On the third day, the ground and water are separated, and the Earth is covered with vegetation. One would expect enough residual moisture to keep vegetation alive on the fourth, fifth and sixth days. There is no place in traditional explanations to fit an Earth with no plant life.

In addition to no plant life, an act of God would have been required to halt evaporation, condensation and precipitation over the entire Earth (as some believe) until Noah built the ark. Scripture nowhere directly supports such a condition. Traditional explanations of this verse completely fall apart with a careful analysis.

C. John Collins in commenting on this verse says, the ESV translates the barren land without rain as applying to "some unnamed place where these conditions held."[97]

Exemplar creation explains this verse as applying to the land of the Garden of Eden, before it was planted on the sixth day. There were no bushes or plants of the field on this piece of ground. The spot chosen for the garden was essentially without rain and probably a desert. We can hypothesize it was in the Arabian desert on the Arabian Peninsula, which includes present-day Saudi Arabia, Oman, Yemen, Kuwait and Qatar (all near the Cradle of Civilization). Since Adam and Eve initially wore no clothing, a location with a hot, dry climate would have been a logical choice. The statement, "there was no man to cultivate the ground" connects readily with Genesis 2:15 where Adam was to cultivate the garden.

• This is the first occurrence of the Hebrew word *tsâmach* (Strong's 06779), here translated "grow. *Tsâmach* occurs 33 times in the Old Testament and is translated "grow" 13 times, "spring forth" 6 times, "spring up" 4 times, "grow up" 2 times, "bring forth" 2 times, "bud" 2 times and "spring out" 2 times, in addition to other miscellaneous words in the KJV.[98] In the first two chapters of Genesis, *tsâmach* occurs in Genesis 2:5 and 2:9.

2:6A. And stream/rise/from the earth/and water/all surface the ground. (LHEIB)
2:6B. And mists/rise up/from the land/and water/all surface the ground.
2:6C. And mists[0108] rose up[05927] from[04480] the land[0776] and water all[03605] [the] surface[06440] [of] the ground[0127] [of the garden].
2:6D. And mists rose up from the land and watered all the surface of the ground *of the garden.*

2:7A. And form/Yahweh/God/the man/dust/from the ground/and blow/in nostril/breath/life/and be/the man/as creature/living. (LHEIB)
2:7C. And Yahweh God form[03335] the man [of] dust from[04480] the ground[0127] and blow in [his] nostril [the] breath [of] life[02416]. And the man be[01961] as [a] living[02416] creature[05315].
2:7D. And Yahweh God formed the man *of* dust from the ground and breathed in his nostrils the breath *of* life. And the man existed as a living creature.

This verse is traditionally translated:

- And the LORD God formed man of the dust of the ground, and breathed into his nostrils the breath of life; and man became a living soul. (KJV)
- Then the LORD God formed the man of dust from the ground and breathed into his nostrils the breath of life, and the man became a living creature. (ESV)

The Hebrew verb *hâyâh* (Strong's 01961) here translated "existed" is a stative verb (rather than a dynamic verb showing action) and means primarily to "be" or "exist."

Genesis 2:7 uses two dynamic verbs when discussing the creation of man and one stative verb to explain the end result. The creation of Adam is described using the dynamic verbs "formed" and "breathed," and the final product is depicted by the stative verb "exist." Prior to this event there was only dust, and now there is Adam the living creature. Animal hominids did not evolve over millions of years to *become* Adam the man. There is no hint bipedal beasts changed over time and *became* Adam the human.

The word "formed" is often used in the Old Testament to describe a potter forming a clay pot with his hands. The elemental materials for Adam came from the dust of the ground, and God arranged them to make his body. The creation of the first human was an awesome display of intelligence and power. God is truly the potter and man is the clay.

In Genesis 2:7, "formed" is translated from the Hebrew word *yâtsar*, and has no relation to the expression "without form" in Genesis 1:2 in the KJV. As discussed in Chapter 5, the translation of "without form" in Genesis 1:2 is illegitimate.

- This is the first occurrence of the Hebrew word *yâtsar* (Strong's 03335), here translated "formed." It occurs 62 times in scripture and is translated "form" 26 times, "potter" 17 times, "fashion" 5 times, "maker" 4 times, "frame" 3 times and "make" 3 times, in addition to other miscellaneous words in the KJV.[99]

2:8A. And plant/Yahweh/God/garden in Eden/in east/and put/there/the man/who/ form. (LHEIB)
2:8C. And Yahweh God plant [a] garden from [the] east in Eden, and there [he] put the man who [he had] formed[03335].
2:8D. And Yahweh God planted a garden from the east in Eden, and there *He* put the man whom *He had* formed.

2:9A. And cause to grow/Yahweh/God/from the ground/every kind of tree/be pleasing/to sight/and good/for food/and tree/the life/in middle/the garden/and tree/the knowledge/good/and evil. (LHEIB)
2:9B. And grow/Yahweh/God/from the ground/every tree/desirable/to sight/and good/for food/and tree/the life/in middle/the garden/and tree/the knowledge/good/ and evil.
2:9C. And Yahweh God grow[06779] from[04480] the ground[0127] every[03605] tree [that was] desirable[02530] to [the] sight[04758] and good for food; and the tree [of] life[02416] in [the] middle[08432] [of] the garden and the tree [of] knowledge [of] good and evil.
2:9D. And Yahweh God grew from the ground every tree *that was* desirable to the sight and good for food, and the tree of life in the middle of the garden and the tree of knowledge of good and evil.

This verse is traditionally translated: "And out of the ground made the LORD God to grow every tree that is pleasant to the sight, and good for food; the tree of life also in the midst of the garden, and the tree of knowledge of good and evil." (KJV)

Genesis 2:9 describes God growing trees from the ground in the Garden of Eden. In the KJV, the word "made" has been added and is not in the Hebrew text. Growing trees in one day was not an issue with God and was no more difficult than causing Aaron's staff to sprout, bud, blossom and produce almonds in one night as described in Numbers 17:1-8. God was fully capable of growing mature trees for the garden in one day.

- This is the first occurrence of the word *châmad* (Strong's 02530), here translated "desirable." It occurs 21 times in the Old Testament and is translated "desire" 11 times, "covet" 4 times and "delight" 2 times, in addition to other miscellaneous words in the KJV.[100]

2:15A. And take/Yahweh/God/the man/he put/in garden Eden/to cultivate/and to keep + she. (LHEIB)
2:15C. And Yahweh God take the man [and] put he in [the] garden [of] Eden to cultivate[05647] and to keep[08104] it.

2:15D. And Yahweh God took the man and put him in the Garden of Eden to cultivate and to keep it.

This verse is translated in the Septuagint as: "And the Lord God took the man whom he had formed and put him in the orchard to till and keep it."[101] The Septuagint adds the expression, "whom he had formed," which is not in the original Hebrew text.

Adam was put in the Garden of Eden to cultivate it. Genesis 2:5 mentions that no man was available to *cultivate* the ground, followed by Genesis 2:15 which says man was put in the garden to *cultivate* it. The Hebrew word *abad* (Strong's 05647) occurs in Genesis 2:5, 2:15, 3:23, 4:2 and 4:12 and is used in these verses to describe cultivating or tilling the ground.

2:16A. And command/Yahweh/God/to the man/say/from all/tree the garden/eat/eat. (LHEIB)
2:16B. And command/Yahweh/God/to the man/say/from any/tree the garden/eat/you eat.
2:16C. And Yahweh God command[06680] to the man say, "From any[03605] tree [of] the garden, you [may] eat[0398]-eat[0398]."
2:16D. And Yahweh God commanded to the man saying, "From any tree of the garden, you *may* eat-eat."

2:17A. And from tree/the knowledge/good/and evil/not/eat/from + he/because/on day/eat + you/from + he/die/die. (LHEIB)
2:17C. And from [the] tree [of] the knowledge [of] good and evil, [do] not eat from, because on [the] day[03117] you eat from it [you shall] die[01491]-die[04191].
2:17D. "And from the tree of the knowledge of good and evil, *do* not eat from, because on the day you eat from it *you shall* die-die."

Word repetition was commonly used by writers of the Old Testament to stress a point. A form of the expression "eat-eat," found in Genesis 2:16, occurs several times in the Old Testament. God wanted Adam to eat fully of the garden and repeated the words for emphasis.

The expression "die-die," found in Genesis 2:17, occurs more than twelve times in the Old Testament. Repetition of the word "die" was used to emphasize the certainty of Adam's death if he ate from the tree of the knowledge of good and evil. In the KJV, this Hebrew expression is typically translated "surely die." The following three examples contain the identical expression found in Genesis 2:17 and show how it was used by those in authority to emphasize the certainty of death as a punishment.

Abraham misled the king of Gerar by telling him Sarah was his sister. As a result, Abimelech took Sarah to be his wife. God appeared to him in a dream and said to return her to Abraham. Genesis 20:7 says, "Now therefore restore the man his wife; for he is a prophet, and he shall pray for thee, and thou shalt live: and if thou restore her not, know thou that thou shalt surely die, thou, and all that are thine." (KJV)

When David was fleeing from King Saul, Ahimelech the priest gave David food and Goliath's sword. When Saul was told of this act of kindness, he was angry at Ahimelech. His response is recorded in 1 Samuel 22:16 when he says, "And the king said, 'Thou shalt surely die, Ahimelech, thou, and all thy father's house.'" (KJV)

King David's son, Absalom, and his followers attempted to take the throne by force. As David fled from Jerusalem during the initial conflict, he was cursed by Shimei, the son of Gera. David regained his throne and eventually his son Solomon became king. He confined Shimei to Jerusalem as punishment for his disrespect. In 1 Kings 2:37, King Solomon decreed, "For it shall be, on the day you go out and cross the Brook Kidron, know for certain you shall surely die; your blood shall be on your own head." (NKJV)

• This is the first occurrence of the Hebrew word *âkal* (Strong's 0398), here translated "eat." It occurs 810 times in the Old Testament and is translated "eat" 604 times, "devour" 111 times and "consume" 33 times, in addition to other miscellaneous words in the KJV.[102] The similar expression "eat-eat" found in Genesis 2:16 occurs in 1 Sam. 14:30; Ezek. 3:1 and Ezek. 16:13.
• This is the first occurrence of the word *mûwth* (Strong's 04191), here translated "die." It occurs 835 times in the Old Testament and is translated "die" 424 times, "dead" 130 times, "slay" 100 times, "death" 83 times, "surely" 50 times and "kill" 31 times, in addition to other miscellaneous words in the KJV.[103]
• The identical expression in Genesis 2:17, containing the two Hebrew words die (מות) and die (המות) occurs in 12 verses in Gen. 20:7; 1 Sam. 14:44, 22:16; 1 Kings 2:37, 2:42; 2 Kings 1:4, 1:6, 1:16; Jer. 26:8; Ezek. 3:18, 33:8, 33:14.[104] A similar expression occurs in Judges 13:22; 1 Sam. 14:39; 2 Kings 8:10; 2 Sam. 12:14.

2:18A. And say/Yahweh/God/not good/be/the man/alone/make for + he/helper/as counterpart + he. (LHEIB)
2:18C. And Yahweh God say, "[It is] not good [that] the man be[01961] alone. For he [I shall] make[06213] as his counterpart[05048] [a] helper.
2:18D. And Yahweh God said, "*It is* not good that the man be alone. For him *I shall* make as his counterpart a helper."

This verse is traditionally translated:
• And the LORD God said, It is not good that the man should be alone; I will make him an help meet for him. (KJV)
• And Jehovah God saith, "Not good for the man to be alone, I do make to him an helper—as his counterpart." (Young's Literal Translation, 1898[105])

The Hebrew word *neged* (Strong's 05048), here translated "counterpart," occurs in Genesis 2:18 and 2:20 when referring to Adam's helper. The Hebrew text is essential to identify this word. *Strong's Exhaustive Concordance* defines *neged* to mean a "counterpart" or "mate." Brown-Diver-Briggs Hebrew lexicon says *neged* can mean "corresponding to" or "opposite to."[106]

Because of its descriptive nature, *The Lexham Hebrew-English Interlinear Bible* rendering of *neged* as "counterpart" is retained in the *exemplar* translation. In many respects, woman is sexually and emotionally the "opposite" or "other side" of man. Their strengths and weakness may be viewed as complementary in marriage.

- This is the first occurrence of the Hebrew word *neged* (נגד), here translated "counterpart." It occurs approximately 170 times in the Old Testament and is translated "before" 124 times, "opposite" 34 times, "front" 5 times, "nearby" 3 times, "counterpart" 2 times and "side" 2 times in the *The Lexham Hebrew-English Interlinear Bible.*
- *Neged* is often used in describing someone or something that is opposite. Joshua 5:13 says, "And it came to pass, when Joshua was by Jericho, that he lifted his eyes and looked, and behold, a Man stood opposite[05048] him with His sword drawn in His hand. (NKJV)
- *Neged* is often used in describing someone or something that is on the other side. 2 Kings 3:22 says, "Then they rose up early in the morning, and the sun was shining on the water; and the Moabites saw the water on the other side[05048] as red as blood." (NKJV)

2:19A. And form/Yahweh/God/from the ground/all animal/the field/and/all bird/the heaven/and bring/to the man/see/what call to + he/and all/what/call to/the man/creature/living/that/name. (LHEIB)
2:19B. And form/Yahweh/God/from the ground/each animal/the field/and/each flier/the sky/and bring/to the man/to see/what call to + he/and all/what/call to/the man/creature/living/that/name.
2:19C. And Yahweh God form[03335] from the ground[0127] each[03605] animal [of] the field[07704] and each[03605] flier[05775] [of] the sky[08064] [within the garden]. And [God] bring [them] to the man to see what he [would] call[07121] to [them] and what[0834] all[03605] the man call[07121] to the living[02416] creature[05315], that [was its] name[08034].
2:19D. And Yahweh God formed from the ground each animal of the field and each flier of the sky *within the garden*. And *God* brought *them* to the man to see what he *would* call them and whatever the man called the living creature, that *was its* name.

The animals God formed in Genesis 2:19 were specifically for the Garden of Eden. Animals of the field and fliers of the sky are mentioned, yet no marine creatures are specified. If this verse were talking about all the Earth, marine life should be included. Since the garden was located on arid land, the absence of marine animals in the garden is not unexpected and suggests there were no large bodies of water in the garden as a habitat.

2:20A. And call/the man/name/to all the domestic animal/and to bird/the heaven/and to all/animal/the field/and for Adam/not find/helper/as counterpart. (LHEIB)
2:20B. And call/the man/name/to all the beasts/and to flier/the sky/and to all/animal/the field/and for Adam/not find/helper/as counterpart + he.
2:20C. And the man call[07121] name[08034] to all[03605] the beasts[0929] and to [the] flier[05775] [of] the sky[08064] and to all[03605] animal [of] the field[07704] [within the garden]. And for Adam [was] not find as he counterpart[05048] [a] helper[05828].
2:20D. And the man called names to all the beasts and to the fliers of the sky and to all animals of the field *within the garden*. And for Adam *was* not found as his counterpart a helper.

2:21A. And cause to fall/Yahweh/God/deep sleep/upon the man/and sleep/and take/one/of rib + he/and close up/flesh/in place of. (LHEIB)
2:21B. And cause to fall/Yahweh/God/deep sleep/over the man/and sleep/and take/one/from rib + he/and close up/flesh/under.
2:21C. And Yahweh God cause [a] deep sleep to fall over[05921] the man and [during] sleep and take from one [of] he rib and close up[05462] under[08478] [the] flesh.
2:21D. And Yahweh God caused a deep sleep to fall over the man and *during* sleep took from one of his ribs and closed up under the flesh.

Genesis 2:21 describes the creation of Eve. It should be obvious that God did not need to take anything from Adam to create Eve. He could have created her out of dust from the ground just as He did man, but apparently wanted to show a symbolical relationship.

From a biotechnology perspective, the biblical description of the making of Eve is entirely plausible. It was easier to take Adam's DNA and modify it to make Eve, rather than beginning completely anew using elemental materials extracted from the ground. God could have used DNA from any cell in Adam's body, but chose a rib. Bones contain stem cells in their marrow, which from a human viewpoint are easier to manipulate. God did not need an entire rib, but only material from a rib. A single stem cell would have been sufficient. The Hebrew text supports the explanation of taking material from one of Adam's ribs.

In reference to this verse, Matthew Henry said that the woman is "not made out of his [man's] head to rule over him, nor out of his feet to be trampled upon by him, but out of his side to be equal with him, under his arm to be protected, and near his heart to be beloved."[107]

Ideally, woman is a counterpart to man in the marriage relationship. Their roles are different, yet important. Conceptually a man and woman bring different physical, mental and emotional attributes to marriage, and these traits complement each other to achieve success. Because of Eve's sin in eating of the tree of knowledge of good and evil, God changed woman's original role in marriage according to Genesis 3:16.

• The Hebrew word *tachath* (Strong's 08478), here translated "under," occurs 544 times in the Old Testament, although only a few occurrences are mentioned in most concordances.[108] *The Lexham Hebrew-English Interlinear Bible* translates *tachath* primarily as "under" and "in place of." In the first two chapters, *tachath* occurs in Genesis 1:7, 1:9 and 2:21.

2:22A. And build/Yahweh/God/the rib/which take/from the man/into woman/and bring + she/to the man. (LHEIB)
2:22C. And Yahweh God build[01129] which [he] take [of] the rib from the man into [a] woman, and bring she to the man.
2:22D. And Yahweh God built—which *He* took *of* the rib from the man—into a woman, and brought her to the man.

This verse is traditionally translated:
• And the rib, which the LORD God had taken from man, made he a woman, and brought her unto the man. (KJV)
• And the Lord God built the rib which he took from Adam into a woman: and brought her to Adam. (Douay-Rheims, 1899)

The KJV mistranslates the Hebrew word *bânâh* (Strong's 01129) in this verse. *Bânâh* means "build" rather than "made," and there is no need to hide its meaning because of theological influence. Noah *built* an altar in Genesis 8:20, Abraham *built* an altar in Genesis 22:9, Jacob *built* an altar in Genesis 35:7, the children of Israel *built* cities for Pharaoh in Exodus 1:11 and God as the Master Builder *built* Eve from Adam's rib in Genesis 2:22. The Douay-Rheims Bible correctly translates *bânâh* as built.

• The Hebrew word *bânâh* (Strong's 01129), here translated "built," is used throughout the Old Testament to describe building altars, cities and houses. *Bânâh* is also used in the sense of building children in Genesis 16:2 and Genesis 30:3. Words based on the root *bânâh* occur 376 times in the Old Testament and are translated "build" 340 times, "build up" 14 times, "builder" 10 times and "built again" 2 times, in addition to other miscellaneous words in the KJV.[109]

2:23A. And say/the man/this/the time/bone/from bones + I/and flesh/from flesh + I/

for this/be called/woman/for/from man/be taken this. (LHEIB)
2:23B. And say/the man/this/now/bone/from bones my/and flesh/from flesh my/for this/called/woman/because/from man/taken this.
2:23C. And the man say, "This[02063] [is] now[06471] bone from my bones and flesh from my flesh; for this[02063] [shall be] called[07121] woman, because this[02063] [was] taken from man."
2:23D. And the man said, "This is now bone from my bones and flesh from my flesh; for this *shall* be called woman, because this *was* taken from man."

2:24A. Therefore/leave man/father + he/and mother + he/and cling to/to wife + he/ and be/as flesh/one. (LHEIB)
2:24B. Over so/leave man/father + he/and mother + he/and hold fast/on wife + he/ and be/as flesh/one.
2:24C. So[03651] over[05921] [this reason a] man [will] leave he father and he mother and hold fast[01692] on he wife. And [they shall] be[01961] as one flesh.
2:24D. So over *this reason,* a man *will* leave his father and his mother and hold fast on his wife. And *they shall* be as one flesh.

God's original plan was for one man to be married to one woman for life, but man quickly violated this arrangement. Genesis 4:19 records Lamech as the first man to have more than one wife. The practice of having two or more wives at the same time, or having a succession of wives through marriage and divorce was not God's plan. This is emphasized by Jesus' teaching in Matthew 19:7-9 and by the requirements for elders (bishops) and deacons in 1 Timothy 3:1-13 and Titus 1:5-9. In the New Testament church, leaders were expected to live in accordance with God's original plan for marriage.

The Old Testament is filled with stories of jealousy and infighting because of the practice of polygamy. Wives fought over their husband's affection and a desire to be the favorite. There was fighting among the children for power, inheritance and the right of succession. Jealousy was a continuous issue. The problems Jacob and David had with their wives and children are a prime example of this conflict.

In addition to the above, sexual relations outside of marriage violate God's plan and have led to a host of sexually transmitted diseases. Diseases transmitted only by sexual contact could be eliminated within one or two generations by strict adherence to a one-man, one-woman, permanent relationship.

2:25A. And be/two/naked/the man/and wife/and not/be ashamed. (LHEIB)
2:25B. And be/two/naked/the man/and wife/and not/ashamed.
2:25C. And [the] two[08147] be[01961] naked[06174], the man and [his] wife, and [were] not ashamed.
2:25D. And the two were naked, the man and *his* wife, and *were* not ashamed.

Conclusion

In the biblical debate over origins, theology should be evaluated based on how well it accurately reflects the Hebrew text. Theology should be assessed on how well it respects the Bible as truthful, trustworthy and inspired of God. Explanations should be judged on how well they harmonize special and general revelation. This is the standard by which all viewpoints of the Genesis story must be measured.

Traditional translations of Genesis Chapters 1 and 2 show the creation story to be illogical, incomprehensible and incorrect on a number of issues. The creation of the heavens and Earth and their description in Genesis 1:1-2 do not easily fit the rest of the creation account. The making of the Sun on the fourth day appears to be out of chronological order with the creation of light on the first day. In addition, the creation of all stars on the fourth day strongly disagrees with physical evidence in regard to time. Alleged statements of fiat creation on the second, fourth, fifth and sixth days are repetitive. God speaks and instantaneously creates, and then repeats himself for good measure. In Genesis 1:20 in the KJV, the statement "let the waters bring forth abundantly the moving creature" appears to support the evolution of life from a primordial soup. In Genesis 2:5, a description of the entire Earth as dry and barren conflicts with the abundant vegetation described on the third day. In Genesis 2:20, the naming of all animals on Earth by Adam in one day is unreasonable. The creation of the entire universe in six days approximately 6,000 years ago irreconcilably conflicts with empirical science in regard to time.

On the other hand, *exemplar* creation confirms the Genesis account as a sophisticated story of beginnings. Considering the limitations of the Hebrew language, the writer of Genesis was a master wordsmith. The narrative in Genesis 1:3-31 occurs during a sequence of six 24-hour days. During the first five days and part of the sixth, God describes what He made and created in the undefined past prior to Genesis 1:3. He describes the completed, working product. On the sixth day God begins creating again. He creates Adam and Eve, plants the Garden of Eden and forms a small number of animals for the garden. Genesis Chapter 2 is a detailed description of events on the sixth day related to man and the garden. This account easily harmonizes with empirical science in regard to time.

Exemplar creation is based on a credible interpretation and translation of several key verses in Genesis Chapters 1 and 2. Sixteen items that distinguish the *exemplar* translation from the KJV are listed below in a descending order of importance. (Appendix C gives a side-by-side comparison of the ESV and the *exemplar* translation.) The treatment of Genesis 1:3, 6, 9, 11, 14, 20, 24 as statements rather than commands substantially exceeds all other items in significance.

1. In Genesis 1:3, 6, 9, 11, 14, 20, 24 alleged commands following, "And God said" are replaced by descriptive statements.
2. Genesis 1:6-7, 14-17, 20-21, 24-25 are not repetitive commands. Rather, a

descriptive verse is followed by one acknowledging God as maker, creator or giver.
3. In Genesis 1:7, 16, 21, 25 the verbs "make" and "create" refer to events prior to Genesis 1:3 as a result of God's actions over a period of time.
4. In Genesis 1:2 the illegitimate term "without form" is replaced by "vacant."
5. In Genesis 1:9 the illegitimate term "let be gathered together" is replaced by "wait." The word "appear" is replaced by "seen" as a more common translation.
6. In Genesis 1:10 "gathering together" is replaced by "collection" as a more accurate term.
7 In Genesis 1:4, 14, 18 "divide" is replaced by "distinguished" as a more specific term.
8. In Genesis 1:10, 11, 12, 20, 22, 24, 26, 28, 29, 30, 2:5, 6 "earth" is replaced by "land" as a more specific term.
9. In Genesis 1:1, 2, 15, 17, 25, 26, 28, 2:1, 4 "earth" is replaced by "Earth" (to indicate the planet) as a more specific term.
10. In Genesis 1:17 "set" is replaced by "gave" as a more common term.
11. In Genesis 1:31, 2:2, 3, 4 "made" is replaced by the more general "did."
12. In Genesis 2:22 the incorrect "made" is replaced by "built."
13. In Genesis 1:14, 15, 16 "lights" are replaced by "luminaries" as a more specific term.
14. In Genesis 1:20 the incorrect phrase, "Let the waters bring forth abundantly the moving creature that hath life" is replaced by, "The waters swarm *with* swarms of living creatures."
15. In Genesis 1:21 "great whales" are replaced by "great reptiles" as a more accurate term.
16. In Genesis 1:5 "the first day" is replaced by "day one" to reflect a cardinal rather than an ordinal number.

All current English translations of the Bible are seriously in error regarding the Genesis creation story. This error has been driven by traditional young-Earth creation theology for more than twenty-two hundred years. Starting in the 1700s, science began to challenge traditional translations of the Genesis creation story. Physical evidence supported a growing belief that the Earth was significantly older than a 4004 BC creation date. This data has grown to the point where the traditional Genesis account is in irreconcilable conflict with general revelation in regard to time, in turn affecting the credibility of the entire Bible.

Sound hermeneutics and exegesis now challenge traditional interpretations of the Genesis creation story. At the heart of the problem is the Septuagint translation and to a lesser extent Masoretic vocal marks. Through their work, Hebrew scribes embedded traditional young-Earth creation theology into the Old Testament. They treated ten imperfect verbs as jussive verbs and transformed seven descriptive statements into commands. This in turn forced the illegitimate translation and poor translation of several related words. These errors have resulted in a translation of Genesis that is indefensible based on the original text of the Bible and a study of general revelation.

Most Hebrew scholars and theologians are aware of some of the issues raised in this chapter. The great majority view the Genesis creation account with absolute puzzlement over its conflict with science in regard to time, yet believe current interpretations are correct. A radical shift in mindset is needed. This begins by realizing the Septuagint translation is: (1) not inspired, (2) contains a number of errors and (3) has traditional young-Earth creation theology embedded within it. The Septuagint has been the primary driver behind incorrect explanations of the creation account and resulting English translations. This in turn has forced the layman to choose between rejecting the overwhelming evidence of empirical science or rejecting the integrity of scripture in regard to the creation account.

A concentrated study rather than casual reading is required to fully understand this chapter, Chapters 1 and 5, and Appendix B. A complete understanding of the information contained herein provides a firm basis for accepting *exemplar* creation as the original creation story that God inspired Moses to record. When properly interpreted and translated, the Genesis creation account is an abbreviated, non-technical, historical account that can be accepted as truthful and trustworthy without apology. Although written thousands of years ago, the creation story is logical, accurate, carefully structured and sophisticated when judged by twenty-first century standards. The Genesis account stands significantly above *all other* religious and philosophical explanations of origins, whether recent or ancient.

There is no intent in this chapter to imply English translations of the Bible are filled with errors. Once past the Genesis creation story and the account of Noah's flood, a significant amount of controversy over the Old Testament disappears. While there is room for improvement in current English Bibles (the more consistent translation of Hebrew words and the identification of added English words), no significant beliefs are misrepresented.

The fear of the LORD is the beginning of knowledge, but fools despise wisdom and instruction. (Proverbs 1:7) KJV

Chapter 7

Design Theory

Evidence of intelligent cause in the world is often subtle, yet clearly evident when the limitations of natural cause are understood. For example, nearby is a park with the normal amenities of trees, grass, bushes and wide open spaces. A casual observer strolling through on weekends might conclude the park is completely natural, yet this is not the case. The grass is always kept between two and three inches tall. There are no weeds, thorns, poison ivy, wild vines or thistles. There are no dead trees due to disease or age, nor are there any broken or twisted tree limbs from high winds or tornadoes. Uniformly spaced rows of trees, all of the same age, size and type grow parallel to some of the walking paths—as opposed to a random pattern. Rather than a layer of decaying leaves, seeds and twigs under each tree from years of accumulation, healthy green grass surrounds the trunk. Close examination reveals a park significantly different from a true natural environment found in a typical forest.

In order to determine intelligent cause or influence, we must understand the effects of totally natural cause. For example, a person cruising on the Missouri River might consider it natural, especially along stretches where there is no evidence of human habitation for miles. If we removed the obvious signs of human influence such as dams, wing dams, levees and revetments, the river would still not be natural. The Corps of Engineers has dredged the river to ensure a guaranteed channel depth to aid river navigation. In addition, the river is virtually free of imbedded trees, debris piles, floating logs and limbs (a big problem for early riverboat traffic), again courtesy of the Corps of Engineers. The present state of the Missouri River is considerably different from that observed by Lewis and Clark when they began their historic journey up the river in 1804.

Discovering the influence of intelligent cause is similar in some respects to a skilled tracker following a person though a forest. The tracker must carefully distinguish between natural and human cause. Footprints are one sure sign of human presence. The remains of a campfire and cut wood might be another. A pile of stones stacked in a pyramid shape, arranged in a row or placed in a circle would be strong evidence. A discarded chewing gum wrapper, cigarette butt or soda container is proof of human presence. Other evidence might be more subtle such as crushed blades of grass, broken twigs, scuffed moss, disturbed stones or bent limbs. With knowledge of both intelligent and natural cause, human passage through a forest can be determined by careful observation.

As the above illustrates, intelligent cause can be inferred without directly observing the causal agent. Just as we can determine air is a real substance by noticing the effects of wind, we can determine intelligent cause by studying its effects. For example, a round brilliant cut diamond with 58 facets can only be attributed to intelligent cause—there is no other plausible explanation. Even if uncovered while mining deep within the earth, found laying on the ground in a remote jungle, encapsulated inside a meteor, or discovered on a distant planet, we would attribute the gem to intelligent cause. Even though natural cause is capable of turning carbon into diamond under the right conditions of heat and pressure, it is totally incapable of forming a brilliant cut diamond with its precise angles and shapes. Likewise, when we see objects about us in the physical world that are totally beyond the ability of natural cause based on a detailed knowledge of the structures, mechanisms and materials involved, we can attribute them to intelligent cause. This is not blind faith, but a product of rational investigation and logical inference.

The three known categories of *cause* can be arranged in a descending order based on their known capabilities. First is intelligent cause responsible for design of the first life. The smallest single-celled organism displays a high level of information and functional complexity many orders of magnitude above the achievements of our brightest engineers. Second is intelligent cause evidenced in man-made products. These exhibit a high level of complexity far beyond the ability of natural cause. Third and ranked dead last is natural cause, capable of building only the simplest of structures such as snowflakes, mineral crystals, concretions, freeze-thaw rock circles, sand dune formations, dried mud crack patterns, eroded structures, clouds, sand particles and salt grains.

Definition 7.1 – Design

Design is intelligence directed change and has a factual and theoretical component. Design as fact is the observation that human designers have created information and complex functional objects not found in nature. Design as theory proposes that intelligent cause is primarily responsible for the information, complex functional systems, mathematically improbable structures and fine-tuned arrangements, not of human origin or influence, found in nature and the universe at large.

A consideration of intelligent cause leads naturally to a discussion of design. A study of *human design* aids in recognizing *generic design* found in living things and the universe at large.

How should design be defined in general terms? In Chapter 2, evolution is defined as being both a fact and a theory. The same can be said of design. Design is intelligence directed change and has a factual and theoretical component. Design as fact is the observation that human designers have created information and complex functional objects not found in nature. Design as theory proposes that intelligent cause is primarily responsible for the information, complex functional systems, mathematically improbable structures and fine-tuned arrangements, not of human origin or influence, found in nature and the universe at large. Natural cause is severely limited as a constructive mechanism.

Human Design Phenomena

Science and technology have radically improved the human condition within the last two centuries. Applied science in the form of medicine, agronomy, pharmacology, transportation and communications has significantly improved the way we live. Human design has made it possible for mankind to have the highest standards of transportation, communication, housing, clothing, food, medical care, entertainment and general comfort the world has ever known. Human design has made possible many new products beneficial to mankind, not found in nature, and in fact completely beyond the ability of natural cause.

Human design is solidly based on real-world experience and our knowledge of physical phenomena. Human design is founded on empirical science and its investigation of the materials, energies, forces, structures and relationships found in nature. Dependable knowledge about the physical world is used by engineers to create new products.

Science helps us understand and control physical phenomena. Knowledge of heat and temperature show that mercury can be frozen, water can be boiled and steel can be melted. Experiments prove that a spark can ignite gasoline to create an explosion, the basis of the internal combustion engine. Personal experience shows that coal, wood, petroleum and diamonds can be burned in the presence of oxygen. Gases cool when they expand and become hot when compressed, the basis for air conditioning. Pressure drops when velocity increases across the top of an airfoil, the basis for aircraft flight. These and a host of other observations and experiments show how natural phenomena can be controlled to achieve specific results in the form of technological products.

Science helps us discern the capabilities and limitations of undirected change. If elemental materials can be randomly mixed in a test tube to create microscopic life, then that is how nature works. If mineral crystals can spontaneously grow into computer chips, then that is how nature functions. If a volcano can spew out molten material that self-forms into chisels, wrenches and drill bits, then that is how nature operates. However, we have never observed these events, no record exists of their occurrence in the past and experiments show they will not happen. This is how nature really works.

From a design perspective, a comprehensive understanding of physical phenomena is of paramount importance. New discovery is the carrot that drives scientific research. The discovery of new materials, processes and methods permits the creation of new designs to benefit humankind. On the other hand, understanding the limitations of phenomena prevents repetitive attempts at doing the impossible. Engineers try not to reinvent the wheel, which includes repeating past mistakes.

Human design is evidenced by the number of products that scientists, engineers and technicians have made. Through the selection, control, sequencing and processing of physical materials, humans have built things not found in nature and in fact completely beyond the ability of natural cause. Human intelligence has directed the laws of chemistry and physics to construct a host of manufactured goods found in industrialized nations.

The accomplishments of human designers can be broadly grouped into four categories, illustrated in Table 7.2. Human design phenomena include (1) information, (2) complex functional systems, (3) mathematically improbable structures and (4) fine-tuned arrangements.

1. Information – A Human Design Phenomenon

The first major category of human design is information. Although an exact definition of information is difficult, it is nevertheless real. Some of its different aspects include design information, information languages and information systems.

Humans can create *design information* in the form of requirements and instructions for building an airplane, car, computer, watch or cell phone. Design information specifies the materials, processes, shapes, sizes, fits, arrangements and sequences for making things. Good design information is specific, integrated, understandable, self-contained and portable. In the information hierarchy, design information is at the very top.

Information, other than design information, exists in many forms and has a multitude of uses. Information can represent accumulated scientific and historical knowledge. Information can sequence the timing of traffic lights and route telephone calls over a network. Information can program a computer to accomplish complex calculations or make a video game possible. Information can represent human art, music, literature, thoughts and ideas. Information can be found in written material (e.g. books, magazines and newspapers) that can be read, stored, copied and exchanged or in audio recordings, video recordings and photographs.

The storing of information has been a pursuit of humans for thousands of years. Early in history, people recognized that if information is not physically retained in some manner it is easily lost. Writing, clay tablets, parchment, papyrus, paper, the printing press, books, copy machines, binary code, computers and compact electronic storage devices represent steps in the development of information storage.

After stored information became commonplace, humans recognized the need to retrieve and copy information, which led to the development of information systems.

A human information system can be defined as having five major components represented by (1) the information itself, (2) the information language, (3) the physical storage media, (4) a means of retrieving information and (5) a method of copying. For example, information might be the story of Tom Sawyer. English is the information language and the storage media is a book in a library. A retrieval system is a person to find the book, while a duplicating system consists of a person and a copy machine.

Computers have revolutionized the way information is handled. For example, information consists of the story of Tom Sawyer in an electronic file. The information language is binary code and the storage media is a computer hard drive, floppy disk, CD or flash drive. A retrieval system consists of software and hardware capable of reading a computer file, while a copy system can duplicate an electronic file or make a paper facsimile.

2. Complex Functional Systems – A Human Design Phenomenon

The second major category of human design is the creation of complex functional systems, defined as an integrated arrangement of discrete parts working together to produce a unique end result not possible by any single part acting alone. For example, mousetraps, light bulbs, electronic calculators, DVD players and laptop computers represent different complex functional systems. A light bulb is made of individual parts, each having unique characteristics, when integrated together contribute toward total system performance. Both discrete parts and complex systems may be characterized as follows:

(1) Unique material properties – Discrete parts in human designs often require unique materials for their function. Two unique materials in a mousetrap are the wood of the base compared to the steel spring. The wooden base must be soft to allow the insertion of staples while the spring must be elastic to store mechanical energy.

(2) Unique shape – Discrete parts in human designs typically require a unique shape for their particular function. Unique shape in a mousetrap is the configuration of the spring which is radically different from the catch, both required for their respective functions.

(3) Unique size – Discrete parts in human designs usually require a unique size (or size range) for their function. Unique size in a mousetrap is illustrated by the size of the hammer, the part that kills the mouse. If the hammer is too large or too small, it will not work as intended.

(4) Unique fit – Discrete parts in human designs commonly require the unique shape and size of one part to fit another. In a ballpoint pen, this is represented by the precision fit of the ball with the tip of the ink stem. If the fit is too tight, the ball will not roll and ink will not flow properly to permit writing. On the other hand, if the fit is too loose the ball may fall out or the ink will leak excessively.

(5) Unique function – Discrete parts in human designs usually perform a unique function not performed by any other part in the system. The clip on a ballpoint pen can hold the pen onto a shirt pocket. No other part on the pen can perform that function.

(6) Similarity – Complex systems of human design often contain similar discrete parts. The reuse of proven design concepts results in similarity. For example, bolts, batteries, switches, insulating wire, light bulbs, transformers, gears, bearings and wheels are used in a multitude of human-designed systems. Human designers have found that certain designs work well and as a result use those designs repetitively.

Discrete parts used in different systems are typically similar rather than identical. For example, an automobile, copy machine and refrigerator contain many similar parts. They all have wheels, bearings, switches, electrical wires, electric motors, screws and hinges. Though it is doubtful any of the components are identical and thus interchangeable, they are similar.

Proven design concepts modified for size, shape and material allow greater applicability. For example, hinges may be used on a house door, kitchen cabinet, refrigerator, microwave, laptop computer, car door or on the landing gear doors of an F-18 fighter aircraft. All of these hinges are based on the same concept, but have been modified for a specific application.

Parts are often standardized to make them identical rather than similar. Small parts (fasteners, washers, light bulbs, clips, terminals, etc.) are often made to a standard drawing, with the goal of creating identical parts that are fully interchangeable for a number of different applications.

In general, similarity is evidence of a common designer, a common design philosophy, the intelligent reuse of successful designs or a design iteration improvement series.

(7) Controlled sequences – Human designed manufacturing systems use controlled sequences to assemble products. For example, all well-engineered assembly lines for tanks, trucks, televisions, torpedoes, toasters, toothpaste and other technological products have a well-defined *critical path.* Parts identified as being on the critical path must be assembled in an exact sequence.

In addition, complex systems of human design often function in a controlled sequence. For example, when the thermostat in a natural gas heating system senses a preset temperature, a switch is closed, in turn activating a relay which sends current to the furnace igniter. After the igniter becomes hot, current passes through a solenoid, opening a valve and allowing gas flow to the burners. As a last step, electric current to the igniter is turned off.

(8) Feedback and control – Complex systems of human design often use feedback and control. For example, a thermostat combined with a furnace is a feedback and control system. The thermostat senses heat and turns the furnace on and off to keep a constant room temperature.

(9) Redundancy – Complex systems of human design often use redundancy to improve dependability. For example, automobiles have two independent hydraulic brake systems to prevent a complete loss of braking ability if a single brake line fails.

(10) Locking mechanisms – Complex systems of human design often use locking mechanisms to prevent dangerous actions, prevent theft or unwanted change. For example, if an automobile is moving, a safety interlock prevents the trunk from being opened, while another interlock prevents key removal from the ignition unless the shift lever is in park. In an airplane, a safety interlock prevents retraction of the landing gear while supporting the aircraft's weight. Locks are used on houses, cars and file cabinets to thwart thieves and vandals. Locks are used on floppy disks, flash drives and computer files to permit information to be accessed and copied, but not changed.

(11) Predetermined response – Complex systems of human design often use sensors to trigger a predetermined response to changing conditions. For example, some automobile headlights automatically turn on when sensors detect a reduced light level. On some cars, when rain is sensed the windshield wipers are activated.

(12) Embedded processing – Complex systems of human design often used embedded processing, made possible by the development of electronic computer chips, to allow sophisticated control of a product. Embedded processing permits stored information to be accessed, sensor data to be monitored and inputs to be processed for real-time response. Embedded processors are used extensively in portable products. For example, embedded processors in an automobile engine (the electronic control module) permit complete control of the fuel, air and ignition. Embedded processors have increased the sophistication of many household appliances such as cooktops, microwave ovens, dishwashers and washing machines. Embedded processors are used extensively in jet engine controllers, automatic flight control systems, fighter aircraft, helicopters, cruise missiles, unmanned air combat vehicles and smart bombs.

3. Mathematically Improbable Structures – A Human Design Phenomenon

The third major category of human design involves mathematically improbable structures, best illustrated by mass production. Henry Ford is normally considered the father of mass production. He introduced the Model T in 1908 and in 1913 installed a moving assembly line in his factory. By 1927, fifteen million Model T automobiles had been manufactured. Mass production reduced manufacturing costs significantly, allowing Ford to undercut his competition and dominate the automobile industry.

The mass production of automobiles requires part consistency. With the metals and machining technology that existed in 1908, it was relatively easy to make a single lug bolt. The difficulty was making thousands of lug bolts exactly the same, which became one of the major challenges Ford faced in producing the Model T and succeeding automobiles. In his assembly line, Ford needed lug bolts and hundreds of other parts manufactured to the same configuration so they would assemble correctly and perform dependably. The control of variation in mass production has always been a challenge because *nature favors randomness*. Identical mass-produced parts are unnatural and may be viewed as mathematically improbable.

Successful mass production requires the control of randomness. Manufacturing begins with a design drawing that describes the requirements for each part. For example, the drawing for making an automobile lug bolt must specify all dimensions with proper tolerances. The steel material, heat treatment, surface finish and any coatings must be specified. A design drawing controls the configuration of the lug bolt, and any variation from drawing requirements is unacceptable. If threads on a lug bolt do not match those of the mating nut, assembly is prohibited. Parts that do not fit at assembly are an obvious indicator of a design or quality problem. Successful mass production requires firm control of the manufacturing process to reduce the natural tendency of parts to vary.

Industry is acutely aware that mass production requires the control of randomness (i.e. variation). Configuration control, quality control, defect reduction, variability control and statistical quality control are some of the names given to programs with the goal of reducing variation and improving quality in mass-produced products. For example, the strict control of variation permits all cans of soup to have the same weight, ingredients and taste. Controlled variation allows all computer chips to have the same structure and performance. The dependability of automobiles, cell phones, computers, televisions and other consumer products are due in part to the relentless pursuit of quality through reduced variation.

By reducing variation, two objectives are achieved for mass-produced products. First, things become interchangeable. A ball bearing used in an automobile can be used interchangeably with any other ball bearing of the same type. A battery of a given type can be used in any flashlight or hearing aid. Design standards exist for hundreds of items from bolts to batteries. Common design standards and controlled variation make them interchangeable.

A second benefit of reduced variation is consistent performance. Once a product has been developed to a given level of performance, engineers prefer that all subsequent items be the same. Consumers want their radios, compact discs, MP3 players and DVD recorders to perform consistently. Uniform performance is achieved by reducing variation in the manufacturing process. All the steel for balls in a bearing must be of the same composition and hardness within defined limits. All balls must have the same diameter, roundness and surface finish within tight tolerances. Only by reducing variation in the manufacturing process to a small amount, can ball bearings of consistent quality be produced.

4. Fine-Tuned Arrangements – A Human Design Phenomenon

The fourth major category of human design is fine-tuning. Engineers and technicians can fine-tune multiple variables in a complex system to achieve optimum performance. For example, in an automobile engine timing of the spark that ignites fuel during the compression stroke represents fine-tuning. If timing varies by more than a few degrees of crankshaft rotation, performance can be dramatically affected. The voltage, amperage and duration of the spark are additional variables. In the same engine, timing for the opening and closing of exhaust and intake valves can affect fuel economy and performance. The fuel to air ratio and fuel injection sequence impact fuel economy. Advances in performance, economy and smoothness of the automobile engine during the last fifty years are primarily due to the precise control of multiple variables, and represents fine-tuning.

Categorizing Human Design

Human design can be grouped into three basic categories: original design, design iteration and design by menu selection. The use of all three design forms, illustrated in Table 7.2, has permitted human designers to create a multitude of products available in our modern world.

(1) Original design – Original design is the most difficult type of design. Development of the first working steam engine, cotton gin, threshing machine, gasoline engine, light bulb, ballpoint pen and airplane was difficult and came about through a combination of mechanical ingenuity, intense motivation and hard work. A new invention requires the designer to conceptualize the integration of a number of detailed parts to make the completed working product. This is followed by the painstaking design, fabrication and assembly of all the required pieces. Unanticipated problems or errors in any step of the process result in failure. Original designs typically require a significant amount of research, development and testing (intelligent directed trial and error).

(2) Design iteration – Design iteration follows original design. In most instances, the first working product of a new invention is quite crude and refined through a series of improvement cycles. Typically, a single design iteration improvement cycle consists of the design, manufacture and sale of a product, followed by improvement. Ideas for improvement come from customer feedback, testing and general technological advancement. Engineers take the old design, slightly modify it and begin the cycle all over again. For example, the telephone represents a design iteration improvement series. The first hand-crank telephone was an original design that developed into the rotary dial phone. It subsequently changed into the touch tone phone, which became the cordless phone. This progressed to the basic cell phone, which morphed into the camera cell phone with Internet capability. Essentially all mass-produced, complex functional products reflect design iteration improvement, illustrated by automobiles, airplanes, motorcycles, televisions, microwave ovens and refrigerators. The ultimate goal of design iteration is to improve efficiency, robustness, miniaturization, capability or sophistication.

Table 7.2 – The Realm of Human Design

Empirical science
- Dependable knowledge about the physical world
- A comprehensive understanding of physical phenomena
- Discerning the capabilities and limitations of undirected change

Design phenomena
- Information
- Complex functional structures
- Mathematically improbable structures
- Fine-tuned arrangements

Discrete part characteristics
- Unique material properties
- Unique shape
- Unique size
- Unique fit
- Unique function

System characteristics
- Similarity
- Controlled sequences
- Feedback and control
- Redundancy
- Locking mechanisms
- Predetermined responses
- Embedded processing

Design categories
- Original design
- Iteration improvement
- Design by menu selection

Design traits (human)
- Product expansion

In a design iteration improvement series, three characteristics are often exhibited. First, most changes in the series will be small, but some will be large and represent a leap in technology. For example, the change from rotary dial phones to touch tone phones was made possible by development of the transistor and printed circuits. The second characteristic is commonality among products, which leads to the third characteristic of common quality defects. Because of common parts among products, they may contain the same defects. For example, automobiles are often recalled because of quality defects. Most recalls affect several different models over multiple years of production because of common defective parts.

(3) Menu selection – Design by menu selection chronologically follows original design and design iteration. Once a specific technology has advanced sufficiently, parts and pieces become standardized. This permits designing new products by selecting from a menu of available parts. Some tailoring may occur, but the new design is primarily an integration of existing components. For example, most houses are designed from available products consisting of shingles, roof trusses, floor beams, studs, siding materials, plumbing fixtures, lights, windows, doors, kitchen cabinets and floor coverings. Virtually all items that go into a house are standard. These can be mixed, matched and tailored to build a unique house.

Another example of menu design is the personal computer, capable of being constructed from an array of off-the-shelf parts including chips, mother boards, hard drives, wiring harnesses, power supplies, cooling fans, CD readers and burners, cases, monitors and keyboards. Design by menu selection normally occurs in a mature technology.

Product Expansion – A Human Trait

Product expansion is a trait of humans and human design. To meet consumer demands, engineers have designed dozens of different kinds of flashlights, hats, cookware, shoes, electric toothbrushes and ballpoint pens. Creative design engineers can take a single product and through small changes make hundreds of related products. There is no rhyme or reason for the diversity of products available today, other than people like variety. A desire for variety, which in turn drives product expansion, appears to be a trait inherent within human nature.

Product expansion is best illustrated by the growth of consumer products in America within the last two centuries. Product expansion is represented by the less than fifty food items in a small town general store in 1800 compared to the 35,000 items typically available today in a large supermarket.[1] Product expansion is represented by the five hundred items or less in a small town mercantile store in 1800 contrasted to the 120,000 items found in a typical Wal-Mart store.[2] If we include clothing, pharmaceuticals, home supply and toys found in large retail chains that specialize in these goods, consumers in a large metropolitan area have access to more than 250,000 items. The number and variety of products available in industrialized countries is a testament to human creativity and represents massive product expansion.

Design Theory

As pointed out in Definition 7.1, design is both a fact and theory. The factual part of design is represented by the accomplishments of human designers. Humans have shown themselves to be quite creative and have built a multitude of structures and products completely beyond the ability of natural cause. Human design aids in the study of generic design. With human design as a solid foundation, design theory becomes a logical explanation for information and functional complexity not of human origin.

Definition 7.3 – Design Theory

Design theory is an inferred explanation that intelligent cause is primarily responsible for the information, complex functional systems, mathematically improbable structures and fine-tuned arrangements, not of human origin or influence, found in nature and the universe at large. Natural cause is severely limited as a constructive mechanism, while intelligent cause, represented by human achievement, has demonstrated considerable ability. Undirected change is rejected as a complete explanation for nature and the universe at large.

Design theory is an inferred explanation that intelligent cause is primarily responsible for the information, complex functional systems, mathematically improbable structures and fine-tuned arrangements, not of human origin or influence, found in nature and the universe at large. Natural cause is severely limited as a constructive mechanism, while intelligent cause, represented by human achievement, has demonstrated considerable ability. Undirected change is rejected as a complete explanation for nature and the universe at large.

Design theory is based upon (1) knowledge of phenomena in the physical world about us, (2) acknowledging the limitations of undirected change and (3) human design.

(1) Knowledge – The first leg of design theory is founded on a thorough knowledge of physical phenomena based upon observation, measurement and experimentation. Design theory relies on verified empirical data and empirical science. Design theory rests on a comprehensive understanding of nature and the universe at large—both its capabilities and limitations. Scientific research and increasing scientific knowledge are the foundation upon which design theory is built.

(2) Limitations of natural cause – The second leg of design theory is based on acknowledging the limitations of natural cause as a constructive mechanism. Undirected phenomena are capable of building simple structures represented by snowflakes, mineral crystals, concretions, freeze-thaw rock circles, sand dune formations, dried mud crack patterns, eroded structures, clouds, sand particles and salt grains. Natural cause *has not* demonstrated the ability to build information or complex functional objects, and belief to the contrary is completely faith-based. The severe limitations of natural cause compel the consideration of design theory.

(3) Human design – A third leg of design theory is based on understanding *human design* as an aid to recognizing *generic design*. Beginning with the industrial revolution two centuries ago, human design has made dramatic advances and is responsible for the explosion of technical products beneficial to mankind. By a careful analysis of human design, its phenomena, characteristics, categories and traits can be identified. Items that distinguish human design in Table 7.2 are not hypothetical or theoretical, but represent reality in the world of engineering, technology and manufacturing. Human design has identifiable features separate from natural cause.

Design theory proposes a design agent or agents other than human. No presupposition is made regarding the origin, identity, character, moral nature or competency of the agent(s). Design theory does not require the identification of a specific agent. By necessity, any design agent(s) must have originated from outside the limitations of the known universe and have the prerequisite intelligence and power. Design as a scientific theory does not appeal to the Bible, other religious books or religious beliefs in an attempt to identify a designer. Design theory is religiously neutral.

Design theory is not an all-encompassing explanation of origins. Theory is specifically limited and primarily focuses on identifying evidence of design. Design theory is a means of rationally explaining information and functional complexity as an alternative to natural cause. In regard to how the universe began, the origin of basic matter, basic energy and space may be considered beyond the purview of design theory. There is no requirement to explain the origin of the design agent(s).

Design and time have no firm relationship, yet natural cause as an alleged constructive mechanism requires extensive time. Design can occur quickly or over great periods of time. In human terms, the Great Pyramid, Taj Mahal, Great Wall of China, Alaska Pipeline, English Channel Tunnel and Petronas Tower of Malaysia represent great design achievements. The time taken to build these structures represents the resources, priorities and self-imposed schedule of the builders. The complex nature of these structures, not the time, is what identifies them as products of design. Whether these structures were built in a day or a million years is irrelevant. If built over a million year span of time, they would still be a product of design. Undirected change is totally incapable of building these structures, even given billions of years. Time is not a magic wand capable of curing the severe limitations of natural cause.

The Limitations of Natural Cause

Design theory accepts the limitations of natural cause. Undirected change as a destructive mechanism can wear away, deteriorate and randomize things. As a weak constructive mechanism, unguided phenomena have demonstrated the ability to produce simple effects, elementary arrangements and mathematically probable structures.

In contrast, philosophical naturalism requires natural cause to be a powerful constructive mechanism when given significant time. Philosophical naturalism requires natural cause to exceed man's design accomplishments, represented by the first microscopic living organism. Natural cause allegedly created single-celled life, something human biologists and biochemists cannot do.

The severe limitations of natural cause are a key issue in discussing philosophical naturalism and supporting evolutionary ideas. Undirected phenomena have not exhibited the ability to produce anything comparable to the music, art, literature and technology humankind has created. Random change has not shown any ability to make double-hung windows, faucets, gas furnaces, water heaters, chandeliers, dishwashers, garbage disposals or central air conditioners. The unguided laws of chemistry and physics have not revealed any ability to duplicate what human designers have accomplished.

Human-produced products can be examined to evaluate the limits of natural cause. For example, undirected phenomena have not demonstrated the ability to create a Phillips screwdriver consisting of a hardened steel shank and plastic handle. Undirected mechanisms cannot take raw iron ore, deoxidize it, remove the impurities and add a small amount of vanadium, carbon and chromium as the ingredients for tool steel. Undirected change cannot melt, shape and harden the steel into a shank with the precise Phillips screwdriver shape on one end. Undirected phenomena cannot combine carbon, oxygen, hydrogen and nitrogen in the proper structure to form Nylon 66 resin. Uncontrolled force cannot shape resin around the steel shank to create a handle. Unless dogmatically driven by some ulterior motive, no scientist, engineer or technician would propose that a screwdriver can occur because of strictly natural cause. Any person who believes otherwise is technologically and scientifically ignorant. A screwdriver is the product of selected materials and processes directed by human engineers and technicians. A screwdriver illustrates the power of intelligence to sequence and control phenomena found in nature, and to control application of the laws of chemistry and physics.

Single-celled life found in nature can also be used to evaluate the limits of natural cause. *Mycoplasma genitalium* is often considered to be the smallest, independent, single-celled, living organism known, yet it is enormously complex and significantly more sophisticated than the space shuttle. No undirected phenomena have been identified remotely capable of creating *Mycoplasma genitalium* from carbon, hydrogen, oxygen, nitrogen, other elements and simple molecules. Extensive research and experiments have failed to discover a plausible natural explanation for this organism or more complex microscopic organisms believed to represent the first life on Earth. If the present is the key to the past, then natural cause completely and unequivocally fails as an explanation for the origin of life.

If we accept the belief that natural cause is a powerful constructive mechanism, then philosophical naturalism is salvaged. However, the problem remains of how to explain the evolution of a Phillips screwdriver and, by logical extension, evolution of the first single-celled life. To do this, evolutionists appeal to missing information, unknown phenomena and undiscovered laws. Their philosophical position forces a shameless appeal to natural metaphysical cause outside of human experience.

In some illogical way, proponents of philosophical naturalism consider natural cause superior to intelligent cause as a constructive mechanism. This is a major point in the design versus evolution debate. This is a significant issue in legal rulings by American courts that evolution is science and that appeals to intelligent cause are faith-based, inherently religious and outside the realm of science.

Generic Design in Nature

Design theory uses human design as an aid in recognizing generic design found in nature and the universe at large. Knowledge of human designers and their tools, methods, thought processes and products helps in understanding generic design. The four human design phenomena illustrated in Table 7.2 are strongly observed in generic designs as follows:

1. Information – A Generic Design Phenomenon

Design theory proposes that intelligent cause is responsible for the *design information* found within the DNA of living things. Design information in DNA specifies how to assemble proteins, cells, structures and systems in living things. Information in DNA specifies amino acids and their exact sequences in protein molecules. DNA information controls the growth, regulation and maturation of the human body. DNA builds the systems that control testosterone, estrogen, thyroid hormone, human growth hormone, blood sugar and temperature. Although not well understood, DNA contains information about cell differentiation, development and placement to create nerves, muscles, bones, tissues and organs. Discovering and comprehending all of the information within DNA represents an enormous challenge to scientists.

Design theory proposes that intelligent cause is responsible for the *DNA-based information system* found within living things. This system consists of (1) information, (2) an information language, (3) physical storage, (4) a means of retrieving information and (5) a means of copying information. *Information* stored in DNA is represented by the sequence of amino acids used to make a protein. *Language* consists of the genetic code capable of specifying one of twenty possible amino acids used to make a protein. *Physical storage* is the DNA molecule and the arrangement of base pairs within its structure. *Retrieval* occurs when information within DNA is used to make a protein molecule. *Copying* occurs when a new DNA molecule is created prior to cell division.

Sophistication in an information system is measured by speed, accuracy, durability, miniaturization and power requirements. Using these criteria, the DNA information system found within living things is significantly superior to human designs. The approximately 3 billion base pairs of DNA found in a single human cell are equivalent to a 1,000,000 page book with 3,000 characters per page.

2. Complex Functional Systems – A Generic Design Phenomenon

Design theory proposes that intelligent cause is responsible for the complex functional systems found in living things. The twelve characteristics previously identified in human design are also found in living things and strongly indicate a design influence.

(1) Unique material properties – Discrete parts in human designs often require unique materials. Within a biological system, unique materials include right-handed deoxyribose sugar molecules used to build DNA and left-handed amino acids used in building proteins.

(2) Unique shape – Discrete parts in human designs typically require unique shapes. Within living things, a distinctive shape may be one of twenty possible amino acid molecules available for building a protein. Unique shape is also represented by the curl of a protein molecule.

(3) Unique size – Discrete parts in human designs may perform a function based on their size. Within the cochlea of the human inner ear, the minuscule size of hair cells permits them to sway easily and allows the sensing of extremely low pressure sound waves acting on the ear drum. A person has more hair cells in their inner ear than hairs on their head.

(4) Unique fit – Discrete parts in human designs often require the unique shape and size of one part to fit another. Within living things, a single enzyme can fit the shape and size of a single molecule. For example, the enzyme catalase fits the hydrogen peroxide molecule and can break it apart into water and oxygen.[3]

(5) Unique function – Discrete parts in human designs often perform a unique function. Within mammals, proteins in hemoglobin found in red blood cells temporarily store and release oxygen, performing a function not possible by any other proteins in the body.

(6) Similarity – Complex systems of human design often share similar features. Within animals, similar features include feathers, scales, skin, blood, nerves, bones, genes, proteins and the genetic code.

(7) Controlled sequences – Complex systems of human design must often function in a specific sequence. In living things, the construction of the AMP molecule involves thirteen steps and twelve enzymes, and occurs in an exact controlled sequence.[4]

(8) Feedback and control – Complex systems of human design often use feedback and control. In humans, the control of body temperature at 98.6° F represents a precise feedback and control system. The regulation of thyroid hormone in the body represents a sophisticated feedback and control system.

(9) Redundancy – Complex systems of human design often have redundant features to prevent complete failure. In the human body, there are two hemispheres of the brain, two lungs, two kidneys, two eyes, two ears, two nostrils, two thyroid lobes, two hands, two feet and two testes or two ovaries.

(10) Locking mechanisms – Complex systems of human design often have locking mechanisms to prevent change from occurring. In living things, genes and other DNA segments have been identified that are invariant (show no variation). This may be the result of a locking mechanism that allows copying but not change.

(11) Predetermined response – Complex systems of human design often have designed responses to changing conditions. In animals, the heavy fur coat a dog develops during cold weather is a predetermined response. The ability of humans to build a tailored antibody to kill an invading virus is a predetermined response.

(12) Embedded processing – Complex systems of human design often use embedded processing to allow sophisticated control of specific functions. Embedded processing uses stored information and logic to control a function. For example, within humans DNA controls cell division and growth within a zygote. DNA additionally controls cell differentiation, reproduction, placement and death (apoptosis) during fetal development.

The complex functional systems discussed above are significantly different from *complex nonfunctional structures* (discussed further in Chapter 9). A complex nonfunctional structure could be a cell phone without a battery or a computer with a faulty motherboard—both incapable of performing their intended function. A complex nonfunctional structure can also be a pail of sand or the collection of stars into a galaxy. None of these four objects are capable of producing a unique end result due to their discrete parts working together in an integrated fashion. For example, although a spiral galaxy consisting of billions of stars represents a fine balance between centrifugal force and gravitational attraction, it does not produce any unique end result. The light produced is the simple collection from all stars in the galaxy.

3. Mathematically Improbable Structures – A Generic Design Phenomenon

At a very simple level, things in nature can be identical. At the *atomic level,* atoms of hydrogen, helium and the other elements exist primarily as exact copies of one another (natural isotopes are exceptions). At the *molecular level*, water, hydrogen peroxide, methane, ammonia, amino acids and other simple molecules form identical structures.

Beyond small molecules, nature strongly favors randomness. Above the small molecular level, the number of possible arrangements for atoms and molecules increases dramatically. Rather than being identical, things become unique. For example, simple structures such as snowflakes, specks of dust, grains of salt or particles of sand are all unique. No two are identical owing to randomness. Even though two snowflakes may appear alike to the eye, the mathematical probability of having the same number of atoms and water molecules in an identical structure is extremely small. The number of possible combinations due to random arrangement is simply too great.

As objects get larger, the probability of two or more identical structures decreases even farther. No two pieces of gravel in the world are the same. No two natural diamonds, rocks, natural gems, mineral crystals or chunks of coal are the same. In nature, these structures arrange themselves according to the laws of chemistry and physics in an uncontrolled fashion. These objects may be viewed as complex nonfunctional structures that are statistical unique.

Because nature favors randomness, the occurrence of two or more identical objects of sufficient size above the basic molecular level can be viewed as unusual and evaluated statistically. As the number of molecules in an item increases, the mathematical probability of two or more identical objects occurring decreases when the molecules are allowed to freely combine. Objects and object populations can be evaluated to determine if they are probable due to natural or intelligent cause.

Design theory proposes that intelligent cause is responsible for the process that makes *one chiral version of a molecule* found in living things. Chirality is the characteristic of a molecule to occur in both a left-handed and right-handed variety. The left-hand version is a mirror image of the right-hand. Nineteen of the twenty amino acids found in the proteins of living things are left-handed chiral molecules (glycine is the exception and is neither right-handed nor left-handed). Sugar molecules found in living things are also chiral, but are right-handed, and include deoxyribose sugar found in DNA and ribose sugar found in RNA. When these amino acids and sugars are made in the laboratory, they occur in statistically equal amounts of the left-handed and right-handed varieties. These experiments have been repeated many times and the results are always the same—chiral molecules form in a random fashion. The occurrence of only left-handed amino acids and right-handed sugars in the first microscopic life can only be attributed to intelligent cause.

Design theory proposes that intelligent cause is responsible for the process that makes *identical protein molecules* found in living things. Within a single plant or animal, there are hundreds of proteins more than 100 amino acids long with the same sequences. These proteins are exact copies at the molecular level—their structures are absolutely identical. The presence of exact copies of proteins in the first single-celled living organism can only be rationally explained by intelligent cause.

4. Fine-Tuned Arrangements – A Generic Design Phenomenon

Design theory proposes that intelligent cause is responsible for the *fine-tuned conditions* on Earth favorable to life. The Earth and Sun may be viewed as two carefully matched components. Earth's orbital distance and current atmosphere are just right for the amount and type of energy received from the Sun. The distance from the Earth to the Sun is approximately 149,600,000 km (93,000,000 miles). If this distance was thirty percent smaller or larger, the amount of radiant energy hitting Earth would be significantly different. Provided all other conditions remained the same, life on Earth as we know it would not be possible. The Sun is a stable, G2 star, representing only five percent of the stars in its neighborhood.[5] The Sun does not have large fluctuations observed in other stars that would make it deadly to life on Earth. Our Sun gives off a spectrum of light highly favorable to plant life.[6] The size, temperature and radiated energy of the Sun are fine-tuned to Earth's orbital distance.

There are a multitude of additional features that illustrate how the Earth is fine-tuned for life. These include the low level of harmful gases and the ratio of nitrogen and oxygen in Earth's atmosphere. Our planet's nitrogen, carbon and water cycles are crucial to living things. The abundance of water and its properties are favorable to life. Earth's strong magnetic field and ozone layer protect life on the planet's surface. When all possible conditions are considered, Earth is unique.

Determining Natural or Intelligent Cause

Unknowingly, design theory has been used for many years to determine natural cause as distinguished from intelligent cause. With knowledge of the *limitations of natural cause*, it can be ruled out as probable during criminal, accident and scientific investigation. With knowledge of the *capabilities of intelligent cause*, it can be accepted as probable during investigation. Both science and forensic science routinely consider intelligent cause in contrast to natural cause as follows:

(1) Suspicious death – Coroners can determine if death occurred by natural cause from a stroke, accident and heart attack. Death may also be attributed to intentional human action from poisoning, suffocation, blunt trauma, choking or electric shock.

(2) Arson investigation – Fire marshals often decide if a fire started due to natural cause from lightning or frayed wiring. Fire can also be the result of arson and the use of accelerants.

(3) Infectious disease – Medical investigators can determine if widespread disease is the result of a naturally occurring virus or bacteria. Disease may also result from terrorists intentionally spreading agents such as anthrax, smallpox or riacin.

(4) Accident investigation – Engineers acting as legal consultants routinely investigate failed machinery to determine if failure is due to natural cause. Failure may also be attributed to bad design, inadequate quality control, poor maintenance or improper operation.

(5) Human disease – Researchers often evaluate if human disease such as breast cancer is due to a hereditary genetic defect. Disease may be caused by exposure to human-developed agents such as cigarette smoke, asbestos, pesticides, herbicides, benzene, isocyanates or other non-natural chemicals.

(6) Human disorder – Doctors determine if natural cause has resulted in a heart, liver or kidney disorder. Disorder may also be attributed to side effects from taking a strong prescription drug.

(7) Cryptography – Cryptologists analyze code to decide if it is a random combination of symbols or contains a secret message.

(8) Computer error – Computer technicians determine if a computer malfunction is because of a natural hardware failure or an overlooked software error. Malfunction may also be the result of an intentionally designed computer virus.

(9) Ancient artifacts – Archaeologists sometimes examine artifacts to determine if they are naturally occurring products. An artifact may also be a tool, weapon or work of art made by an ancient civilization.

(10) SETI – Scientists examine electromagnetic signals from outer space to determine if they are naturally occurring random noise or if they contain information from extraterrestrial intelligence.

Design Theory versus Biblical Belief

Although design theory has theological implications, it is not a theological enterprise. In like manner, ideas of evolution are not a theological enterprise. However, both design theory and evolution theory have definite theological and philosophical implications.

Bible students and Christians should be aware of the distinct differences between design theory and biblical belief. Critics of design theory attempt to connect it with traditional young-Earth creation theology and the Bible. By making this connection, they discredit design theory as a scientific idea with scientific motives. This connection is the foundation for legal attacks (based on the separation of church and state) against design theory to keep it from being taught in public education and other public venues. Some of the major differences between design theory and biblical teachings are as follows:

(1) Explanatory scope – Design theory is a means of explaining the information and functional complexity found in the universe as a rational alternative to natural cause. Design theory focuses on evidence of design and is specifically limited. The Bible presents God as an all-encompassing explanation of origins. He created the heavens and Earth.

(2) Belief basis – Design theory is based upon (1) knowledge of phenomena in the physical world about us, (2) acknowledging the limitations of undirected change and (3) human design. The Bible is God's message to humankind and is the basis for belief in God.

(3) Agent identity – Design theory has no presuppositions as to the origin, identity, character, moral nature or competency of a design agent or agents. The Bible identifies God as the creator (sole design agent) of the universe. He is eternal, powerful, benevolent, knowledgeable, intelligent, wise, just, loving and holy.

(4) Religious appeal – Design theory does not appeal to the Bible, biblical theology or any religious belief. The Bible is inherently religious and identifies God as Creator, Lord and loving Father who sent his son Jesus Christ to Earth to die for the sins of humankind.

(5) Believer constraints – Design theory may be fully accepted by pagans, atheists, infidels and agnostics. A full acceptance and belief in the entire Bible as the truthful, trustworthy, inspired word of God is incompatible with beliefs held by the pagan, atheist, infidel and agnostic.

(6) Design issues – Design theory proposes that information, complex functional systems, mathematically improbable structures and fine-tuned arrangements within nature and the universe at large are primarily the result of design. The Bible is not a scientific book and does not address these detailed issues.

(7) Moral issues – Design theory does not address sin and moral issues. One of the main purposes of the Bible is to address human sin, God's condemnation of sin, the loving grace of God, personal redemption through belief in Jesus Christ and instructions for living a holy life.

An appeal to intelligent cause does not require identification of a designer, nor does it necessitate a religious entity. For example, we can firmly say a tricycle is due to intelligent cause. Based on the severe limitations of natural cause, we can reject evolution as a plausible explanation. Given billions of years of time, undirected change cannot create a child's tricycle. On the other hand, we know for a fact that humans can build tricycles. However, if the NASA Exploration Rover were to find a tricycle on Mars, it would not be of human origin. Some might attribute it to God while others would cite another deity or alien influence. We could argue indefinitely over the identity of the maker, but few would doubt its existence due to intelligent cause.

Conclusion

The examples of human design used in this chapter and summarized in Table 7.2 are factual rather than theoretical or hypothetical. They are real examples taken from personal experience in the world of engineering, technology and manufacturing. The products developed by human engineers give us insight into the characteristics of designed systems and establish a solid basis for design theory.

A hierarchy of cause shows that intelligent cause, responsible for design of the first single-celled living organism, ranks first. The smallest microscopic life is incredibly complex and completely dwarfs all human engineering achievement. Intelligent cause, represented by man-made products, ranks a distant second. Third and extremely limited is natural cause, capable of building only the simplest of structures.

Natural cause as a weak constructive mechanism has the ability to produce simple effects, elementary arrangements and mathematically probable structures. As a destructive mechanism, natural cause can wear away, deteriorate and randomize things. Undirected change best explains stellar ageing, stellar novae, meteor impacts and planetary cooling. Undirected change best explains oxidation, crumbling rocks, soil formation, water erosion, sediment hardening, accumulated marine sediments, volcanoes, earthquakes, natural fires, continental drift, tsunamis, hurricanes and floods. Evolution best explains naturally occurring amino acids, simple molecules, point mutations, defective genes, protein isoforms, chromosome abnormalities, genetic disease, cancer, aging, limited species divergence, new viral strains and the spread of an antibiotic-resistant bacterial population.

The best explanations in science arise when design theory is allowed to explain information and functional complexity due to intelligent cause, and evolution is allowed to explain undirected changed due to natural cause. A combination of these two ideas is fully capable of plausibly explaining all that is found in nature and the universe at large.

With him is strength and wisdom: the deceived and the deceiver are his. (Job 12:16) KJV

Chapter 8

The Myth of Chemical Evolution

When Charles Darwin proposed his theory of species evolution, the first microscopic life was assumed and believed to be composed of some kind of simple material. The first single-celled life just happened. Spontaneous generation was a popular idea at the time and could account for the first life. With microscopic life as a beginning, Darwin explained the naturalistic development of all subsequent species on Earth. Little did he know that explaining the first living organism would become the greatest challenge to evolution. In time, this issue by itself would show the bankruptcy of natural cause as an all-encompassing explanation of origins.

Ideas about the natural origin of life can be divided into chemical and biological evolution. Chemical evolution is the development of non-living material into the first living organism. Biological evolution (Darwinian and neo-Darwinian evolution) is the development of the first living organism into the multitude of Earth's plant and animal species. The mechanisms of chemical and biological evolution are different, though they both rely on undirected change. Chemical evolution may have occurred anywhere in the universe and is not restricted to Earth. Life may have come to Earth in the form of spores or bacteria. A completely naturalistic view of life requires that chemical evolution occurred first, followed by biological evolution.

As defined in Chapter 2, evolution has both a factual and theoretical component. Chemical evolution loosely falls under the theoretical component. Chemical evolution proposes that the first microscopic life in the universe developed from non-life due to natural cause. Known materials and mechanisms available for chemical evolution include elemental materials, simple molecules, random molecular combinations and other undirected phenomena acting over significant time. The first life may have occurred on Earth or within the universe at large.

There is no accepted theory of chemical evolution. It is simply an idea. Neither Charles Darwin nor anyone since has developed a credible explanation for the natural origin of life. No undirected phenomena have been identified to remotely account for the information and functional complexity found in microscopic life. However, supporters of chemical evolution firmly believe it happened. The National Academy of the Sciences in its book *Science and Creationism: A View from the National Academy of Sciences* says, "For those who are studying the origin of life, the question is no longer whether life could have originated by chemical processes involving non-biological components. The question instead has become which of many pathways might have been followed to produce the first cells."[1] This statement represents an entrenched philosophical position. Not a single plausible pathway has been identified to explain the naturalistic origin of life.

Definition 8.1 – Chemical Evolution

An idea that the first microscopic life in the universe developed from non-life due to natural cause. Known materials and mechanisms available for chemical evolution include elemental materials, simple molecules, random molecular combinations and other undirected phenomena acting over significant time. The first life may have occurred on Earth or within the universe at large.

Philosophical naturalism is the primary driver behind chemical evolution. Philosophical naturalism requires the "fact" of chemical evolution. Every single step in the development of life must have a strictly natural explanation. Therefore, all appeals to intelligent cause are categorically rejected, including design theory.

Chemical Evolution Overview

A strictly natural explanation for life requires the occurrence of chemical evolution. It is not an option. Chemical evolution is a required precursor to biological evolution. Proponents absolutely believe chemical evolution occurred and have developed the following eight concepts in support of their convictions.

(1) Evolution is a fact – Philosophical naturalism considers the chemical evolution of life a fact. Life exists, therefore natural cause created the first living organism. Regardless of the impossibility of life evolving, nature somehow overcame all obstacles. Chemical evolution happened—there are no alternatives. A compelling lack of scientific evidence in support of this belief is dismissed as irrelevant. This position represents dogmatic faith in philosophical naturalism.

(2) Evolution needs no proof – People who accept philosophical naturalism often treat chemical evolution as a "given" and focus on biological evolution as an explanation for how life evolved. This belief is stated in McLean v. Arkansas Board of Education (1982 U.S. District Court Case). Because of the difficulty in attempting to explain how the first living organism came about by natural cause, it is easier to ignore the issue.

During Darwin's time, the prevailing theory of spontaneous generation explained the appearance of the first life on Earth. Simple life just happened and could be easily "proven" by a number of popular scientific experiments. Broth in a sealed glass container would grow microorganisms after several days, evidence interpreted as proof of spontaneous generation.

In 1860, a French scientific organization came forward to help resolve the spontaneous generation issue. A year after Darwin published *The Origin of Species*, the French Académie des Sciences offered the Alhumbert Prize of 2,500 francs for the best attempt by experimentation to throw new light on the question of so-called spontaneous generation. Louis Pasteur became interested and began experimenting with bacteria. He showed that sterilized broth inside a swan-necked flask would not grow microorganisms. Spontaneous generation would not occur in a sterile environment. He wrote a paper describing his experiments and was awarded the Alhumbert Prize in 1862.

Even though Louis Pasteur won the prize, spontaneous generation continued to be a popular idea for many years. Félix Archimède Pouchet, Director of the Museum of Natural History in Rouen, France, was a staunch supporter of spontaneous generation and rejected Pasteur's conclusions. Pouchet obtained different results than Pasteur with his sterile culture experiments. In 1876, Pasteur debated Dr. Henry Charlton Bastian, a physician and professor of pathological anatomy at the University College in London. Bastian was a strong advocate of spontaneous generation and had also obtained different experimental results than Pasteur.

Between 1862 and 1876, knowledge of bacteria slowly emerged through the work of Ferdinand Cohn, Robert Koch, John Tyndall and others. They helped establish the existence of highly heat-resistant bacterial endospores. During this time, Pasteur struggled to develop a comprehensive understanding of aerobic and anaerobic bacteria found in a variety of substances. The presence of bacteria in air was an especially controversial subject.

Pasteur began to suspect that the liquids used by Pouchet, Bastian and others in their sterile culture experiments were not truly sterile. Pasteur and his associate, Charles E. Chamberland, discovered that certain materials, notably milk, cheese, egg yolks, dried meat, urine and infusions of hay would not remain sterile when boiled. They discovered that some liquids had to be heated to 115° C for twenty minutes to kill heat-resistant microbial life, accomplished by a pressure steam sterilizer (autoclave) invented by Chamberland.

With a better understanding of germs and improved sterilization techniques, sterile culture experiments became more consistent and failures could be explained. In 1878 Pasteur gave a lecture at the French Académie de Médicine on germ theory. Criticism and doubt about Pasteur's experiments and ideas were slowly laid to rest around this time, after more than sixteen years of debate.[2] The theory of spontaneous generation, accepted for more than two thousand years, was finally discredited and represents one of the major reversals of scientific thought in the history of science.

Chemical evolution should be subject to scientific investigation and analysis. Philosophical naturalism requires the undirected, uncontrolled, random origin of life. However attractive this belief, it should be subject to the scrutiny of empirical science. If no plausible mechanisms can be identified, chemical evolution should be rejected along with spontaneous generation. Just as *negative experiments* discredited spontaneous generation, they must be allowed to discredit chemical evolution.

(3) Life evolved due to unknown laws – Some individuals believe chemical evolution occurred because of yet undiscovered laws of chemistry and physics. Known laws do not support chemical evolution. Extensive scientific research has not brought to light any natural phenomena capable of plausibly explaining single-celled life. The book *Biochemical Predestination* by Gary Steinman and Dean H. Kenyon released in 1969 gave hope to discovering a strictly naturalistic mechanism for life, but this has not occurred. Any appeal to unknown phenomena is strictly an unsupported, faith-based exercise.

(4) Life evolved by special effects – It has been speculated that chemical evolution occurred because of special effects during the distant past. Proponents say that self-emergent, complex functional systems developed the first life. No one can identify these mechanisms or explain why they are not observed today. One of the basic principles of science is the uniformity of nature. We live in an orderly universe where the laws of chemistry and physics observed today on Earth are consistent throughout time and space. A plea to special effects appeals directly to unknown phenomena and natural metaphysical cause.

(5) Life evolved elsewhere – This belief proposes that life came to Earth on an asteroid, meteor, comet or space dust. Chemical evolution happened somehow, somewhere in the vast universe. Panspermia (belief that the seeds of life came from outer space) and speculative concepts in astrobiology encourage this belief. Alleged Mars rocks found in Antarctica and evidence of surface water on Mars give hope to this idea. The discovery of other planets orbiting distant star systems expands the scope of possible chemical evolution. Billions of worlds are now hypothetically available as birthing grounds for life. Chemical evolution has more time to occur and is not limited by the age of the Earth, its atmosphere or available resources.

On the surface, belief that life evolved somewhere on a distant planet appears impressive; however, the issues are formidable. An initial problem involves a rock achieving escape velocity. Rocks do not fly into space of their own volition. Hypothetically, this occurs when a large meteor impacts a planet with weak gravity. Impact results in the ejection of fine dust and small particles into space. Another problem involves the ability of life to survive from thousands to billions of years in space during transit to Earth. An object moving at the speed of comet Hyakutake at 156,000 km/hour (43.3 km/sec.) would take 31,000 years to travel from the nearest star, Alpha Centauri. During this time, ultraviolet radiation, high energy particles, vacuum and extreme cold would kill any organism. The probability of an unguided object striking the Earth from another planet five light years away or more is extremely remote. Another problem includes the ability of an organism to survive the heat of reentry. Small rocks burn up during entry into Earth's atmosphere. Putting these problems aside, there is no reason to believe another planet would be more conducive to chemical evolution than Earth. The impossible task of life developing by natural means is not diminished by moving the event to another location in the universe, distant in time and space. This belief is fraught with more problems than having life evolve on Earth.

(6) Life is a result of alien influence – Some people believe aliens seeded the Earth with life. Rather than life arriving by chance, it was intentionally brought to this planet by intelligent beings with a means of space travel. This conclusion subtly concedes that life is too complex to have developed on Earth by natural means. With this conclusion, the next question is obvious, "Where did the aliens come from?" This forces the issue back to a discussion of chemical evolution elsewhere in the universe.

Several organizations currently promote the search for life beyond Earth. The Search for Extraterrestrial Intelligence (SETI) Institute is a major force in the hunt for intelligent life beyond our solar system. Through radio and optical telescopes, and millions of dollars in public and private support, believers in extraterrestrial intelligence look to the heavens to validate their belief. SETI is considered serious scientific research. Major promoters of SETI include the late Dr. Carl Sagan (professor of Astronomy at Cornell University and author of 17 books), The Planetary Society (with a membership of more than 100,000), The National Radio Astronomy Observatory and the NASA Astrobiology Institute. Not a single example of intelligent life has been found.

(7) Multiverse theory can account for life – Some scientists speculate that our universe is one of many universes. Physical laws and materials in other universes are allegedly different. Based upon this idea, the origin of life is no longer constrained by the resources of our universe. On the down side, scientists have failed to find any direct evidence to support multiverse theory. It is complete speculation to believe microscopic life could find its way between two universes and travel for billions of years across the vast space among galaxies to Earth. Given the estimated age of the universe, it is not old enough for an object traveling at low sublight speeds to go from the edge of our universe to Earth. Multiverse theory represents speculative, unsupported, philosophical musings and is a veiled appeal to natural metaphysical cause.

(8) Life is not supernatural – Some believe that if man can create life, it will prove life did not originate due to supernatural cause. At present, scientists are unable to make life using any known technique, process or technology. The smallest known organism is too complex to build starting from small biomolecules. Scientists are learning how to manipulate existing life by cutting, splicing and duplicating DNA. Genes have been modified and transplanted between species, while plants and animals have been cloned. These activities are substantially different from making life.

To create life from non-living molecules is an incredibly challenging task. If perhaps five hundred years from now man has the technology to make a microscopic living organism starting from basic molecules, it will only confirm that an intelligent entity can make life. The technical difficulty of such an accomplishment will further discredit the origin of life due to natural cause. The making of a microscopic organism will show that humans can take physical materials and processes and build a non-natural, extra-natural and in fact supernatural living organism beyond the ability of natural cause. Such an accomplishment will show that design theory is the only rational scientific explanation for the origin of life.

Scientific Thought Favorable to Chemical Evolution

A review of scientific effort in favor of chemical evolution helps to provide perspective. The book *Genesis: The Scientific Quest for Life's Origin* by Robert M. Hazen, PhD, gives a comprehensive discussion of recent research and thought in support of chemical evolution. Dr. Hazen received his MS degree from Massachusetts Institute of Technology and his PhD from Harvard University. He conducts astrobiological research at the Carnegie Institute in Washington, DC. Hazen has performed personal research in support of chemical evolution, appears favorable to the idea and is an authority on the subject.

Hazen attempts to structure the origins debate by saying, "Barring divine intervention, life must have emerged by a natural process—one fully consistent with the laws of chemistry and physics."[3] By barring divine intervention, philosophical naturalism and natural cause are subtly established as the only viable explanation.

The debate over the origin of life can be framed in several different ways. The debate can be framed between science and faith, between science and religion or between science and biblical belief. By implication, anything not science is false. Debate can be more equitably framed between philosophy and religion, between philosophical naturalism and biblical belief, or between natural cause and intelligent cause. The most objective and meaningful approach is to frame the debate between *evolution theory* and *design theory*, with science as the arbitrator. Science is a search for dependable knowledge based on physical evidence and its logical inferences. Natural cause is a biased assumption of science rather than a dogmatic requirement.

As pointed out in Definition 2.9, evolution theory uses natural cause over time to explain things that are clearly not of intelligent origin or influence. Philosophically, some scientists believe evolution is a powerful constructive mechanism when given significant time. Therefore, somehow somewhere the first single-celled life evolved.

Based upon Definition 7.3, design theory rejects natural cause as an explanation for life because of its severe limitations. A design agent is responsible for the information, complex functional systems, mathematically improbable structures and fine-tuned arrangements found in microscopic life. The first single-celled life in the universe was due to intelligent cause.

When the origin of life debate is defined between evolution theory and design theory, then real progress is possible. Scientists can conduct research to determine if undirected phenomena can create life or if the limitations of natural cause preclude its development. As a contrasting idea, scientists can evaluate the possibility of creating life through intelligent cause. Humans, as surrogate design agents, can manipulate amino acids, proteins, DNA, genes and other biological components in an effort to create life. When framed in this manner, scientists can evaluate whether natural cause or intelligent cause is a more rational explanation. Contrary to what some have said, an appeal to intelligent cause does not inhibit science or stop research.

Most arguments about chemical evolution discuss the Second Law of Thermodynamics. Hazen points out that according to this law, heat always flows from warmer to colder regions and never the other way.[4] Heat energy loses its ability to do work. Structured systems deteriorate and become unstructured. Things tend to become randomized. Disorder and entropy increase. The Second Law of Thermodynamics goes against the natural formation of organized structures necessary for chemical evolution to occur.

Proponents of chemical evolution say that while the Second Law of Thermodynamics is a total system effect, local effects permit functional complexity to emerge.[5] Oft-cited examples of organization arising in a natural system include ripple structures due to blowing sand, dried mud crack patterns, concretions and freeze-thaw rock circles.[6] These examples of simple organization are extrapolated to support the idea that complex functional systems can naturally emerge. This extrapolation is completely speculative and unsupported by empirical science. No “law of emergence” or “principle of self-organization” based strictly on natural cause has been identified capable of creating a complex functional system.

The Miller-Urey experiment is considered a major benchmark for chemical evolution. Hazen discusses the good and bad of this experiment, first published in the May 15, 1953, issue of *Science*. Stanley Miller and Harold Urey put methane, ammonia, hydrogen gas and water vapor inside an enclosed chamber and subjected them to an electric spark.[7] Molecules from the combination of these gases were collected in a water trap. Some of the twenty essential amino acids in living things were found among the many non-biological amino acids and molecules produced. The basic Miller-Urey experiment has been repeated many times with different gas mixtures, ultraviolet radiation and powered minerals.[8] Some gases used in these experiments, such as concentrated hydrogen cyanide, do not occur naturally. Through the careful selection of gases and other ingredients, many of the basic molecules for life have been formed to one extent or another.

The Miller-Urey experiment is often described as a breakthrough experiment. Some consider it compelling synthesis research. In reality, this experiment fails to solve a host of problems dealing with chemical evolution. Hazen points out that a Miller-Urey type experiment cannot build the complex structures and macromolecules (large biomolecules such as proteins, enzymes, DNA and RNA) necessary for life as we know it. He says, "Macromolecules tend to fragment, rather than form, when subjected to the energetic insults of lightning and the Sun's ultraviolet light."[9] Hazen reinforces this point again in referring to Miller-Urey type experiments when he says, "Interesting molecules inevitably emerge, but those energetic processes effectively prevent the formation of essential macromolecular structures, including the polymers and membranes required by all known life-forms."[10] An effect in reverse of what is needed occurs. Large molecules break apart into smaller ones rather than growing into bigger, more complex structures.

In the exploration of Earth, scientists have found life growing under unusual circumstances. Ocean floor volcanic vents have been discovered to contain organisms that thrive in this hot environment. Thermophilic bacteria have been identified living in water near these vents at temperatures of 200° F and more. Volcanic vents are typically rich in iron, copper, zinc and sulfur compounds. Worms and bacteria in this environment are unique since they live without sunlight while under great pressures and elevated temperatures. Their main source of energy appears to be hydrogen sulfide. Some scientists believe that life evolved around these vents. Stanley Miller emphatically rejects this idea.[11] The temperature of a volcanic vent environment is too high for amino acids and essential molecules to form because heat is destructive. Hot hydrothermal zones are antithetical to life.[12]

Meteorites from space can contain amino acids.[13] For example, the Murchison meteorite that fell near Melbourne, Australia, in September 28, 1969, contained amino acids. Even though the presence of amino acids in meteors is a mystery, it does not prove chemical evolution. No proteins, DNA or life have ever been found on a single meteor. Despite the media hype about rocks from Mars containing life, this has not been confirmed. Hazen and other scientists are highly skeptical of the Mars rock claims.[14]

Hazen discusses lipid molecules as support for chemical evolution. Molecules of this type have shown the ability to form tiny spheres or vesicles.[15] Lipid molecules can form self-organized lipid layers and are speculated to have formed the first cell membranes. Even though lipid molecules are interesting, Hazen says they are substantially simpler than cellular membranes found in living things and are not equivalent.

Clays and their effect on amino acids and RNA nucleotides are often considered supportive of chemical evolution. Clay material can cause both amino acids and RNA nucleotides to link up and form short chains.[16] Some scientists have speculated that clays provided a scaffold which directed the organization of the first living organism. Hazen points out that the structure within clay is essentially random and cannot account for the information and precise sequencing found within proteins, RNA and DNA.

Exhaustive research in support of chemical evolution has reached a dead end. After 246 pages of exploring scientific thought about the chemical evolution of life, Hazen brings *Genesis* to a close. Although presented in a favorable light, a careful reading of his book reveals that evolution of life research has failed to explain the formation of macromolecules necessary for life. In reality, science is now farther away from showing the possibility of chemical evolution than it was in 1953 when the Miller-Urey experiment was announced. The more we have discovered about single-celled life, the less plausible chemical evolution has become. Hazen discusses a multitude of evolutionary ideas and experiments in his book, but all have failed to come to fruition under the harsh reality of empirical science. Hazen is to be commended for his thorough (although one-sided) presentation of this subject. While his book closes in an upbeat manner with a solid commitment to chemical evolution, no significant evidence is presented in support of this belief.

In *Darwin's Black Box*, Dr. Michael Behe sums up the current state of affairs in regard to chemical evolution by saying, "Scientists working on the origin of life deserve a lot of credit; they have attacked the problem by experiment and calculation, as science should. And although the experiments have not turned out as many hoped, through their efforts we now have a clear idea of the staggering difficulties that would face an origin of life by natural chemical processes."[17] Speculation that life originated due to natural cause has thus far been thoroughly discredited by extensive scientific research.

As a result of the many failed experiments and discredited ideas, the search for the origin of life appears to be moving beyond the boundaries of Earth. The hunt for life has moved outside the constraints of Earth's resources and the domain of empirical science. Astrobiology is now the great hope of chemical evolution. Other planets and moons within our solar system must have the answer. If not there, perhaps elsewhere in the Milky Way Galaxy. Beyond our galaxy lie billions of other galaxies. Proponents of philosophical naturalism absolutely believe that somehow, somewhere, someway life evolved due to natural cause.

Seven Ideas that Discredit Chemical Evolution

There are at least seven major ideas that discredit chemical evolution. These ideas primarily address the lack of a viable mechanism to explain how life could evolve from non-life. Experiments and extensive investigation conclusively show that the undirected combination of elemental materials and simple molecules cannot create the macromolecules necessary for life. Our detailed knowledge about the complexity of single-celled life and the severe limitations of natural cause strongly discredit chemical evolution.

(1) Based on present knowledge of single-celled life, scientists have failed* to *develop a plausible explanation for the naturalistic origin of the first viable living organism.

No likely explanation has been developed for the evolution of life. *Life-The Science of Biology* by authors Purves, Sodava, Orians and Heller is a science textbook.[18] Chapter 25, "The Origin of Life on Earth," discusses the chemical evolution of life from basic elements and molecules. The eight pages of material presented in favor of this idea are embarrassing. The authors struggle to make this theoretical wasteland appear scientific and respectable. Other than the Miller-Urey experiment, no evidence is given in support of chemical evolution. Discussion consists primarily of conjecture and speculation.

The resources available for chemical evolution are quite limited. Resources include billions of years of time, elemental materials, simple molecules, random molecular combinations and other undirected phenomena. Currently no accepted hypothesis exists to explain how these resources can account for the information and complex functional structures found in single-celled life. Time is not a magic wand capable of curing the severe limitations of natural cause.

The gulf between non-life and the smallest single-celled independent organism is too large to be bridged by natural cause. There is no continuum of small steps between the two. Dr. Michael Denton in *Evolution: A Theory in Crisis* states, "Between a living cell and the most highly ordered non-biological system, such as a crystal or a snowflake, there is a chasm as vast and absolute as it is possible to conceive."[19] This statement is repeated by Hazen when he says, "The great mystery of life's origin lies in the huge gap between molecules and cells."[20] No scientific hypothesis has yet emerged capable of rationally explaining natural cause as a means of bridging this gap.

The Achilles' heel of natural cause is its inability to form complex functional systems and create information. Complex functional systems require a number of unique parts to be arranged in a specific relationship. This is true in regard to radios, cell phones, computers, automobiles, cruise missiles and smart bombs. It is ludicrous to believe a personal computer with its hardware and software could evolve from basic elements and simple molecules over billions of years. It is likewise absurd to believe the first life—millions of times more complex than a computer—could occur because of undirected change. As Hazen says, "The simplest living cell is intricate beyond imagining, because every cell relies on the interplay of millions of molecules engaged in hundreds of interdependent chemical reactions."[21] This level of functional complexity defies an evolutionary explanation.

It is important to realize that natural selection does not play a role in chemical evolution. Biological evolution proposes that a random beneficial mutation can occur to give an organism a survival advantage. Natural selection can act upon a beneficial mutation and, combined with heredity, spread it throughout a species. Hypothetically, by repeating this process over and over a species can improve. However, before the first microscopic life existed, there was no heredity. Before the first life there was no DNA, no DNA information, no hereditary mutations, no hereditary laws, no genes, no chromosomes and no sexual reproduction. Natural selection cannot act on non-hereditary features in non-living things to build complexity in a step-by-step series.

Without random mutation and natural selection working together, chemical evolution is forced to rely exclusively on *random combination.* Mindless, undirected change as a mechanism for creating complex functional structures absolutely does not work. This is why we do not find light bulbs, calculators, MP3 players and laptop computers in the fossil record. Undirected change cannot remotely explain these non-living complex functional structures, nor can it plausibly explain single-celled life that is millions of times more complex.

(2) Scientists have failed to demonstrate by experimentation the natural formation of biomolecules found in life today and probable in the first viable living organism.

As previously mentioned, scientists Stanley Miller and Harold Urey at the University of Chicago conducted an experiment to simulate conditions for the evolution of molecules necessary for life. When the results of their experiments were released by the new media in 1953, they rocked the world. Some of the basic components of life had been created by a natural process. This event was heralded as virtual proof of chemical evolution. News coverage was so positive that some people were led to believe life itself had been created in the laboratory. Research within the last fifty years has considerably dampened the initial enthusiasm generated by the Miller-Urey experiment. Knowledgeable scientists now recognize this experiment and subsequent variations do not remotely explain the origin of life.

When all of the evidence is considered, the Miller-Urey experiment is powerful support against chemical evolution. The Miller-Urey experiment has been repeated many times and with different materials. Although presented in glowing terms in biology textbooks, the results have been extremely disappointing. When all of the raw ingredients for life are made available for this experiment (under quite unnatural conditions), they are capable of only producing the most basic biomolecules in small amounts. This experiment cannot form the *macromolecules* found in microscopic life.

The gases used in the Miller-Urey experiment are not representative of the atmosphere on early Earth. The gases are a highly reactive mixture of methane, ammonia, hydrogen and water vapor. Scientists have since come to realize that Earth's early atmosphere was probably nitrogen and carbon dioxide.[22] As Hazen points out, when nitrogen, carbon dioxide and water vapor are used in a Miller-Urey type experiment, nothing of interest is produced.[23] These three gases cannot be used to "synthesize very many useful biomolecules."[24]

Oxygen is deadly to the formation of biomolecules and must be artificially excluded in any Miller-Urey type experiment. Some scientists believe Earth has always had an atmosphere containing oxygen.[25, 26] Approximately 46 percent of Earth's crust is composed of oxygen combined with other elements and the deepest strata contain oxides as evidence of an oxygen atmosphere.

The Miller-Urey experiment has led to the belief that life emerged from a primordial soup. Hypothetically, early Earth was covered by water filled with biomolecules. Scientific investigation has shown that Earth's strata lack evidence of the prebiotic molecules formed in the Miller-Urey experiment. A lack of confirming evidence strongly discredits the existence of prehistoric oceans of amino acids, proteins, lipids and macromolecules. The missing molecules are a predictive failure of this idea.

In addition to the above, the primordial soup idea is plagued with basic chemistry issues. Hypothetically, amino acids suspended in water randomly joined together to form complex protein chains necessary for the first life. Behe points out, "The presence of water strongly inhibits amino acids from forming proteins."[27] The major problem is that amino acids readily dissolve in water. Any water molecule attached to an amino acid must be removed before it can be used in a working protein chain. No water molecules are allowed. Experiments show the impossibility of functional proteins developing by natural means in a solution of water.

Behe discusses the formation of nucleotides found in DNA as a major issue in evolution experiments. The four DNA nucleotides are built from adenine, guanine, thymine, cytosine, deoxyribose sugar and phosphate molecules. The processes necessary to form these different molecules are chemically incompatible. When made in the laboratory, these components are formed under separate conditions and then combined together. Efforts to produce these molecules using a common process have been without success. This problem is commonly referred to as "the prebiotic chemist's nightmare."[28] Experiments strongly discredit the ability of naturally occurring conditions to make the biomolecules found in DNA.

(3) Scientists have failed to identify a natural mechanism capable of separating chiral amino acids and sugars found in living things.

The chiral amino acids and sugars found in living things defy a natural explanation. As previously mentioned, chirality is the characteristic in which a molecule occurs in both a left-handed and right-handed variety. Chiral molecules are mirror images of each other. Except for glycine, all of the twenty amino acids found in proteins are left-handed chiral molecules. Deoxyribose sugar (found in DNA) and ribose sugar (found in RNA) are right-handed. Amino acids are used in the building of proteins and enzymes, while sugars are an integral part of the DNA and RNA macromolecule. When chemically synthesized in the laboratory, both amino acids and sugars occur in approximately equal amounts of the left-handed and right-handed variety. The presence of essentially only left-handed amino acids and right-handed sugars in life has no natural explanation. As Hazen says, "In spite of a century and a half of study, the origin of this biochemical 'homochirality' remains a central mystery of life's emergence."[29] The failure to identify a credible natural mechanism capable of separating chiral molecules in the first living organism strongly discredits chemical evolution.

Some natural phenomena show that limited chiral separation can occur. Meteorites from outer space have been found to contain naturally occurring amino acids with a slight excess of one chiral variety over the other. In addition, scientists have discovered that the mineral calcite, found in limestone, will cause slight chiral separation between amino acids. The minor chiral separation in these instances cannot be extrapolated to explain how it could occur in a churning ocean of primordial soup.

(4) Scientists have failed to identify a natural mechanism capable of controlling the combination of small biomolecules to produce working proteins, enzymes, membranes, DNA and RNA observed in life.

Essentially, only twenty kinds of amino acids are used in the proteins of living organisms, and their selection is a major problem for chemical evolution. A number of amino acids not found in life have been produced by the Miller-Urey experiment and other origin of life experiments. Non-biological amino acids are commonly found in nature. For example, the Murchison meteorite that fell in Australia in 1969 was analyzed and found to contain approximately 50 non-biological amino acids.[30] *The Practical Handbook of Biochemistry and Molecular Biology* lists more than 300 amino acids.[31] No undirected mechanism has been identified capable of selecting only the twenty amino acids found in proteins when a multitude of non-biological amino acids would have been present in a hypothetical ocean of primordial soup. As Hazen says, "But what most books don't mention is that for every potentially useful small molecule in that prebiotic environment, there were dozens of other molecular species with no obvious role in modern biology."[32] He goes on to say, "So the real challenge turns out to be not so much the making of molecules, but the selection of just the right ones and their assembly into the useful structures we call macromolecules."[33]

To build macromolecules of working proteins, enzymes, membranes, DNA and RNA, the controlled combination of small biomolecules is required. A hypothetical ocean of primordial soup filled with amino acids, sugars, phosphates, lipids and a host of non-biological molecules has no mechanism to control their combination. For example, there is no mechanism to control when a protein should stop growing, continue growing, select one of twenty correct amino acids or discard a non-biological amino acid. The incorporation of incorrect chiral amino acids or sugars into a macromolecule is incredibly destructive. The inclusion of a single non-biological molecule in a macromolecule severely degrades or destroys function. Lack of a natural mechanism capable of bringing order out of chemical chaos strongly discredits the natural formation of working macromolecules.

(5) Scientific investigation and mathematical analysis point out that the amino acid sequences found in the proteins necessary for the smallest known single-celled organism would not occur as a result of random combination over billions of years.

Proteins and enzymes are required for life as we know it. Proteins are molecules used by the cell to give it structure and perform a host of chemical functions. An enzyme is a specialized protein whose main function is that of a biochemical catalyst.

Enzymes are the catalysts of life. Enzymes occur not only in our stomach to digest food, but throughout the body to make it function. They exist in muscle, brain, nerve and blood cells. Enzymes cause the combination or separation of molecules at extremely high rates and can easily speed up a chemical reaction by more than one million times. Without enzymes, some chemical reactions would not occur or occur so slowly that life would not be possible.

In living things, enzymes appear to be custom designed to facilitate specific chemical reactions. Each enzyme type acts on one kind of molecule or a small group of molecules. For example, the enzyme carbonic anhydrase is found in red blood cells and speeds up the combination of one water molecule with one carbon dioxide molecule. This enables red blood cells to transport carbon dioxide from cell tissues to the lungs where it is released. One enzyme of carbonic anhydrase can process between ten thousand and one million molecules of carbon dioxide per second.[34] One enzyme of catalase can break apart more than a million molecules of hydrogen peroxide (a normal by-product of metabolism) per second into water and oxygen.[35] Without these enzymes, we could not live. The Brenda Enzyme Database lists more than 4,780 enzymes identified in living things.[36] Enzymes represent an extremely sophisticated level of chemical engineering.

Proteins are made from a selection of twenty amino acids. A protein may be visualized as a long chain, with each link being one of twenty possible amino acids. Sometimes proteins consist of two or more chains twisted together. The sequence and number of amino acids give a protein its properties. Information carried by genes within DNA determines the exact sequence of amino acids within a protein. If left to combine naturally, amino acids join with each other in a completely random order.

Mathematically it can be shown that any structure can be formed due to random combination if unlimited resources are available. Hypothetically, microchips, radar detectors and GPS locators can originate from the unlimited random combination of atoms and elementary molecules. However, resources are always limited—even if we appeal to the entire universe. Based on (1) a limited amount of material in the universe, (2) a limited amount of time for the universe and (3) a limited number of combinations per second, a probability limit can be established. This is called the *universal probability boundary*.

Small probability and the limited resources of the universe can prevent a chance event from happening. Mathematical analysis and the universal probability boundary are discussed in *The Design Inference: Eliminating Chance Through Small Probabilities* by mathematician William Dembski, PhD. The universal probability boundary is calculated to be 1×10^{-150}. The negative sign in the exponent makes this expression equivalent to a decimal point followed by 149 zeros and the number one. This is a very small number. If the mathematical probability of an event is below this number, then the event is unlikely to occur.

Scientists and mathematicians have identified resources available in the universe upon which a probability limit can be determined. As mentioned, they are the number of atoms in the universe, the age of the universe and the possible number of combinations per second. There are an estimated 1×10^{80} atoms in the known universe. There are an estimated 1×10^{25} seconds in the age of the universe from its origin until it dies of heat death in the distant future. Planck time is estimated at 1×10^{-45} seconds. Planck time hypothetically limits the number of combinations of a single atom to a maximum of 1×10^{45} combinations per second. When these three numbers are multiplied together, the number of 1×10^{150} is obtained. So if the probability of an event happening is less than one chance in 1×10^{150}, then it is unlikely to occur. Normally a minus sign is placed in front of the exponent to make it a number smaller than one. This leads to the statement that if the probability of an event is smaller than 1×10^{-150} it is unlikely to occur.

What is the mathematical probability of a single protein forming by chance from a mixture of twenty essential amino acids? First of all, chain structures such as proteins can be easily analyzed. The probability of any linear sequence of random letters, numbers, objects or molecules from a defined set can be mathematically determined. Secondly, according to the UniProt/Swiss Protein Knowledgebase, 98 percent of identified proteins in biological life are composed of a chain of 50 amino acids or more. The size of the average protein found in living things is 367 amino acids. The largest known protein contains 34,350 amino acids.[37] For sake of illustration, assume each amino acid must be in an exact sequence within a chain to form a specific working protein. Also assume a purified primordial soup exists containing only the twenty essential amino acids, equally divided between left-handed and right-handed varieties. A single 100-amino acid protein, arising from an equal mixture of chiral amino acids, has a probability of 40^{100} or 1.6×10^{-160} of occurring by chance. This number is substantially smaller than the universal probability boundary. A 100-amino acid protein requiring an exact sequence of amino acids is unlikely to form by chance in our universe according to mathematical analysis.

Proponents of chemical evolution often bring up the issue that a protein can work with several variations, which increases the probability of a protein occurring by chance. Robert Sauer, Department of Biology at the Massachusetts Institute of Technology, investigated the amount of variation possible in proteins before they ceased to function.[38] By using viral proteins and substituting amino acids one at a time, he was able to determine which amino acids could be changed before a protein stopped working. Some amino acid positions were critical and could not be changed without a loss of protein function. Many positions in the protein chain worked with three or four different amino acids while some positions worked with up to fifteen. His experimental work concluded that a 100-amino acid protein with possible amino acid combinations of 1.27×10^{130} could be reduced to 1×10^{65} combinations (A single 100-amino acid protein arising from a mixture of *only left-handed amino acids* has a probability of 20^{100} or 1.27×10^{130} of occurring by chance). Rather than only one protein, a family of related proteins is capable of a particular function.[39] Regardless, each member of this protein family represents a mathematically improbable combination.

The smallest known single-celled organism illustrates the problems of life arising by chance. *Nanoarchaeum equitans* Kin4-M, a tiny heat-resistant microbe, has 536 protein-encoding genes and 490,885 base pairs of DNA in its genome.[40] It is believed to have an obligatory symbiotic relationship with a larger organism.

The smallest, single-celled, free-living organism is generally considered *Mycoplasma genitalium* G-37 which has 482 protein-encoding genes and 580,076 base pairs of DNA in its genome.[41] *Mycoplasma genitalium* is a parasitic bacterium (it causes a sexually transmitted disease in humans) and is dependent upon its host.

Prochlorococcus marinus CCMP1378 MED4 (blue-green algae) is a small cyanobacterium with 1,713 protein-encoding genes and 1,657,990 base pairs in its genome.[42] *Prochlorococcus marinus* is capable of photosynthesis, is robust and can live in an unprotected natural environment using inorganic molecules, water and sunlight. This organism represents the most reasonable candidate for the first life on Earth.

Regardless of which of these three single-celled organisms we accept as a model for the first viable life on Earth, they represent a formidable challenge to chemical evolution. The probability of 482 or 1,713 different proteins occurring by undirected combination substantially exceeds the universal probability boundary. According to mathematical analysis, the proteins necessary for single-celled life as we know it absolutely would not occur by chance in our universe.

Mathematical probability analysis is useful in evaluating the chance formation of the proteins necessary for life. Unfortunately, analyses such as the above ignore important information about chemistry and physics. These issues are difficult to quantify in a mathematical manner; nevertheless, they are real and further eliminate the probability of life occurring by natural means. Additional major issues are:

• More than 99 percent of the visible mass in the universe exists as plasma within stars. Life cannot chemically evolve within hot plasma.

• The majority of components for life and their assembly would most plausibly occur on a single planet. This dramatically limits the materials, environments and resources available.

• Proteins, DNA and RNA break apart and deteriorate unless carefully protected. They go from being organized to being random, rather than in the opposite direction. These macromolecules degrade in the presence of heat, oxygen, water, ultraviolet light and high energy particles.

• All planets and moons in our solar system are dominated by one or more conditions hostile to the chemical evolution of life. These conditions include high temperatures, vacuum, oxygen, chlorine, water, ultraviolet light, solar radiation and extreme low temperatures.

• Non-biological amino acids, incorrect chiral amino acids and non-biological molecules are abundant in nature. In any conceivable evolutionary environment, they would be present. Lack of a means to exclude these molecules prevents the formation of working proteins necessary for life.

• Molecules in laboratory primordial soup experiments combine and eventually approach chemical equilibrium based on temperature, mixing and available resources. The chemical reaction process increasingly slows. Molecules do not randomly recombine indefinitely at the speed of 1×10^{45} combinations per second as assumed in the above probability calculations.

(6) Scientific investigation and mathematical analysis indicate that the information within DNA necessary for the smallest known single-celled organism would not occur as a result of random combination over billions of years.

Experiments have shown that random change is incapable of creating significant information. The random arrangement of alphabetical characters can produce simple words but little else. Chance cannot produce lengthy meaningful sentences. Undirected change within information is always destructive. For example, information in a book is quickly destroyed when random change occurs. Random change does not know which words to alter in a sentence when selecting from 200,000 or more words in the English language. Undirected change does not know which letter to alter within a word when selecting from a possible twenty-six letters in the English alphabet. Indiscriminate change always adds, deletes or replaces letters and words that need no change. The ultimate effect of extensive, undirected, random change in any book is complete nonsense.

Professor Werner Gitt in his book *In the Beginning Was Information* has proposed that information is evidence of intelligence. Information may be represented by symbols, selected from a set of symbols and sequenced in a specific manner. The genetic code within DNA meets these criteria. Three base pairs of DNA can make a codon. There are 64 possible codons, each of which can specify one of twenty possible amino acids. DNA contains the information for building functional proteins and therefore is a result of intelligence.

DNA is a macromolecule that may be compared to a spiral staircase. Each step of the staircase is called a base pair. In turn, each base pair can consist of four possible nucleotide pairs. These nucleotide pairs contain the information for life. Mathematical analysis shows that 250 base pairs of DNA information can form 4^{250} or 3.27×10^{150} possible combinations. Therefore, the universal probability boundary is exceeded by any random sequence of DNA more than 250 base pairs long.

As previously mentioned, the smallest free-living organism is often considered *Mycoplasma genitalium* with approximately 580,076 DNA base pairs. Most of its DNA is believed to be used for encoding proteins. This is equivalent to a 193 page book containing 3,000 characters per page. No random process can create this amount of information.

Proponents of evolution often say that given enough time a monkey bashing away at random on a typewriter could produce all the works of Shakespeare.[43] To the ignorant and gullible this statement sounds impressive; however, it is completely false. The first seven lines of dialogue in Hamlet, Act I, Scene I contain 138 alphabet characters (actor names, spaces and punctuation are excluded). With a possible twenty-six alphabetical characters, this equates to a total of 26^{138} or 1.8×10^{195} possible combinations which substantially exceeds the universal probability boundary. Therefore, even given the entire resources of the universe, the first seven lines of Shakespeare's Hamlet cannot be reproduced by chance.

(7) Scientific experiments prove the inability of undirected mechanisms to assemble basic biomolecules and create life.

Discrete parts do not make a living organism—assembly is required. Experiments show that undirected change cannot assemble molecules to create a living organism. When large quantities of biomolecules such as left-handed amino acids, right-handed deoxyribose sugar, right-handed ribose sugar, adenine, guanine, thymine, cytosine, phosphates and lipids are mixed together, life does not arise. When all of the known basic biomolecules for life are made available and all unnecessary non-biological molecules are excluded, life does not happen. Basic molecules cannot self-assemble to create life. Based on current scientific knowledge, life will not occur due to undirected change even given oceans of purified basic biomolecules and billions of years of time.

Design Theory as an Alternative to Natural Cause

Design theory proposes that the functions and features found in life today are representative of the first microscopic life on Earth. The information, complex functional systems, mathematically improbable structures and fine-tuned arrangements in the first microscopic life were due to intelligent cause. Scientific knowledge about single-celled life and the severe limitations of natural cause exclude evolution as a means of creating the first life.

Design theory recognizes the possibility of life beyond Earth. Life may exist elsewhere in the solar system or the universe. Owing to the functional complexity of any conceivable biological life, its origin is the result of intelligent cause. Design theory makes the following predictions:

(1) As our knowledge of microscopic life increases, it will become more information rich, functionally complex, integrated and fine-tuned rather than simpler. Life will have less probability of occurring by natural cause.

(2) A substantial gap will remain between the most complex inorganic structures and the smallest single-celled living organism. No step-by-step series will be found between the two.

(3) No new laws of chemistry and physics will be discovered to make chemical evolution plausible. A law of "emergence" or "self-organization" capable of creating single-celled life will not be discovered.

(4) No inorganic (non-living) source for DNA nucleotides, the DNA macromolecule and DNA information will be discovered.

(5) No experiments that accurately represent Earth's early atmosphere (consisting primarily of nitrogen, carbon dioxide and water vapor) will produce all of the biomolecules found in microscopic life which include amino acids, lipids, sugars, phosphates, nucleotide bases, proteins, cellular membranes, RNA and DNA.

Conclusion

In March of 1989, Dr. Stanley Pons and Dr. Martin Fleischmann at the University of Utah announced they had conducted an experiment where cold fusion was observed. The experiment consisted of a battery that passed an electric current through two palladium electrodes immersed in a beaker of heavy water. Electrolysis of the water allegedly resulted in the fusion of hydrogen atoms and the production of excess energy. Immediately after the announcement, debate raged for years over the validity of their experiment and the promise of cold fusion.

Cold fusion theory has strong environmental and financial appeal. However, theory cannot be demonstrated and is dying after millions of dollars in investment, multiple experiments, eight international conferences and volumes of discussion. The primary problem is that the alleged results cannot be verified by experiment. Empirical science does not support theory. No excess energy has ever been produced and the alleged process does not work. Believers in cold fusion say this does not disprove the theory, but only that it cannot be duplicated. Regardless of the arguments, the burden of proof lies squarely upon the shoulders of its supporters. Negative experiments have discredited this theory. Without verifiable results, cold fusion is an attractive idea and nothing more.

The past state of cold fusion is similar to the present state of chemical evolution. Although emotionally appealing to some, this idea has reached a dead end. Chemical evolution cannot create the small biomolecules necessary for life when constrained by natural conditions that exist on Earth. Scientists and mathematicians discount the ability of undirected change to create the proteins, nucleotides and information necessary for life. Regardless of the clever ideas, arguments and excuses, chemical evolution cannot be duplicated by any experiment that approximates natural conditions. Because of negative results, chemical evolution is destined for the great trash bin of ideas discredited by empirical science.

Although evolution may be viewed as both a fact and theory, chemical evolution is neither. There is no accepted theory of chemical evolution. *The Problems of Biology* by John Maynard Smith examines the challenges of evolution in detail. Smith says, "He [Darwin] did not claim to account for the origin of those first living things. This part of Darwin's theory, sometimes referred to as the 'fact' of evolution, is almost universally accepted by biologists today."[44] The "fact" of chemical evolution, accepted by most scientists and taught in schools, is not supported by empirical science. Belief in chemical evolution is driven by a refusal to accept the evidence of science and the influence of philosophical naturalism.

Philosophical naturalism requires that chemical evolution be true and promotes wishful thinking about its occurrence. Intelligent, motivated people, absolutely convinced of the truth of this philosophy, have developed clever arguments in support of their belief. Speculative ideas describe how the first microscopic life may have evolved. Block diagrams describe how DNA, RNA, metabolism, reproduction and other functions may have developed. Simple cartoons describe how molecules may have assembled. A lack of detail and empirical evidence underscores the implausibility of these ideas, which are nothing more than a mental exercise in creative speculation. Believers in philosophical naturalism have no rational mechanism for chemical evolution, and the burden of proof lies squarely upon their shoulders.

A discussion about the origin of life can be framed as a choice between evolution theory and design theory. One idea proposes that life evolved due to undirected phenomena, in clear disregard of the severe limitations of natural cause. The contrasting idea is that life originated due to intelligent cause. Humans, as surrogate designers, can create information and complex functional structures and manipulate the molecules found in life. The likelihood of humans creating microscopic life far exceeds that of natural cause by many orders of magnitude. Design theory and intelligent cause provide a logical explanation for life.

Philosophical naturalism is the primary driver behind chemical evolution. Proponents of this philosophy have misused science to support their errant convictions. This belief has driven scientists to appeal to missing information, unknown phenomena and undiscovered laws. This idea has forced scientists to appeal to natural metaphysical cause, while rejecting intelligent cause as an explanation.

The evidence against chemical evolution is overwhelming. Unless the scientific community is willing to critically examine this idea, any discussion of more complex issues is essentially futile. Without a willingness to recognize the limitations of natural cause and the legitimacy of design theory, it is pointless to discuss other issues with less clarity. Evidence convincingly shows that chemical evolution, above all other ideas of natural cause, is a scientifically bankrupt idea. For people knowledgeable of biology and science, belief in chemical evolution represents a resolute, faith-based commitment to philosophical naturalism.

Scientists in general avoid use of the word *impossible* as it denotes finality and a strong sense of decisiveness. It almost sounds dogmatic. Regardless, we can say that based on the observed constraints of the universe, the known laws of chemistry and physics, mathematical probability, the severe limitations of natural cause, exhaustive scientific investigation and our knowledge of microscopic life, chemical evolution is impossible.

Science is a rational search for dependable knowledge about the world based on physical evidence and its logical inferences. Natural cause is a biased assumption of science rather than a dogmatic requirement. If errant deep-seated belief in philosophical naturalism and its role in science can be exposed, chemical evolution will die. Regardless of how entrenched an idea, without a viable engine, it will eventually slow and stop. Philosophical inertia will only drive an idea so far. If scientific thought can remain free in our schools and universities, a relentless search for truth will eventually discredit chemical evolution. Science will again show itself to be a self-correcting process.

Lead me in thy truth, and teach me: for thou art the God of my salvation. (Psalm 25:5) KJV

Chapter 9

The Illusion of Biological Evolution

Charles Darwin is considered the father of biological evolution. His adult career began when he graduated from Christ's College at Cambridge, England in early 1831 where he had studied theology in preparation for entering the ministry. However, his interests took him in a different direction and in December of that year boarded the HMS *Beagle* as its unpaid naturalist. The ensuing five year trip took him around the world. Among notable places visited were the coasts of South America, the Galapagos Islands, Australia and Cape Town, South Africa. He returned to England on October 2, 1836 and for the next two years edited notes taken during his trip. His book *Journal and Remarks, 1832-1835* (which later became known as *The Voyage of the Beagle*) was published in 1839. During the next twenty years Darwin did minor research, revised the *Journal,* wrote and eventually married. His experiences aboard the *Beagle* were not forgotten and formed the basis for *On the Origin of Species by Means of Natural Selection* published in 1859.

Charles Darwin's theory of species evolution was developed during a time of great ignorance and was a leap in the dark. For example, Darwin and a majority of scientists in his time believed in spontaneous generation, discredited by Louis Pasteur in 1862. Darwin did not understand the genetic basis for heredity founded on the work of Gregor Mendel, first published in 1866 but essentially lost until rediscovered in 1900. Darwin knew little about meiosis (sexual reproduction), first observed and described by the German biologist Oscar Hertwig in 1876. Darwin was ignorant of chromosomes (called chromatin initially) and their description by Walther Flemming in 1882. Darwin was unaware that genes are carried on chromosomes, discovered by Thomas Hunt Morgan in 1910. Darwin knew nothing about the double-stranded helical model for the DNA molecule proposed by James Watson and Francis Crick in 1953. Darwin was ignorant of DNA information, protein-encoding genes and point mutations.

In 150 years since Darwin's book, much has been discovered about biology. Research within the last fifty years, when properly understood, seriously discredits Darwin's theory of evolution. His simplistic ideas of macroevolution are challenged by our current knowledge of heredity, mutations, DNA, genes and proteins.

Darwin theorized that the first life on Earth developed into all subsequent species (i.e. common descent) because of small changes and natural selection over many generations. Chemical evolution occurred first and then biological evolution took over. Charles Darwin described biological evolution as, "Descent with modification."[1] Today, geneticists often describe biological evolution as a process that results in heritable changes in a population spread over many generations.

Concise definitions for biological evolution vary. There is no inherent disagreement with the above two definitions; however, they fail to mention certain key items and may be either too simple for some or overly technical for others. The theory of biological evolution is an inferred explanation that undirected change is responsible for the development of all species on Earth, whether living or extinct, starting from single-celled life. Development of new DNA, features, functions and species occurred by the hereditary accumulation of genetic material within species populations from generation to generation over time. The primary mechanisms of biological evolution are random beneficial mutations within hereditable DNA and natural selection acting upon those mutations. The term "biological evolution" is used in the National Academy of Sciences book *Science, Evolution and Creationism.*[2]

Definition 9.1 – Theory of Biological Evolution

An inferred explanation that undirected change is responsible for the development of all species on Earth, whether living or extinct, starting from single-celled life. Development of new DNA, features, functions and species occurred by the hereditary accumulation of genetic material within species populations from generation to generation over time. The primary mechanisms of biological evolution are random beneficial mutations within heritable DNA and natural selection acting upon those mutations.

Proponents of biological evolution believe random variation and natural selection are a simple, yet elegant means nature has provided for creating new information and functional complexity. When given several choices, natural selection always picks the most advantageous change. By this means, nature can build new DNA, genes, proteins, chromosomes, features, functions and species in a step-by-step series. The concept is amazingly simple, always make the right choice when confronted with a number of options and anything is possible. While an appealing idea, it is simplistic and ignores a host of other factors.

Similarity as Evidence of Evolution

Any comprehensive discussion in biology textbooks normally presents comparative anatomy, homology, analogous structures and related physiology—summarized as similarity among species—as major evidence in support of biological evolution. Similarity consists of DNA information, molecules, proteins, genes, pseudogenes, features and functions common among different species. Or as described by John Maynard Smith in *The Origins of Life: From the Birth of Life to the Origin of Language,* "The structural similarity between the human hand, the flipper of a seal and the wing of a bat has long been familiar: it is one of the decisive reasons for accepting the theory of evolution."[3] Similarity among different species can be interpreted as support for common descent due to evolution.

Similarity among biological species is a fact. Similarity is observed in the *functions* of sexual reproduction, placental birth, digestion, sight, hearing, balance, smell and taste. Similarity is found in *features* such as hair, claws, flippers, fins, feathers, scales, legs, eyes and organs. Similarity has been discovered on the *molecular level* by comparing DNA sequences, genes, proteins and pseudogenes. Genes and proteins are powerful evidence of similarity, as pointed out in *The Making of the Fittest* by Sean Carroll. He says, "We find about 500 genes that exist in all domains of life."[4] These 500 genes are not necessarily identical, but they code for proteins that are similar among Earth's species. The 39 Hox genes found in humans and all other mammals which control the formation of body parts are evidence of similarity.[5]

Similarity among species can be plausibly explained by either biological evolution or design theory. For example, virtually all cars are similar because they have four wheels, an engine, transmission, steering wheel, windshield, springs and headlights. Some parts are identical and directly interchangeable. In addition, some automobiles contain the same defects that extend over many years and product lines. If similar features are the only criteria evaluated, it is difficult to determine whether the automobile is due to evolution or design. In this instance, we know for a fact that similarity is because of design. Automobiles represent a design iteration improvement series where similarity is based on the development of good designs, standardization, economics and the requirements of mass production. Less desired features of similarity include common defects and repetitive quality problems.

Design theory explains similarity as evidence of a common designer, a common design philosophy, the intelligent reuse of successful designs or a design iteration improvement series. Similarity represents the development of good designs and the widespread reuse of those designs. Similarity represents the ability to modify good designs for size, function and variety for use in different applications.

Similarity undeniably exists among species, and so the real issue is whether it is due to evolution or design. From a Darwinian perspective, similarity is attributed to the evolution of all species from a common ancestor (i.e. common descent) over millions of years. From an opposing viewpoint, similarity would also be strongly evident if all species were due to a common designer or a common design philosophy.

Design theory easily explains both *similar* and *unique features* in living things, whereas evolution only explains similarity. Some genes, enzymes, proteins, features and functions are quite unique and appear in only few species. For example, the star-nosed mole (*Condylura cristata*) has an unusual nose that is only found in its species.

Two species of gastric brooding frog in Australia give birth by swallowing their fertilized eggs and allowing tadpoles to develop in their stomachs. The archerfish is unique among fish, being able to shoot down an insect up to six feet in the air by squirting it with water. The sea horse, hermit crab, sea cucumber, starfish, electric eel, squid, octopus, kangaroo, giraffe, elephant, platypus, woodpecker, turtle and long-beaked echidna all have unique traits. Some entire orders or families such as bats, butterflies, nonplacental mammals and bombardier beetles have unique characteristics. Unusual features strongly discredit biological evolution. Unique features cannot be rationally explained due to evolution because of their complexity, dissimilarity and lack of a step-by-step series connecting them with other species.

Fossils as Evidence of Evolution

Evidence from paleontology is often presented as supportive of evolution. Fossils show a record of life from small multicellular organisms to complex vertebrates. Fossils are firm evidence of the existence of dinosaurs and other extinct animals. A sequence of fossils from the oldest to the newest does not exist in a continuous stratigraphic sequence at one location. Fossil sequences are built from many different locations, and the animals and plants found therein are arranged in an ascending series.

The substantial lack of transitional fossils in the fossil record is a major predictive failure of evolution. Biological evolution requires a step-by-step series of small changes from the first single-celled organism to large warm-blooded vertebrates. These step-by-step changes should be substantiated by corresponding evidence in the fossil record. With millions of years of alleged evolution, fossils in support of this theory should be abundant. The actual fossil record lacks the vast majority of transitional fossils. There are gaps among major species. Animals and plants appear suddenly without any close ancestors. Based on current knowledge, it is doubtful these missing fossils will ever be found.

The fossil record is an embarrassment to Darwinian evolution. As Darwin himself wrote, "Natural selection can act only by taking advantage of slight successive variations; she can never take a leap, but must advance by the shortest and slowest steps."[6] The small step-by-step changes required appear to be permanently missing. Alleged examples of fossils in support of evolution are pathetically small. Based on an expected twenty or more intermediates between major species, biological evolution is not well supported. The fossil record shows abrupt jumps between major species and overwhelmingly fails to support Darwinian evolution.

Several prominent scientists have proposed hypotheses to explain the steps, jumps and gaps in the fossil record. Richard Goldschmidt, PhD, was a prominent American geneticist and member of the National Academy of Sciences. In *The Material Basis of Evolution* released in 1940, he attempted to integrate genetics, fossils and evolution theory. Goldschmidt proposed large abrupt changes as an explanation for evolution rather than slow step-by-step changes. Large genetic changes between a parent and its offspring were proposed as the only way to explain jumps in the fossil record. His idea became known as the "hopeful monster" hypothesis. Critics say his idea permits a reptile to lay an egg capable of hatching into a bird.

Niles Eldredge, paleontologist at the American Museum of Natural History, and Stephen Jay Gould, professor of geology and zoology at Harvard University, proposed the hypothesis of punctuated equilibrium. Their first joint article on the subject appeared in 1972 in the book *Models in Palaeobiology*. Subsequent articles appeared in *Natural History*, *Nature* and *Paleobiology*. They pointed out that the fossil record does not show the development of animals and plants in small step-by-step increments over millions of years as proposed by evolution. Life appears in sudden jumps with long periods of time between each jump. They called these sudden jumps punctuated events. Long periods of time between jumps, or periods of no change, were called equilibrium. From this, they derived the term punctuated equilibrium.

Eldredge and Gould hypothesized that evolution occurred in evolutionary bursts, with rapid undocumented changes in the fossil record. Their idea was based on the lack of intermediate fossils showing step-by-step change. The largest and most prominent example of missing fossils is the sudden appearance of large complex life in the Cambrian layer without any significant previous fossils. Cambrian fossils consist of sponges, jellyfish, trilobites, mollusks, jawless lampreys and chordates. Only microfossils exist below the Cambrian layer.

The hypothesis of punctuated equilibrium was severely criticized because it lacked a demonstrable mechanism. Observations of periodic jumps and equilibrium in the fossil record are essentially correct, but no evolutionary mechanism exists to explain its occurrence. Biological evolution over millions of years has enough problems without proposing rapid change over thousands of years.

Goldschmidt, Eldredge and Gould are intelligent, educated, accomplished individuals, knowledgeable of biology and paleontology. Their ideas reflect an attempt to explain the huge discrepancy between step-by-step Darwinian evolution and evidence in the fossil record. The scientific establishment may reject and ridicule their ideas, but no one has yet solved this problem.

In some respects, the Ford automobile may be compared to the fossil record and interpreted as an evolutionary series. As previously mentioned, Henry Ford began producing the Model T in 1908, which subsequently developed into the many Ford products today, represented by a sequence of one hundred vehicles leading up to the 2008 product line. At first glance, this can be interpreted as an impressive evolutionary series. This is indisputable classic Darwinian evolution that substantially exceeds any plant or animal fossil sequence in existence. However, on closer examination, evolution has trouble explaining the large steps in this series and the lack of intermediates. Evolution has no mechanism to explain the sudden appearance of electric starters, air conditioning, disc brakes, cruise control, overhead cam engines and front wheel drive. In front wheel drive cars, the drive mechanism does not slowly move from the rear to the front on an annual basis. Change appears suddenly in one step. All of the major technical changes mentioned appear abruptly with no intermediates. Evolution lacks a mechanism capable of creating large steps of complex change.

In contrast to evolution, design theory easily explains the Ford automobile. Abrupt appearance of the first Model T was a significant improvement over the closest four-wheel carriage of the day and represents an *original design.* From that point onward, a *design iteration improvement series* is represented by both small and large changes in each yearly model. Through feedback from testing and customer complaints, each yearly model was incrementally improved. The original design of new components caused large step changes, represented by the battery, generator, hydraulic brakes, windshield wipers, power steering, alternator and fuel injection. Small incremental changes plus large step changes have resulted in the current level of technical sophistication. *Product expansion* is represented by the Model T that has developed into the 2008 full line of Ford products which includes light pickups, full sized pickups, commercial trucks, over the road trucks, convertibles, sports cars, compact cars, family sedans, luxury sedans, minivans, full sized vans and sport utility vehicles available in a variety of colors and options.

A theorized evolutionary series and a design iteration improvement series are similar, yet significantly different. Theoretically, an evolutionary series is characterized by small step-by-step changes. As defined by Darwin, evolution does not advance in large steps. However, a human design iteration series, characterized by the Ford automobile, contains both small and large changes.

The fossil record documents the sudden appearance of new plants and animals. Large steps are present and transitional fossils are missing among major species. The fossil record may be explained as a series of events over extensive time where new species were introduced (product expansion due to intelligent cause) with evidence of original design, design by menu selection and design iteration improvement.

The fossil record is a predictive failure of Darwinian evolution and can be viewed from three possible perspectives. (1) Animals and plants developed in a step-by-step series, but their record of development is permanently missing from the fossil record. Many plants and animals were not fossilized. With this explanation, biological evolution appeals to missing evidence. (2) The fossil record contains a complete record of step-by-step development, but that record is yet to be discovered. Gaps in the fossil record will eventually be filled in by additional exploration. Biological evolution again appeals to missing evidence. (3) Permanent gaps exist in the fossil record and the known fossil record accurately reflects the appearance of life on Earth. New animals and plants appear suddenly in jumps and steps with no intermediates. Biological evolution cannot explain these steps and is discredited by their presence.

Microevolution as Evidence of Macroevolution

Discussions of biological evolution often present *microevolution* as supportive of macroevolution. Small undirected changes have been well documented by the scientific community in support of microevolution. However, small changes cannot automatically be extrapolated to support large changes required by Darwinian evolution. Some of the categories of microevolution are:

(1) Random mutation – Mutations in heritable DNA are primarily due to copying errors or mutagenic agents. Mutations are a fact of biology and are a documented phenomenon, but the vast majority are neutral or deleterious. Neo-Darwinian evolution requires beneficial mutations, which are nonexistent or extremely rare. Within the human genome, sickle cell is one of the few mutations identified with some type of beneficial effect. It helps a person survive in a malaria-infested environment. Under normal conditions, sickle cell trait (one defective gene) is essentially neutral while sickle cell disease (two defective genes) is quite harmful.

The human genome has been studied more than any other and gives the best information about the effects of random mutations. There are a number of documented mutations in the human genome. For example, the UniProt/Swiss Protein Knowledgebase lists 15,796 known protein isoforms in humans which cause 1,941 different genetic diseases.[7] The Online Mendelian Inheritance in Man database lists 18,715 genetic disorders.[8] The Human Gene Mutation Database has identified 79,078 mutations in 3,000 genes in the human genome.[9] Mutations in the human genome are essentially either neutral or deleterious and account for a substantial number of genetic diseases. Truly beneficial mutations (under all environmental circumstances) in the human genome are nonexistent or extremely rare. The total effect of all known human mutations is overwhelmingly destructive rather than constructive. Random mutations cannot be legitimately extrapolated to support macroevolution.

(2) Natural selection – The action of natural environment on species characteristics is called natural selection. For example, if a mutation causes an animal to be faster, smarter, healthier, have more offspring or in some way survive better than other animals, then that mutation can spread. The increasing occurrence of the mutation is due to natural selection. In *The Making of the Fittest,* Sean Carroll, PhD, provides four well-documented examples of natural section. Evidence of natural selection includes the dark and light pepper moths of England and population ratio changes due to predation and environmental factors.[10] Natural selection is shown in the different survival rates of white rump feathered pigeons versus gray rump feathered pigeons from peregrine falcon predation in Davis, California.[11] Natural selection is evidenced by the change in population ratios of stickleback fish in Loberg Lake, Alaska. Over a twelve year period, stickleback fish with a full row of side armor plating decreased in the lake while stickleback fish with a short row of plating increased in number.[12] Natural selection is shown by the population of different colored rock pocket mice living in the Pinacata Desert of Arizona. Dark colored mice are the majority living among dark colored lava flows while light colored mice are the majority living on light colored sandy ground.[13] As a fifth example in Carroll's book, it is easy to visualize how the different coloration in the Northwestern garter snake might provide a survival advantage under certain predation and environmental conditions.[14]

In addition to those mentioned by Carroll, other examples of natural selection are commonly known. Antibiotic-resistant bacteria can increase in population size within a sterile hospital type environment. Natural selection can cause an increase in pesticide resistant bugs that destroy farm crops. It is well known that strong, healthy animals survive better than the sick, deformed and crippled. Certain plants under specific growing conditions based on temperature, soil condition, wind, moisture and sunlight survive better than others. A white rabbit living in a snowy environment will evade predators better than a brown rabbit because of its camouflage. These are all examples of natural selection acting on certain genetic traits.

Natural selection is real, but severely limited. Natural selection cannot create new genetic material and can only act on existing material. Natural selection cannot act upon small genetic changes which give no survival advantage based on predation, reproduction, environment or disease. The few examples of natural selection observed today cannot be legitimately extrapolated to support macroevolution.

Natural selection by itself cannot create new genetic information and may destroy it. When natural selection forces a trait to go extinct, it actually destroys genetic information and decreases genetic diversity. For example, if there are light and dark colored moths, and all the light colored moths are killed off by predators, then genetic information is lost. In some instances, natural selection is a winnowing process that reduces variation.

(3) Species divergence – Changes in a species weight, size, color, mating habits, feeding habits and other features are called divergence. This may occur when a species becomes divided geographically between two different environments. Limited species divergence has been observed. The Rhagoletis fly is an example of species divergence. Different populations of the Rhagoletis fly feed on different fruit (hawthorn plants or apple trees), mature differently and have slightly diverged.[15] Mosquitofish in Hawaii have diverged slightly in length, embryo size and other features.[16] The many breeds of sheep, dogs, cats, cows, pigs and horses are good examples of species divergence (in many cased aided by selective breeding). Based on extensive breeding experiments, minor species divergence cannot be legitimately extrapolated to support macroevolution.

(4) Hereditary variation – When a male and female mate, their offspring vary because of heredity. In sexual reproduction, variation is caused when the genes contributed by each parent are mixed. The different looks of people in the world are a testament to this effect. Eighteen different wing patterns observed in the fruit fly illustrate hereditary variation.[17] The coloration and variety found in tropical fish, butterflies and birds demonstrate variation.[18] Normal hereditary variation cannot be legitimately extrapolated to support macroevolution.

(5) Genetic transfer – At least three different processes permit the transfer of genetic material into bacteria. Genetic recombination occurs by transformation, conjugation and transduction. *Transformation* occurs when a bacteria takes up genes or genetic material from its surroundings. *Conjugation* occurs when two bacteria share genetic material by direct contact. *Transduction* occurs when a virus infects a bacterium and transfers new genetic material into it. The effects of genetic transfer into bacteria cannot be legitimately extrapolated to support macroevolution because of the barrier to genetic exchange among more advanced life.

(6) Adaptive change – Animals and plants can adapt to their environment to a limited extent. Adaptive changes are autogenous (come from within) and are a direct response to a new environment. Darwin's finches are an example of adaptive change. Their beaks slightly change shape and size based on food availability. Minor changes in animal species due to adaptation have been observed to occur rapidly.

Adaptive change is real, but does not appear to be a random process. Dr. Lee Spetner in *Not by Chance* hypothesizes that environmental factors can cause heritable mutations within DNA to result in rapid adaptation.[19] He says that bacteria have mutated and developed the ability to live on totally different food when change is forced upon them. The results have been duplicated multiple times with separate colonies of bacteria.[20] The mathematical probability of this occurring by random mutation is extremely small and strongly suggests that a predetermined response, triggered by environmental influence, is capable of causing the same repetitive genetic mutation. Spetner says evolutionists ignore this evidence, since according to theory, mutations must be random. He also says that environmental factors can turn genes on and off to result in rapid adaptation that is heritable without any DNA changes.[21] The epigenetic code, histone code and DNA methylation are factors in this effect. In general, adaptive changes occur rapidly, can be repeated and are always beneficial within the confines of the new environment. On the other hand, the occurrence of a truly random mutation is based on mathematical probability. If the probability is small enough, a random mutation will not readily repeat itself.

Design theory explains rapid adaptive change as a predetermined response, with the automobile as a good analogy. The basic Model T was transportation and little else. The modern luxury automobile is substantially more sophisticated. If it becomes dark, the headlights turn on automatically. If it rains, the windshield wipers begin operating. As speed and road noise increase, the radio volume is raised to compensate. If it is hot or cold, the climate control system adjusts the car interior to a preset temperature. If a wheel is slipping, torque shifts to the wheel with more traction. If the road is rough, suspension stiffness changes to soften the ride. All of these actions are carefully designed predetermined responses to specific environmental conditions. Sensors detect change and initiate a response. Within living things, rapid species adaptation may be viewed as a predetermined design response to environmental factors to help ensure species survival under adverse conditions.

Scientists currently lack comprehensive information about the details of most rapid adaptation. The cause for swift response to environmental change is not well understood. Based on present knowledge, adaptive changes cannot be legitimately extrapolated as support for the major changes required by macroevolution.

Irreducible Complexity

Irreducible complexity is probably the single most significant idea to challenge biological evolution in the twentieth century. This idea refutes the key mechanisms of evolution—random beneficial mutations and natural selection. Irreducible complexity is explained in *Darwin's Black Box: The Biochemical Challenge to Evolution* by Michael Behe, PhD and professor of biochemistry at Lehigh University in Bethlehem, Pennsylvania.

Biological evolution proposes that significant species modification and improvement can occur through a continuum of small step-by-step changes guided by natural selection. Hypothetically, this process begins as a single random beneficial mutation within one member of a species. A single mutation occurs within one member's heritable DNA, creating a functional advantage. This permits the owner to be smarter, healthier, stronger, faster, have more offspring or in some way survive better than other members of its species (under specific environmental conditions). As generation after generation passes, the genetic material of the superior member spreads throughout the species (within a specific environment). Members with inferior genetic material become extinct. This cycle is repeated to build new genes, proteins, features, functions and species in a step-by-step series.

While this argument sounds impressive, the rationale totally falls apart when there is no functional advantage for natural selection to act upon.

A mousetrap is used by Dr. Behe as an example of an elementary, yet complex functional system. A mousetrap has multiple discrete parts made of different materials that fit together in a precise arrangement to perform a given function (catch mice).

Irreducible complexity can be illustrated by taking parts away from a functioning mousetrap. As parts or pieces of parts are removed, at some point function will cease (the trap will be unable to catch mice). It is possible to remove a single staple or break a corner off the base and still have a working mousetrap. However, if one of the major parts is removed such as the base, spring, catch/bait holder, hammer wire or restraining bar, then function immediately ceases. By taking away parts from the mousetrap, it can be reduced to a point where it will not work.

Irreducible complexity can also be illustrated by building a mousetrap. As assembly progresses, the trap will not function until all the essential components are present and in the proper arrangement. If the alleged assembly process is evolutionary, the mousetrap cannot evolve a base, catch a few mice, evolve a spring, catch a few mice, evolve a hammer wire, catch a few mice, etc. The mousetrap must be complete with all parts in place to catch any mice. Natural selection cannot operate on a partial mousetrap with no function. Step-by-step evolution does not work. The entire system must be complete and functional to catch mice.

The logic of irreducible complexity applied to a mousetrap is also applicable to living organisms. Behe shows that many biochemical processes in animals such as gated transport, AMP synthesis and blood clotting are irreducibly complex. It is estimated there are tens of thousands of irreducibly complex systems at the cellular level within mammal species. In a complex biological system, if any critical part is missing it will not work. Construction that is fully functional at inception must take place for there to be any benefit—compelling a design influence.

Behe uses the blood clotting mechanism as an example of irreducible complexity found in animals. Blood clotting is a cascading chemical response to bleeding, and the extinguishing of that response once certain conditions are met. If blood did not clot, we would bleed to death with a minor cut. If the clotting response was not stopped, our entire blood system would become clogged. The clotting response is started by tissue factor which usually only appears on the outside of cells not in contact with blood. Once the blood-clotting cascade is started, thrombin is activated, which slices off small pieces of fibrinogen to make fibrin. The fibrin has sticky ends which adhere to each other to make a meshwork capable of covering a large area (similar to netting) with a minimum of protein material. Antithrombin and heparin help stop the cascade by inactivating clotting proteins. The total blood-clotting cascade from start to stop involves more than twenty-five proteins, enzymes and molecules.[22] We can hypothesize that blood clotting did not evolve in a step-by-step series, but was made as a completed functioning system.

Behe cites the production of AMP in living things as an example of a complex functional biological system. AMP (adenosine monophosphate) is a nucleotide molecule found in RNA. Synthesis of AMP in a human takes fourteen steps and involves thirteen enzymes in addition to several precursor molecules.[23] This functional system can be destroyed by removing a single enzyme. It can be argued that some AMP might still be produced after removing an enzyme, although at a rate 100,000 times slower than normal. The counter argument is that AMP production is so slow that for purposes of supporting metabolism, the organism will die.

Any complex functional system can be rendered nonfunctional by taking away its parts. Design engineers know that when designing a complex system composed of hundreds of parts, each part contributes in a small way to total system performance. However, the failure (or removal) of a single part can completely destroy all function. The Space Shuttle Challenger is a prime example of this principle. The shuttle exploded immediately after launch in 1986 due to the failure of an O-ring seal in the solid fuel booster. The Space Shuttle Columbia disintegrated over Texas in 2003 due to the loss of a ceramic insulating tile. A single part destroyed both of these shuttles, which were otherwise functioning perfectly. The failure of a single inexpensive capacitor in a Dell personal computer can turn its Pentium IV chip and motherboard into nonfunctioning junk.

Any complex biological system can have parts or pieces of parts removed until function ceases. Below this point, no function exists for natural selection to act upon. Below this point, undirected change is the only mechanism available to build *complex nonfunctional structure* as a precursor to a complex functional system. Undirected change alone is incapable of the task. It cannot tell what pieces to save or reject. It cannot tell if things are going in the right direction. It cannot tell when change should start or stop. The mathematical probability of random mutation building the required complex nonfunctional structure is so small it will not occur—even given billions of years. Time is not a magic wand capable of curing the severe limitations of natural cause.

Definition 9.2 – Principle of Irreducible Complexity

Any complex functional system with an identified function can have individual pieces removed until function is lost. As applied to biology, a system within living things can be reduced by the removal of atoms, molecules, proteins, cells, tissues and other discrete parts until its principal function is lost. Below this point, the building of complex nonfunctional structure as a precursor to the complex functional system cannot be guided by natural selection. Random change alone is incapable of building the required nonfunctional structure.

In his book, Behe presents irreducible complexity as an idea, which in my opinion deserves elevation to the status of principle. Any complex functional system with an identified function can have individual pieces removed until function is lost. As applied to biology, a system within living things can be reduced by the removal of atoms, molecules, proteins, cells, tissues and other discrete parts until its principal function is lost. Below this point, the building of complex nonfunctional structure as a precursor to the complex functional system cannot be guided by natural selection. Random change alone is incapable of building the required nonfunctional structure.

Supporters of biological evolution have attacked the principle of irreducible complexity with a vengeance. Most of these attacks are based on issues of semantics and debate, rather than on the facts of biochemistry and the substance of Behe's idea. For example, critics point out that a mousetrap is not irreducibly complex. One staple can be removed and the trap will still work (staples are not mentioned in Behe's book). Up to fifty percent or more of the wooden base can be carefully removed and the mousetrap will still function. However, if we continue to remove pieces from the mousetrap it will eventually reach a point where it cannot catch mice. The same applies to living things. Removal of a single atom, molecule, amino acid, protein, cell or tissue from a biological system might not cause system failure; however, if we continue to remove pieces, the system will eventually lose its prime function.

Critics of Behe's mousetrap example argue that the trap will still work when parts are substituted. For example, the wooden base can be discarded and all parts attached to the floor on which the mousetrap is resting. Behe makes no claim that another part, capable of performing the same function, cannot replace an existing part. It should be obvious that any part of a mousetrap can be substituted with another slightly different one having the same function. This does not invalidate the principle of irreducible complexity.

Opponents of Behe say that a mousetrap to catch and kill mice can be designed totally different from the traditional mousetrap. Behe does not claim this is the only way to make a mechanical mousetrap, and admits there are other possible designs. The number of potential designs for a mousetrap is a completely different issue. However, regardless of the design it will be irreducibly complex.

Critics of irreducible complexity argue that some biological systems have multiple pathways which produce the same function. If one pathway becomes inoperable, another pathway can still function, which skeptics say invalidates his principle. To address this argument, we need to understand redundancy.

Engineers commonly design redundancy (multiple pathways) into critical systems that must not fail. For example, modern commercial aircraft are designed with a high level of redundancy to prevent catastrophic failure. Every aircraft must have two engines, each one being controlled by a digital electronic engine controller having multiple circuits to prevent a single point failure. All critical instruments and components in the cockpit exist in duplicate or triplicate. An automatic landing system will have one primary system plus redundant systems and power supplies. Suppose an aircraft is making a fully automated landing in the dark of night, with fog, rain and a 20 knot crosswind with intermittent gusts. The aircraft is on final approach at 140 knots and 200 feet above the ground. Lightning strikes the aircraft, trips all the circuit breakers and knocks out all electrical power. In realistic situations such as this, these systems must not fail. The engines, instruments, controls and automated landing system must perform flawlessly without a second of hesitation. Human designers use independent, redundant systems with battery backup power as one method of achieving this level of dependable function.

Within complex living things, redundant systems are to be expected. Each alternate pathway should be treated as a separate system, capable of being reduced to a point where function is lost. Below this point, random mutation is incapable of building the necessary complex nonfunctional structure. Design theory recognizes redundancy as an indicator of sophisticated design with the purpose of providing robust function.

Skeptics have developed the idea of a biochemical scaffold as a counter to irreducible complexity. Hypothetically, a scaffold system developed first and guided the formation of complex molecules, proteins, enzymes and systems. After formation of the necessary parts, the scaffold became obsolete and disappeared. This is an impressive sounding argument; unfortunately, we are faced with the problem of how the scaffold system developed. A scaffold system itself would be irreducibly complex. Unless it had some function, it could not develop using natural selection as a guiding mechanism. A complex scaffold cannot be built by random variation alone. A scaffold or pre-scaffold cannot explain how a complex nonfunctional structure can evolve as a precursor to a complex functional system. In human directed construction projects, scaffold systems are normally quite complex and are often more precise and complicated than the object being built.

Opponents of irreducible complexity have said that the existence of similar parts in other structures invalidates irreducible complexity. For example, some parts of the bacterial flagellum (used by Behe to illustrate an irreducible system) can be found in other systems. Critics say this shows that identical parts can evolve independently in totally different systems. Behe makes no claim that parts used in one system will not be found in another system (design theory predicts commonality and the reuse of parts). Behe does not address commonality or claim it cannot occur among irreducibly complex systems.

Similar or identical parts among complex systems do not invalidate the principle of irreducible complexity. In fact, they help illustrate the inability of a complex nonfunctional structure to develop based solely on random change. For example, the human body contains twenty different kinds of amino acids, approximately 29,000 genes and an estimated 100,000 or more different proteins. Suppose humans did not possess sight of any kind. Without having any function for natural selection to act upon, which of the available genes or proteins should be selected to build a chemical pathway to turn light into a nerve impulse? Random chance alone is incapable of such a task. The fact that a protein may exist in several different systems is irrelevant as far as the principle of irreducibility complexity is concerned.

Critics of irreducible complexity have attempted to use the eye as an example of how a complex functional system can evolve, arguing that a light sensitive spot could have developed naturally. The simplest known eyes are capable of turning light into an electric signal, in turn stimulating a nerve. Behe discusses in detail the biochemistry of turning a photon of light energy received by an eye into an electric signal.[24] Without every single molecule working properly, no signal is produced. The simplest chemical pathway for sight in any living organism on Earth is irreducibly complex.

Opponents of Behe's idea have identified several basic eyes they believe could have evolved. The National Academy of Science in *Science and Creationism* uses the eye of the flatworm as an example of a rudimentary eye.[25] An article in *National Geographic* gives the eye of the limpet (a marine mollusk) as a possible starting point for eye evolution.[26] In *The Making of the Fittest,* the ragworm's eye is mentioned as an example of a simple eye.[27]

Actually all of these eyes in the adult are substantially complex and defy an evolutionary explanation. Development of the ragworm eye from a single cell observed in larva to the full eye of an adult is not evolution, but a growth process controlled by DNA. Drawings of eye evolution are nothing more than speculative cartoons. These articles appear to miss the entire point of Behe's argument. The most rudimentary eye known works because of its complex molecular pathway, which is irreducibly complex. It is inconceivable this biochemical system could have occurred by chance.

Skeptics claim that similarity among species invalidates irreducible complexity, and the protein cytochrome c is used to illustrate their point. The textbook *Life-The Science of Biology* gives the amino acid sequences of 29 different cytochrome c proteins found in 33 species.[28] The cytochrome c protein is slightly different for most species listed. Cytochrome c protein consists of a chain of 104 amino acids and exhibits variation in 69 of those amino acid positions (35 positions are invariant) for the species given. Variation in 69 of the amino acid positions is believed to be neutral because it has no known effect on the species listed. When humans are used as a reference, these 29 cytochrome c proteins can be arranged in a descending series based on increasing variation consisting of humans, gray whales, snapping turtles, tuna fish, baker's yeast and sesame plants.

Irreducible complexity does not address the issue of similarity among the 33 species listed, but rather how the first cytochrome c protein came into existence. Whether this protein is similar among species or first appeared in baker's yeast is immaterial as far as the principle of irreducibly complexity is concerned. The most primitive cytochrome c protein is an irreducibly complex molecule and random mutation alone is incapable of building the necessary nonfunctional structure leading to a functional protein. Other than essentially appealing to a paranormal event, biological evolution has no plausible explanation for the first cytochrome c protein.

In summary, between a single-celled organism and humans, evolution must plausibly explain the first occurrence of each complex functional system when evaluated by the principle of irreducible complexity. There are tens of thousands of molecules, molecular pathways, proteins, genes, tissues, features and functions found in mammal species not evident in microscopic life. A single-celled organism lacks cell differentiation, a circulatory system, a digestive system, a lymphatic system, a skeletal system, hormone systems, sight, hearing, smell, balance, taste, a brain, muscles, skin, teeth, sexual reproduction, bisymmetry and other features. The additional capabilities found in mammals represent a host of irreducibly complex systems which evolution cannot explain when evaluated by Behe's principle.

Point Mutations and the Step-by-Step Evolution of Proteins

Charles Darwin proposed that biological evolution occurred due to small step-by-step changes. He believed that small beneficial changes were acted upon by natural selection to result in new features, functions and species. Darwin did not know the size of the smallest change possible or how it would occur. We now know that a point mutation is the smallest possible structural change in DNA.

There are many different types of heritable DNA changes that can occur within a species. DNA material can be substituted, deleted, inserted, inverted or otherwise changed.[29] Individual genes can be duplicated, and in at least one instance, the entire genome of an organism has been duplicated.[30] A point mutation is the smallest possible structural change and occurs when a single base pair of DNA is altered.

A point mutation alters one base pair by substitution, addition or deletion. As previously mentioned, the DNA macromolecule may be visualized as a long spiral staircase. Each step of the staircase is called a nucleotide (or base pair). The four bases are adenine (A), guanine (G), thymine (T) and cytosine (C) which are paired together to make four possible nucleotides consisting of A-T, T-A, C-G and G-C. A point mutation may add, delete or substitute one nucleotide for another.

Point mutations of interest are those that occur within heritable DNA found in germline cells rather than somatic cells. In animals that sexually reproduce, heritable DNA lies within the sperm of the male and egg of the female. Point mutations may come from several possible sources. A point mutation may occur when DNA is copied during egg production in the female or during sperm production in the male. Point mutations may occur when the egg and sperm combine during fertilization. A mutation may occur as the one-celled zygote grows into an embryo. Regardless of how they occur, point mutations are an observed phenomenon within heritable DNA.

For biological evolution to be true, billions of beneficial point mutations had to occur. Genetic information for proteins found in the first single-celled organism had to develop into the genetic information for proteins observed in all subsequent species. The vast majority of this change had to occur due to point mutations rather than whole genes or DNA segments.

As previously mentioned, *Prochlorococcus marinus* CCMP1378 MED4 (blue-green algae) is a small, free-living, robust, microscopic organism that has 1,713 protein-encoding genes and 1,657,990 base pairs in its genome.[31] It is a strong potential candidate for the first microscopic life on Earth. The human genome contains roughly 29,000 protein-encoding genes and approximately 3,000,000,000 base pairs of DNA, some of which are listed in Table 9.3. For biological evolution to be true, the 1,713 genes represented by *Prochlorococcus marinus* (or a similar first microscopic organism) had to develop into the roughly 29,000 genes found in humans. The additional genes in humans are the basis for additional proteins, in turn making possible additional features and functions not found in blue-green algae.

A point mutation can only be called beneficial if it imparts an advantageous trait at the organism level. Beneficial traits cause an organism to be smarter, healthier, stronger, faster, have more offspring or in some way survive better than other members of its species. For a mutation to be considered beneficial at the genetic level (genotype), it must impart a beneficial trait at the organism level (phenotype). Mutations in genes and pseudogenes are considered one evolutionary pathway.

Nonfunctional pseudogenes cannot cause a step-by-step series of point mutations to occur. Actually some pseudogenes appear to have a function, but for sake of argument assume they are completely nonfunctional. Suppose a point mutation occurs in a nonfunctional pseudogene. Since the pseudogene is nonfunctional, a point mutation cannot impart any beneficial advantage to the organism. Without a beneficial trait, natural selection cannot work. This makes a step-by-step series of point mutations guided by natural selection impossible. Nonfunctional pseudogenes are not a plausible means of developing new genes and proteins within a species.

On the other hand, can functional genes (or pseudogenes) logically account for step-by-step evolution? In this scenario, the first point mutation in a step-by-step series must impart a beneficial trait to get things going. Based on our knowledge of proteins, the probability of any beginning mutation imparting a new beneficial trait is very small, thus preventing the series from starting. In addition, it is impossible to develop a two, three, four, five or six step series leading to a new protein (with a new function) without each point mutation imparting a beneficial trait. As discussed in the textbook *Life-The Science of Biology* about the constraints of biological evolution, "All intermediate forms must work; that is, all modifications must benefit their bearers in every generation."[32]

Allegedly, a step-by-step series of beneficial point mutations in a functional gene can build a new gene. Initially a random point mutation creates a protein imparting a significant beneficial trait to one animal. Natural selection then spreads the first step trait throughout the species (or to all members living within the environment benefited by the mutation). After the first step has spread to 10,000 animals, a large enough population is established so that a second point mutation has a mathematical probability of occurring. A second beneficial point mutation then occurs and subsequently spreads to 10,000 animals. A third beneficial point mutation then occurs and spreads. The final result is a population of 10,000 animals containing the first three steps of a series. Each subsequent step in the continuing series must go though the same cycle. A single step at the genetic level (genotype) which fails to impart some significant new beneficial trait at the organism level (phenotype) will stop the series. A fairly large population (i.e. 10,000 animals) is required to have a reasonable probability of a beneficial mutation occurring.

A protein series, separated by small steps between each protein, would be expected if evolution were true. Based on present information, a continuous sequence of proteins from the smallest to the largest does not exist. The current UniProt/Swiss Protein Knowledgebase contains the amino acid sequences for 267,354 proteins from 5,198 species.[33] While small steps exist within some protein families (such as cytochrome c) and some phylogenetic tree arrangements show structural similarities, the vast majority of proteins have large differences between each other (similar to gaps in the fossil record).

Most mutations are neutral or deleterious, rather than beneficial, and prevent a step-by-step evolutionary series from occurring. The book *Genetic Entropy & The*

Mystery of the Genome by John Sanford, PhD, discusses point mutations. He describes the effect of a point mutation at the genetic level (genotype) and its effect at the organism level (phenotype). He states that point mutations which do not create a beneficial effect at the organism level cannot be acted upon by natural selection.[34]

Sanford uses the example of a television screen and pixels to illustrate the effect of a small random change on a large system.[35] His logic applies equally well to a computer screen. Suppose we have a screen saver that consists of an array 1,152 x

Table 9.3 – Human Chromosomes, Genes and Pseudogenes

No.	Chromosome Length in Base Pairs[1]	Number Protein Encoding Genes[1]	Number Pseudo-genes[2]	Comments[3]
1	247,000,000	2,782	1,861	Last chromosome sequenced
2	243,000,000	1,888	1,887	Density is 1 gene per 128.7 kbp
3	200,000,000	1,469	1,507	Density is 1 gene per 136.1 kbp
4	191,000,000	1,154	1,329	Density is 1 gene per 165.5 kbp
5	181,000,000	1,268	1,903	Density is 1 gene per 142.7 kbp
6	171,000,000	1,505	1,398	Density is 1 gene per 113.6 kbp
7	159,000,000	1,452	1,823	Density is 1 gene per 109.5 kbp
8	146,000,000	984	1,026	CSMD1 gene has 2,056,947 bp
9	140,000,000	1,148	1,017	Average gene length 46,410 bp
10	135,000,000	1,106	850	Widespread overlapping genes
11	134,000,000	1,848	1,321	One-quarter of genes overlap
12	132,000,000	1,370	1,203	Density is 1 gene per 96.4 kbp
13	114,000,000	551	538	Lowest gene density
14	106,000,000	1,275	971	Density is 1 gene per 83.1 kbp
15	100,000,000	945	896	Average gene length 57,960 bp
16	89,000,000	1,109	945	Density is 1 gene per 80.3 kbp
17	79,000,000	1,469	1,006	Density is 1 gene per 53.8 kbp
18	76,000,000	432	454	Average gene length 75,520 bp
19	64,000,000	1,695	1,521	Highest gene density
20	62,000,000	737	416	Density is 1 gene per 84.1 kbp
21	47,000,000	352	324	Density is 1 gene per 133.5 kbp
22	50,000,000	742	603	First chromosome sequenced
X	155,000,000	1,336	1,738	DMD, largest gene 2,220,390 bp
Y	58,000,000	307	586	95% of DNA is male specific
Totals	3,079,000,000	28,924	27,123	

Note: The above information may change with ongoing research. The number of genes and pseudogenes will vary among sources because of the identification method used.
References:
(1) National Center for Biotechnology Information (2006). Human Genome Resources, Build 36.2.
(2) Gerstein, Mark and John Karro (2007). Search 31, Pseudogene.org.
(3) Genomics (2007). Post-Human Genome Project Progress & Resources.

864 points for a total of 995,328 pixels. If the intensity or color of one pixel is changed, what is the overall effect on the screen saver? The change or complete deletion of a single pixel has little effect (is neutral). Many pixels must be changed in an organized manner to achieve a beneficial improvement. The effect of changing one pixel in order of probability is neutral (highly probable), deleterious (small probability) and beneficial (impossible).

A book can also be used to illustrate the effect of small random changes within a large system. Suppose we have a 300 page technical manual that describes how to assemble an automobile engine. What is the effect of randomly changing one character within the manual? Any change would probably be deleterious at a local level. The word "now" might be changed to noa, nor, nod, noc or nox, and affect a single sentence. The total effect on the entire book would most likely be neutral. On the other hand, it is virtually impossible for a single character change to improve the instructions. The effect of changing one character in order of probability is neutral (highly probable), deleterious (small probability) and beneficial (impossible).

Based upon these examples and his knowledge of mutations, Sanford makes the statement that essentially all point mutations are neutral or deleterious.

Evolutionists believe natural selection can act on small changes in a step-by-step manner to produce information. In *The Blind Watchmaker*, Richard Dawkins, PhD, zoologist at Oxford University and leading spokesman for biological evolution, gives an example of how "cumulative selection" can occur. He starts out with a random string of 28 characters and uses a computer program to simulate random change and step-by-step natural selection. In 43 generations, a string of random characters (WDLTMNLT_DTJBKWIRZREZLMQCO_P) is changed into a line from Shakespeare.[36]

Dawkins proposes that when a correct character randomly occurs it imparts some information to the string. He leads people to believe that an increase in information is recognized by the computer and retained, similar to natural selection.

Dawkins' example of random change and natural selection is deceitful and misleading. Suppose the first letter in the string is changed from W to M and the last letter from P to L. Do we have enough of an increase in information to tell if a correct change has occurred and should be retained? Absolutely not! With two, three or four letters changed, it is impossible to determine if they should be retained. They are neutral changes without a discernable benefit. In Dawkins' example, 24 letters out of 28 are changed correctly before a coherent statement begins to emerge. The only way Dawkins' example works is if METHINKS IT IS LIKE A WEASEL is loaded into the computer program at the start, and when a correct letter randomly occurs it is locked in place. This is intelligence in action, not natural selection.

In summary, (1) biological evolution hypothetically occurs in a series of small step-by-step changes, (2) a point mutation is the smallest structural change possible, (3) a beneficial trait is required for each step or the series will stop, (4) essentially all point mutations are neutral or deleterious rather than beneficial and (5) neutral and deleterious mutations will stop a series from starting or continuing. Our knowledge of point mutations strongly discredits Darwinian evolution.

We should realize that a new trait can develop from a single mutation event. One random point mutation in the protein-encoding gene of a light colored mouse can result in a black colored mouse.[37] One random point mutation in *Plasmodium falciparum* (the most common parasite causing malaria) can make it resistant to the anti-malarial drug atovaquone.[38] One random point mutation in the HBB gene of humans causes sickle cell trait. These are all mathematically probable events for a single point mutation. Regardless, these three rare examples cannot be extrapolated to support the formation of tens of thousands of completely new proteins where a step-by-step series of beneficial point mutations are required.

Darwin versus the Human Genome

The human genome has been studied more than any other species in an effort to understand genetic disease. As a result, knowledge of human DNA, chromosomes, genes, proteins and mutations far exceeds that of any other animal species and provides the best current information to evaluate the claims of Darwinian evolution.

Starting with the first microscopic life, Darwinian evolution proposes that small beneficial changes were acted upon by natural selection to result in all of Earth's species. If biological evolution is true and humans evolved, there should be significant evidence of protein evolution in the human genome.

Human proteins and *protein isoforms* (a protein different from the common form due to mutation) have been extensively studied. The Swiss Protein Knowledgebase is a consortium of organizations whose goal is to catalog all known proteins. This includes information about protein mutations in humans. The UniProt/Swiss Protein Knowledgebase lists 31,601 protein isoforms made by 5,778 different genes found in humans. In the Swiss Protein Knowledgebase, each protein isoform represents a normal protein in which one amino acid has been changed. Of the total number of protein isoforms in humans, 15,796 cause some type of disease, 13,833 are neutral and 1,972 are unclassified.[39] In other words, 50% are deleterious, 42.8% are neutral and 6.2% are unclassified.

The actual categorizing of protein mutations shows the majority are deleterious or neutral. This confirms the statement by Sanford when he says, "Geneticists widely agree that essentially all of these [human] mutations are either neutral or deleterious (if any are beneficial, they are considered so rare as to be entirely excluded from consideration)."[40] The effects of all known human protein isoforms (mutations) are a continuum from completely neutral to totally lethal. Some of the lethal mutations include Tay-Sachs disease, cystic fibrosis, hemophilia and Wilson's disease (all are fatal without medical treatment). Based on this information, we can say that 99.9% or more of random mutations affecting proteins in the human genome that have been classified are either deleterious or neutral.

Table 9.4 – Distribution of Human Genes and their Protein Isoforms (Arranged in descending order by protein isoforms per gene)			
Protein Isoforms per Gene	Number of Genes	Protein Isoforms per Gene	Number of Genes
>501	0	21 - 30	90
401 - 500	1	11 - 20	270
301 - 400	1	10	41
201 - 300	4	9	74
101 - 200	26	8	96
91 - 100	4	7	109
81 - 90	11	6	170
71 - 80	11	5	237
61 - 70	13	4	377
51 - 60	25	3	639
41 - 50	27	2	1090
31 - 40	51	1	2410
Note: A total of 5,778 different genes and 31,601 protein isoforms. Reference: UniProt/Swiss Protein Knowledgebase (2006). Release 52.5.			

As shown in Table 9.4, a distribution analysis of human protein isoforms by gene reveals they are not random. For example, 2,410 different genes have one protein isoform per gene, while 270 genes have from 11 to 20 protein isoforms per gene. One gene (the F8 gene that causes hemophilia) has 456 protein isoforms. Analysis shows that 20% of the genes in this table contain 73% of the protein isoforms. Based on this data, the protein-encoding sequences of most genes are conserved, contain few mutations and suggest the presence of a locking or repair mechanism. Contrary to the claims of evolutionists, mutations are far from random.

Evolution (undirected change) is occurring in human proteins, but is not resulting in improvement. Our detailed knowledge of the human genome strongly discredits the idea of beneficial protein evolution in humans and can be summarized as follows:

- Most protein-encoding genes produce either few or no protein isoforms. Most protein-encoding genes appear to be highly conserved (with few mutations) or invariant (with no mutations).

- Protein isoforms (mutations) are not random and are produced by a minority of genes (5,778 out of 29,000).

• More than 99.9% of classified protein isoforms are either neutral or deleterious. The lack of beneficial mutations prevents a series of small step-by-step changes from occurring based on natural selection.

• Protein isoforms demonstrate that evolution in the human genome is overwhelming destructive rather than constructive.

New Alleles but No Improvement

According to neo-Darwinian evolution, new genes and proteins are continuously evolving within species. Through mutation, the genes and proteins in the first microscopic life are hypothesized to have evolved into the tens of thousands evident in species today. A review of actual human genes shows that mutations are occurring; however, these mutations are not creating new beneficial proteins. In the human genome, new beneficial proteins are nonexistent or extremely rare.

Different versions of a gene located at a specific location (locus) in a chromosome are called alleles. These different versions of a gene create variation from one offspring to the next. For example, during sexual reproduction in humans each child inherits one gene from her mother and one from her father. The two inherited genes may be identical or different. If the genes are different, they represent two alleles or versions of the same gene. These alleles determine the features of the child. Some normal human features due to different alleles are eye color, skin color, hair color, curly or straight hair, height, body type, face shape, freckles, attached or detached ear lobes, fleshy or thin lips, dimples, double-jointed thumb, widow's peak and the ability to tongue roll.

A point mutation in a gene can cause a new allele. If the new allele changes an amino acid, in turn affecting protein function in a deleterious manner, a new genetic disease is the result. Some common hereditary genetic diseases are sickle cell disease, Tay-Sachs disease, Marfan syndrome, Wolfram syndrome, Waardenburg syndrome type I, Wilson disease, diabetes, colon tumors, leukemia, liver tumors, hemophilia, cystic fibrosis, breast cancer and age related macular degeneration. Several variations of a genetic disease may be caused by different mutations affecting the same protein. For example, the HEXA gene responsible for Tay-Sachs disease has 44 protein isoforms (produced by 44 different alleles) due to mutation according to the UniProt/Swiss Protein Knowledgebase.[41] While mutations in the original HEXA gene have created new alleles and protein isoforms, they are all deleterious. No new proteins beneficial to humans have evolved in the HEXA gene.

Hemoglobin is often cited as an example of gene and protein evolution. The hemoglobin molecule transports oxygen from the lungs to the cells. Normal hemoglobin molecules in humans are made of two alpha protein chains (alpha subunits) and two beta protein chains (beta subunits) combined together. Each alpha protein chain consists of 141 amino acids and each beta protein chain contains 146 amino acids. The most notable mutation in sickle cell trait or sickle cell disease is caused by a point mutation in the HBB gene which makes one of the hemoglobin proteins (the 146 amino acid beta protein chain). A point mutation in the seventeenth nucleotide of the HBB gene changes thymine to adenine, causing a corresponding change in the sixth amino acid of the beta protein chain from glutamic acid to valine.

The sickle cell mutation affects red blood cells in a deleterious manner. Red blood cells are deformed from their normal donut shape to a crescent shape. Deformed cells cannot pass through blood vessels easily and may cause pain or organ damage. Normally red blood cells last 120 days in the body, while those containing defective hemoglobin may last only 16 days. The body's inability to replace these blood cells can lead to anemia. In areas infested with malaria, the sickle cell mutation can help people survive and is "beneficial" in this unique environment.

Beyond sickle cell, the hemoglobin molecule shows a lot variation due to mutation. The UniProt/Swiss Protein Knowledgebase gives 252 protein isoforms in the beta chain made by alleles of the HBB gene.[42] While these mutations are not as well known as the sickle cell mutation, they exist and are a testament to the many deleterious alleles and their protein isoforms caused by mutation.

In summary, according to biological evolution new genes and proteins have been continuously evolving since the first living organism to create new features, functions and species. The alleles and protein isoforms caused by mutation of the HBB hemoglobin gene are often given as a prime example of evolution. It is a fact that many new alleles have evolved from the original HBB gene; however, these alleles (1) contain no new information, (2) function of the hemoglobin molecule has not improved and (3) no beneficial effects have been observed. The best known example of protein evolution in humans is going nowhere. It is undirected change that is neutral or deleterious and nothing more.

Missing Neutral and Deleterious Mutations

Natural selection acting upon random beneficial mutations within heritable DNA is the primary mechanism of biological evolution. For this mechanism to work, a substantial number of mutations had to occur. Allegedly, natural selection retained a small percentage of beneficial mutations and discarded a vast multitude of neutral and deleterious mutations.

Mutations occur within every species genome and have been measured and extrapolated to calculate various mutation rates. Heritable mutation frequency rates for a given species are debatable; however, mutations occur and are the primary hypothetical mechanism for the evolution of new genes and proteins. Conceptually, low mutation rates slow down the rate of evolution to unacceptable levels while high rates do not allow natural selection enough time to eliminate deleterious mutations.

Supporters of evolution are quick to calculate how mutations can give rise to new features and functions. Sean Carroll in *The Making of the Fittest* shows how a random mutation can create a black colored mouse from a population of light colored mice.[43] A mouse has about 2.7 billion base pairs in its DNA.[44] There are 2 MC1R genes with 10 available sites that can be mutated to get a black mouse.[45] The mathematical probability of one base pair being affected by a mutation in a single mouse is one in 135 million (2.7 billion/(2 x 10)). Based on this example, it would take 67.5 million mutations to have a 50% probability of getting the desired black mutation.

There should be 540 million or more mutations in the mouse genome if estimated mutation rates during reproduction are correct. If we consider one mouse generation to be six months (in captivity mice sexually mature in eight weeks and can produce a new litter every 18 to 28 days), then two complete generations can occur per year. During reproduction, one mutation is estimated to occur per 100 million base pairs of DNA[46] to result in each mouse having 27 mutations (2.7 billion/100 million). Mice have been on Earth longer than humans and can be considered at least 10 million years old. Based on the number of mutations added by a single mouse in each generation, in 10 million years one line of mice would add 540 million mutations to the mouse genome (2 generations per year x 27 mutations per mouse x 10 million yrs.). At present roughly 14 million single-nucleotide polymorphisms have been identified in the mouse genome.[47]

Study of the human genome gives us our best current information about DNA and mutations. Although mutations exist in the human genome, the number is relatively small. The human genome contains approximately 3 billion base pairs of DNA. Of that amount, 99.9% of all base pairs are estimated to be the same among humans. Approximately 1.4 million base pair locations have been identified where nucleotide differences exist.[48] The International HapMap Project estimates there are 10 million single-nucleotide polymorphisms in the human genome.[49] Regardless of which figure we accept, the human genome is relatively homogeneous and conserved.

For biological evolution to be true, a substantial number of mutations in heritable DNA had to occur. The genome of each species should be filled with mutations. One of the major questions in regard to biological evolution is, "What happened to the missing mutations?" This is especially true of mutations within protein-encoding genes. If millions of years of evolution took place, accompanied by hundreds of millions of mutations per species, there should be ample evidence of these mutations today. Every gene of every species should be filled with neutral and deleterious mutations. The mouse genome should have 540 million or more single-nucleotide polymorphisms rather than 14 million. The human genome should have hundreds of millions of point mutations. Allegedly, a vast number of mutations occurred to facilitate evolution and then somehow most disappeared.

Mutations within genes can be classified in several different ways, and up to this point have been deleterious, neutral or beneficial. An additional method of classifying mutations within a gene is by their *recessive* or *dominant* nature. Recessive mutations require two mutated genes in the offspring to show evidence of a trait while dominant mutations require one mutated gene.

Children receive one gene from their father and one from their mother for a specific locus (location) within a chromosome. If one parent has two mutated genes and the other has one mutated and one normal gene, then the following will occur for a recessive monogenic trait caused by the mutation.

- Fifty percent of their children will have two mutated genes. The recessive trait will be expressed in the phenotype (they will have signs of the disease).

- Fifty percent of their remaining children will have one normal gene and one mutated gene. They will be carriers, but show no signs of the recessive trait (they will not have the disease).

According to Mendel's laws of heredity, recessive mutations which become fixed with a species genome cannot be removed by natural selection. As shown above, a single recessive mutation can never affect more than 50% of the offspring in a deleterious manner. The number of people who carry two mutated genes is limited to 50% of the population. Even if the recessive deleterious mutation is one hundred percent lethal (e.g. Tay-Sachs disease) and kills the child, natural selection cannot remove it from the human genome. The only way a recessive mutation can be removed from a species genome is if all carriers of that mutation die without producing offspring. This requires genetic testing and strict control of mating. Elimination of a recessive mutation in this manner is an intelligence directed process.

The unlimited accumulation of recessive deleterious mutations within a species genome over time can have disastrous effects. Hypothetically, recessive deleterious mutations can accumulate within a species genome to the level where each member carries many defective genes. For example, if each member carries five hundred different recessive deleterious mutations, in a small population it is highly likely that each offspring will have a least one genetic disease. A heavy genetic load of this type can hypothetically lead to species extinction.

Human genetic diseases are a testament to the inability of natural selection to eliminate recessive deleterious mutations. There are currently around four thousand known human genetic diseases,[50] most of which are caused by recessive deleterious mutations. The vast majority are rare because they exist in a small percentage of the population. Sickle cell disease, cystic fibrosis, hemophilia, breast cancer, age related macular degeneration and Tay-Sachs disease are the most recognized heritable genetic diseases among Americans. Approximately eight percent of African Americans carry the sickle cell mutation.[51] One in twenty-five Caucasians (four percent) carry a copy of the defective gene causing cystic fibrosis.[52]

According to the known laws of heredity, *neutral mutations* and *recessive dele-*

terious mutations established in a species genome cannot be eliminated by natural selection. Neutral mutations cannot be removed because there are no inferior traits to act upon. Recessive mutations cannot be removed as previously discussed. It is illogical to believe uncontrolled breeding, starvation, disease, predators or natural catastrophes selectively eliminated hundreds of millions of mutations in every species on Earth.

To solve the missing mutation problem, proponents of evolution hypothesize that all species must have passed through a "genetic bottleneck" where most mutations were eliminated. No convincing ideas have been fleshed out to show how this happened to all mammals, birds, fish, reptiles, insects, worms and bacteria within the last 65 million years. This idea requires a Noah's Ark event, except on a much grander scale. Empirical evidence in support of this hypothesis is missing.

Occam's razor is the principle that when explaining a phenomenon, explanations should rely on the fewest assumptions possible. The simplest solution with the least supposition tends to be the best one. The best explanation for neutral and recessive deleterious mutations observed today is that they have continuously accumulated since species inception. The explanation that hundreds of millions of mutations occurred in each species and then somehow disappeared violates Occam's razor. This explanation is a bizarre idea fabricated to support Darwinian evolution.

In summary, if biological evolution is true then hundreds of millions of point mutations had to occur in each species. A vast multitude of neutral and recessive deleterious mutations had to occur for every alleged beneficial point mutation. As a side effect, the protein-encoding genes of every species should be saturated with neutral and recessive deleterious mutations. The absence of a significant number of point mutations in mice and men discredits biological evolution. It is further discredited by lack of a rational explanation for the disappearance of the alleged missing mutations.

Limited Natural Change

Biological evolution is alleged to be a powerful constructive mechanism that is essentially unlimited. In support of this belief, extensive breeding experiments have been conducted in an attempt to create major species change.

Bacteria have been extensively tested under laboratory conditions for evidence of evolution. Bacteria populations have developed resistance to different chemicals, disinfectants, antibiotics and pesticides. They have gained the ability to use different food sources such as sugars, cellulose, chemical waste and hydrocarbons. In all instances, bacteria at the completion of the experiments were essentially the same. Even though some had gained new traits, none had evolved into worms, mites, ants, ticks or significantly different bacteria. Richard E. Lenski, PhD and professor at Michigan State University, has bred twelve different populations of the same bacteria for more than 20,000 generations. Other than random mutations, little has changed.[53]

The fruit fly *Drosophila melanogaster* has been bred extensively over the past eighty years to test for evolution. Flies have been bombarded with X-rays and exposed to mutagenic chemicals. As mentioned in *Genetics and Heredity* by Maurice Caullery, approximately four hundred fly mutations have been identified: legs growing in place of the antenna, eyes on wings, eyes on the tips of antennae, no heads and two abdomens. Flies have developed multi-colored eyes, truncated wings, miniature wings, no wings, body hair or no hair. The most striking mutation has been two sets of wings similar to a dragonfly (the second set of wings is nonfunctional and offers no advantage). Flies have not changed into wasps, bees or hornets. No new non-fly structures (e.g. butterfly wings, a wasp stinger or grasshopper legs) have ever appeared. Flies are still flies and the limits of natural change have never been exceeded.

Breeding experiments strongly discredit the idea of essentially unlimited natural change required by Darwinian evolution. Yet, the above experiments are often discussed in glowing terms as support for evolution. Minor changes are extrapolated in an effort to illustrate how any type of change is possible; however, these extrapolations are false. Real-world experiments using natural selection and artificial selection, natural environments and artificial environments, and natural mutations and artificially induced mutations show there are strict limits to species change.

Genetic Entropy

An increase in entropy occurs when heat loses its ability to do work, normally associated with the Second Law of Thermodynamics. Increasing entropy may be viewed as a randomizing effect (rather than a concentrating or separating effect). In *Genetic Entropy & The Mystery of the Genome,* Dr. John Sanford introduces the idea of genetic entropy.[54] Within genetic systems, information within DNA can become randomized due to naturally occurring mutations. When this happens, genetic entropy increases.

The idea of genetic entropy contradicts biological evolution. According to Sanford, all species genomes are accumulating mutations, and the information within their DNA is being randomized. Natural selection cannot stop the relentless increase in neutral and recessive deleterious mutations. The ultimate fate of every species on Earth is an increasing genetic load and increasing genetic entropy.

Even though there is good support for the idea of genetic entropy, there is much research to be done and other issues to consider. Will genetic research verify that older species have accumulated more mutations in their genome than younger species? Genetic entropy should tend to show this effect. This raises another interesting question. Why is some DNA conserved (few mutations) or invariant (no mutations), yet other DNA shows significant change? Why doesn't genetic entropy affect all DNA the same?

Another intriguing question in regard to genetic entropy relates to the long term survival of some species. How have animals and plants living today survived for millions of years without becoming extinct because of genetic load (the accumulation of deleterious mutations)? Ancient species alive today include the horseshoe crab whose fossils have been found in strata estimated to be 300 million years old. The sea turtle is estimated to be 200 million years old. Scientists have found centipede fossils estimated to be 100 million years old. Tadpole shrimp fossils are found in strata estimated to be 65 to 250 million years old. Cryptobranchid salamander fossils, similar to living salamanders, are estimated to be 60 to 160 million years old. Paleontologists have found coelacanth fish fossils in strata estimated to be 65 million years old. The gladiator insect (Mantophasmatodea) was found in amber estimated to be 45 million years old. An ever increasing genetic load should hypothetically have resulted in the extinction of these species.

Genetic entropy and the long term survival of ancient species raise some challenging questions. A species that has existed for 50 millions years should have its entire genome saturated with neutral and recessive deleterious mutations. So what has happened? Is some of our basic information about mutation rates or the age of these species in error?

Repair genes are one known means of slowing down the number of mutations within somatic cells (cells that form body parts as opposed to germline cells). One hundred thirty repair genes have been identified capable of repairing errors during somatic cell division.[55] Do germline cells have a similar mechanism capable of correcting errors?

Sanford hypothesizes that locking mechanisms must exist to prevent genome degradation over extensive time. He says that most DNA sequences are polyfunctional and hypothesizes they are also poly-constrained[56] similar to a Rubik's Cube or a palindrome (word puzzle). A single feature cannot be changed without destroying the pattern. Poly-constraint limits the number of hereditary mutations and prevents an ever increasing genetic load. Important sections of DNA are locked to prevent the destruction of information over tens of thousands of generations and millions of years. On the other hand, unconstrained DNA is free to accumulate errors.

Invariant and conserved DNA within species discredit the idea of completely random mutations promoted by Darwinian evolution. Current data about human gene mutations and protein isoforms shown in Table 9.4 suggest the existence of locking or conservation mechanisms.

Design Theory as an Alternative to Biological Evolution

Design theory proposes that all original species on Earth are the result of intelligent cause. Provided current interpretations of the fossil record are correct, life on Earth was a progression from microscopic life to the multitude of species existing today over extensive time. The first life on Earth may have been cyanobacteria (blue-green algae) or similar life. The first major event was the Cambrian explosion where a multitude of complex species were introduced. This was followed by the introduction of reptiles, mammals, birds and eventually present-day species.

Earth's many species may be explained as product expansion (the introduction of new species at specific intervals due to intelligent cause) over extensive time with evidence of original design, design by menu selection and design iteration improvement. Product expansion is represented by the first microscopic life on Earth, expanded over time into a multitude of radically different plant and animal species. There is no rhyme or reason for the existence of Earth's many species, other than a design agent(s) wanted to create an enormous variety of unique, diverse living things. Most species were introduced during a series of major events, typified by the Cambrian explosion.

Earth's species in their present state are best explained as the result of design and limited undirected change (microevolution). Each species was given its DNA, chromosomes, genes and pseudogenes at inception. The genome of each original species was due to intelligent cause, but then modified by microevolution over time.

An automobile is a good analogy of design and microevolution. When an automobile rolls off the assembly line, it is one hundred percent the product of human design. Engineers and technicians have precisely controlled a multitude of elemental materials to create the structures and systems representing an automobile. By the controlled application of the laws of chemistry and physics, they have made a complex functional machine. Engineers have used physical materials and processes to build a non-natural, extra-natural and in fact supernatural product beyond the ability of natural cause. As an automobile leaves the assembly line undirected change begins to occur, and after twenty years and 200,000 miles its effects are quite evident. Light from the Sun and oxygen have deteriorated the plastic dashboard and interior components. The piston rings, cylinders, valves and seals in the engine are worn, evidenced by a loss of power, black smoke and leaking oil. The sheet metal body is heavily rusted and its paint is pitted, faded and peeling. Grease and grime have built up on the engine, while lost coins and dirt have accumulated under the seats. Close examination shows that a multitude of undirected changes have occurred to bring this once new automobile to its current pathetic state. This car now represents a combination of design and undirected change. It is undeniably a product of design, yet has been modified by microevolution. While living things are millions of times more complex than an automobile, this simple analogy best explains what is observed in species today.

According to design theory, each original species was given its genes and alleles at inception. Alleles were not inherently harmful and gave variety and robustness to the species. The complex relationship among multiple genes to create polygenic hereditary traits was designed at inception. Original genes have been subject to mutation over time to result in a number of deleterious alleles and their defective proteins.

Design theory proposes that undirected change over time has modified the genome of all species to a limited extent. DNA has been substituted, deleted, inserted, inverted and duplicated. For example, the genome size of the domestic dog (*Canis familiaris*) varies in weight from 2.80 picograms to 3.54 picograms (weight is proportional to the number of DNA base pairs).[57] Variation in size of the dog genome indicates an accumulation or deletion process.

Some scientists have said the human genome is primarily junk DNA. The human genome contains approximately 3 billion DNA base pairs, which some have claimed is 98 percent junk. This conjecture is premature and most likely incorrect, especially since the function of most DNA is not understood. Human development from a fertilized egg throughout the fetal stage and into adulthood is controlled by DNA. Programmed cell death (apoptosis) and cell differentiation during early fetal growth in the womb are controlled by DNA. It controls the placement of all body structures, including the development of vascular pathways to supply every cell in the body with blood. DNA ultimately controls growth, regulation, hormonal changes and development throughout life. With an estimated 100,000 or more different proteins and 100 trillion (1×10^{14}) cells in the human body, the amount of DNA required to control these may be substantial. Scientists are beginning to increase their estimates of useful DNA in the human genome.

However, there is strong evidence that large amounts of junk DNA exist in some species. For example, the Animal Genome Size Database lists the marbled lungfish (*Protopterus aethiopicus*) with a genome weight of 132.8 picograms, the Atlantic deep-sea shrimp (*Hymenodora*) has a genome weight of 38.0 picograms, the common toad (*Bufo bufo*) has a genome weight of 6.6 picograms and the human (*Homo sapien*) genome has a weight of 3.5 picograms.[58] The weight of these genomes corresponds to the amount of DNA they contain. While the exact function of all DNA in these organisms is unknown, it is doubtful smaller and simpler life has more functional DNA than larger, more complex human life. Some genomes appear to contain massive amounts of junk DNA. The total amount of DNA in a species is not necessarily proportional to its size and complexity. It appears the majority of truly junk DNA is due to copying errors that have accumulated over many generations.

Pseudogenes (nonworking copies of a gene) are considered to be junk DNA. For example, the L-gulonolactone oxidase gene (GLUO gene) in mice and other animals codes for an enzyme that is active in making vitamin C.[59] The GLUO gene appears in the human genome as a pseudogene and is not active. This is why humans must have an external source of vitamin C, yet mice can make their own. Primates, guinea pigs and bats are among the few species lacking the ability to synthesize vitamin C.

Design theory explains the GLUO pseudogene better than evolution. Having the ability to make vitamin C is highly advantageous. If primates were once capable of making vitamin C, loss of the ability would have been a deleterious mutation. It appears implausible that natural selection would spread this harmful mutation, which initially occurred in one member, throughout its species. According to biological evolution, deleterious mutations are eliminated and beneficial mutations are spread throughout a species—not the reverse. Evolution does not have a rational explanation for the GLUO pseudogene in primates, guinea pigs and bats. Design theory proposes that the GLUO gene was inactivated at inception in these species, and their inability to make vitamin C was an intentional decision.

From a design perspective, pseudogenes can be explained due to four basic reasons. (1) Pseudogenes may have a function that is not readily apparent. (2) A few pseudogenes may be explained as functional genes inactivated by random mutation. They are remnants of dead genes and are the result of undirected change. (3) Multiple copies of a pseudogene in a species may represent repetitive copying errors. (4) A functional gene in one species corresponding to a pseudogene in another species may be attributed to intentional deactivation at species inception.

The ratio of genes and pseudogenes involved in the function of smell varies among species. A family of more than 1,000 genes which code for olfactory receptors has been identified in mammals.[60] In humans, around 50% appear as functional genes and 50% appear as pseudogenes. In mice, about 80% appear as functional genes and 20% as pseudogenes.[61] The ratio of functional genes to pseudogenes can be attributed to design at inception. Table 9.3 lists pseudogenes in the human genome.

Design theory proposes that within a section of DNA, it may be easier to deactivate a gene (make it a pseudogene) than remove it. This is especially true if a gene is poly-constrained and its removal would affect the operation of other DNA because of its three dimensional configuration. Deactivating a gene is analogous to deactivating a section of computer code. Computer program writers often deactivate code rather than delete it, especially if the code is complex, difficult to write and may potentially be used again.

Design theory proposes that all major species were designed, and uses human design as its model. One human trait and three categories are used to explain design applicable to biological life as follows:

(1) Product expansion – An increase from the first microscopic organism on Earth to the multitude of plant and animal species known today represents massive product expansion. There is no rational explanation for the diversity of life on Earth other than a creative designer who enjoys variety—analogous to humans who like an assortment of products.

(2) Original design – The first living organism on Earth was an original design. Original design is represented by the sudden appearance of species containing new information, DNA, molecules, proteins, genes, features and functions. The first appearance of cellular specialization, bisymmetry, sexual reproduction, nerves, flippers, wings, legs, eyes, kidneys, lungs, gills, intestines, a skeleton, circulatory system, liver, stomach, heart and brain in living things represent original design.

(3) Design by menu selection – Many features in living things have been selected from a menu of standard items. Design by menu selection is represented by similar molecules, genes, proteins, features and functions found among species. The 500 genes which exist in all domains of life represent a menu of standard genes.[62] The more than 1,000 genes which code for olfactory receptors in mammals characterize standard genes.[63] The 39 Hox genes found in mammals that control body part formation represent standard genes.[64] Selecting from a menu permits the rapid, efficient design of basic biological systems, while allowing tweaking and tailoring for unique features.

(4) Design iteration improvement series – Biological parts and systems showing a sequence of changes that increase efficiency, robustness, miniaturization, capability or sophistication over time represent a design iteration improvement series. In a human design iteration improvement series, engineers often take an existing product, errors and all, and modify it to create a new product. This may be evident in some living things, where a species genome, errors and all (without any clean up), was taken and modified to make a new species.

Design theory does not propose perfection or goodness in the design of living things. The measure of perfection is dependent on purpose, which in turn is based on the designer's intent. The fact that primates cannot make vitamin C appears to be an intentional design change rather than a mistake—a question only answerable by the designer.

Evolutionists often criticize design as an explanation for living things because of alleged imperfections or lack of optimization. Because of the complexity of living systems and our inability to do significant biological design, restraint rather than arrogance should be the attitude. Perfection in design is a fuzzy concept and quite debatable. For example, no transportation vehicle appeals to everyone. Some people prefer a small motorcycle while others a full-sized SUV. To a large extent, perfection, utility and desirable features are in the eye of the beholder.

Conclusion

Biological evolution is a theory that natural cause is responsible for all of Earth's species, whether living or extinct, starting from the first microscopic life. Similarity, fossils and microevolution are the primary evidences used to support this belief. Hypothetically, improvement was due to massive random mutation. Natural selection acted on a miniscule percentage of heritable beneficial mutations to build new features, functions and species in a step-by-step series over time. This idea is discredited by the following:

(1) Irreducible complexity – The principle of irreducible complexity proposed by Michael Behe discredits natural selection as a means of building complex non-functional structure as a precursor to a complex functional system. The belief that random mutations (unguided by natural selection) can build the required complex nonfunctional structures is false.

(2) Absent beneficial mutations – The absence of beneficial point mutations, and an abundance of neutral and deleterious point mutations, prevents a step-by-step series from occurring due to natural selection.

(3) Human genome – Protein isoforms in the human genome caused by point mutation are neutral, deleterious or uncategorized. Beneficial protein isoforms (under all environmental conditions) in the human genome are nonexistent or extremely rare.

(4) Unimproved alleles – The HBB hemoglobin gene illustrates undirected change. Mutations in this gene have resulted in at least 252 new alleles and protein isoforms; however, the new alleles do not show improvement. Undirected neutral and deleterious change has occurred, but not beneficial change.

(5) Absent mutations – Evolution requires the occurrence of hundreds of millions of mutations per species. According to the laws of Mendelian inheritance, neutral mutations and recessive deleterious mutations that become fixed within a genome cannot be naturally removed. The alleged occurrence and then disappearance of these mutations is a bizarre explanation.

(6) Limited change – Extensive breeding experiments using natural selection and artificial selection, natural environments and artificial environments, and natural mutations and artificially induced mutations discredit the ability of species to change beyond certain limits.

(7) Genetic entropy – The natural tendency of DNA is to become randomized through the accumulation of neutral and recessive deleterious mutations.

Earth's species only makes sense when explained as a combination of design theory and microevolution. The sudden introduction of new information, DNA, molecules, proteins, genes, features, functions and species at different intervals over the course of Earth's history is best explained by design. After introduction, each species was subject to limited change through random mutation over extensive time. Undirected change best explains point mutations, defective genes, protein isoforms, chromosome abnormalities, genetic disease, cancer, aging, limited species divergence, new viral strains and the spread of an antibiotic-resistant bacterial population.

Charles Darwin presented his theory of evolution in *On the Origin of Species by Means of Natural Selection* in 1859. He attempted to explain complex phenomena he did not understand. Darwin knew little about the fossil record, archaeology and animal physiology. He knew nothing about heredity, meiosis, mitosis, chromosomes, genes, the DNA molecule, DNA information, point mutations, proteins, enzymes, cellular structure and the immense complexity of living things. Darwin's theory of natural evolution was a leap in the dark of ignorance. Although wrong, his theory proved to be immensely popular by making philosophical naturalism acceptable as an alternate to nineteenth century biblical belief.

In the last fifty years, scientific knowledge of biology has seriously discredited Darwin's theory of evolution. Unfortunately, philosophical naturalism has become firmly entrenched within scientific thought, education and popular media. Darwinian evolution is kept alive by mistaken beliefs about science, peer pressure, hubris and entrenched power within the academic and scientific community. Among intelligent, knowledgeable, thinking people, Darwin's ideas can be rejected when several key concepts are accepted.

- Individuals who accept science as a search for dependable knowledge about the world based on physical evidence and its logical inferences—without the constraints of philosophical naturalism—can reject biological evolution.

- Individuals who accept natural cause as a biased assumption of science, rather than a dogmatic requirement, can reject biological evolution.

- Individuals who accept the severe limitations of natural cause can reject biological evolution.

- Individuals who understand philosophical naturalism as a faith-based belief and recognize its influence on biology can reject macroevolution.

Philosophically, natural cause is alleged to be a powerful constructive mechanism and is capable of logically explaining Earth's species. When carefully examined in the light of twenty-first century science, these claims cannot be substantiated and in fact are refuted. This is especially evident when detailed knowledge exists about a species genome. Present knowledge about the human genome shows that evolution is overwhelming destructive, rather than constructive.

For as the heavens are higher than the earth, so are my ways higher than your ways, and my thoughts than your thoughts. (Isaiah 55:9) KJV

Chapter 10

The Origin of Man

A study about the origin of man is a complex subject requiring a diverse collection of information. While a detailed analysis of this subject is beyond the scope of this book, some of the main evidence concerning the origin of man can be discussed. Based on current knowledge, a valid hypothesis can be developed for the origin of man due to intelligent cause.

An objective analysis of human origins requires a strict adherence to verified empirical data. In the following, a determined effort has been made to separate fact from inference. Genes, DNA, skulls, skeletal remains, artifacts and archaeological sites are all empirical evidence. Theories, hypotheses and conjecture are not, and will be clearly identified. Much is unknown about human history, and filling in the gaps with speculative ideas detracts from the search for truth.

Human and Chimpanzee Biology

A study of the origin of man begins with biology and involves a careful examination of human and animal DNA. Biology is definitive information in the search for man's origin. A comparison of humans and chimpanzees shows their DNA to be amazingly similar, even though the two species are significantly different.

Evolutionists use similarity between human and chimpanzee DNA as support for biological evolution. Hypothetically, chimpanzees and humans evolved from a common ancestor roughly six million years or more in the past through random mutation and natural selection. Natural selection allegedly eliminated a vast multitude of neutral and deleterious mutations, while keeping beneficial mutations. A large number of heritable beneficial mutations accumulated over many generations, bringing humans to their present state.

Design theory points out that similarity is an inherent characteristic of design. Similarity among species is an absolute fact. Comparable DNA information, molecules, proteins, genes, pseudogenes, features and functions among species prove this issue. As mentioned in Chapter 7, similarity is evidence of a common designer, a common design philosophy, the intelligent reuse of successful designs or a design iteration improvement series.

DNA is the standard for comparing genetic similarity between humans and chimpanzees. The Chimpanzee Sequencing and Analysis Consortium compared chimpanzee and human DNA and reported the results in the "Initial Sequence of the Chimpanzee Genome and Comparison with the Human Genome" published in the September 2005 issue of *Nature*. This is the most complete report issued to date comparing the two genomes.

A comparison of human and chimpanzee DNA shows them to be similar. The *Nature* article compared the chimpanzee genome (94% sequenced) with the human genome (99% sequenced). Each genome contains approximately three billion base pairs (nucleotides) of DNA. Single-nucleotide substitutions account for a 1.23% difference (35 million nucleotides) between the two genomes. Single-nucleotide insertions and deletions account for approximately a 1.5 % difference in each genome for a total of 3% difference. To obtain the 3% value, five million single-nucleotide insertions and deletions were identified in a portion of each species genome and extrapolated to 90 million.[1] According to their figures, the total difference between the two genomes is 4.23%, representing approximately 125 million single-nucleotide differences between the two species. More complete sequencing of each genome may change this percentage slightly. Beyond any doubt, human and chimpanzee DNA are highly similar.

However, genes and proteins within the chimpanzee and human genome were compared and found to be significantly different. A set of 13,354 genes[2] were compared between chimpanzees and humans (the human genome contains approximately 29,000 protein-encoding genes). The encoded proteins from these genes were identical in 29% of the cases.[3] From an opposing perspective, 71% of the proteins between chimpanzees and humans were not identical (varied by one amino acid or more). Most of these amino acid changes are neutral or deleterious. This information is similar to another study which reported that 80% of the proteins between the two species are different.[4]

A major portion of the *Nature* article discusses divergence, with DNA as the measuring stick. The report is written from the perspective that biological evolution is true and that chimpanzees and humans diverged from a common ancestor in the past. Divergence is measured by the DNA difference observed today between the two species. A small amount of gene divergence shows little DNA difference, while a large amount of divergence shows major differences. Table 4 of the report lists sixteen rapidly diverging gene clusters in humans and chimpanzees and concludes, "A total of 16 such neighborhoods were found, which greatly exceeds random expectation."[5] In other words, random mutation and natural selection cannot account for the significant differences between chimpanzees and humans at these gene clusters.

The *Nature* article also discusses similar disease alleles between chimpanzees and humans, as genetic disease can be used to analyze whether two different species have a common past. If a genetic disease existed in a common ancestor, hypothetically it would be passed along by heredity to each species. Table 7 in the report lists sixteen disease alleles found in both chimpanzees and humans. These disease alleles represent 16 cases out of 12,614 genetic disease variants in 1,384 human genes.[6] The sixteen similar disease alleles may be attributed to random mutation rather than inheritance from a common ancestor. Based on these similar disease variants alone, there is no need to infer chimpanzees and humans are tied together by evolution.

Design theory readily accepts similarity between humans and chimpanzees defined by DNA. Man was not a "clean sheet" design without any previous basis. Man was the last of a large number of species introduced on Earth, starting with the first microscopic organism. Chimpanzees were introduced near humans at the very end. Though humans are relatively recent, much of the DNA from which they are made is similar to DNA found in animals that have lived on Earth for millions of years.

Based on design theory, we can predict that a complete comparison of the chimpanzee and human genome will show the undeniable imprint of design. Comparison will identify significant DNA differences between the two species, incapable of being bridged by random mutation. Additional research will clarify large areas of "divergence" between their genomes, unexplainable due to undirected change. No evolutionary mechanisms will be identified capable of explaining differences between the two species within a six million year time frame. Evidence of this type will be compelling support for design theory as the best explanation.

The HAR1 gene is an example of compelling evidence in support of design. This gene consists of 118 base pairs of DNA. The HAR1 gene does not code for a protein but for an mRNA which is believed to affect brain development. All species from chickens to chimpanzees have an almost identical version of this gene—it varies by only two nucleotides. Hypothetically, 310 millions years of evolution from chickens to chimpanzees changed this gene by only two nucleotides. Between chimpanzees and humans, a huge change occurred represented by 18 nucleotide differences in the human HAR1 gene.[7] There are no intermediates between the two species—the improvement appears in one massive step. No rational evolutionary explanation exists for the sudden change in the HAR1 gene, though steps of this nature are predicted by design theory.

The human genome is relatively conserved (consistent throughout) and does not contain the many mutations required for biological evolution. As previously mentioned, the human genome contains approximately 3 billion base pairs of DNA. Of that amount, 99.9% of all base pairs in humans are exactly the same. Approximately 1.4 million base pair locations have been identified where single-nucleotide differences exist.[8] Relatively speaking, there is little variation in the human genome and few hereditary mutations have occurred. Hundreds of millions of mutations as evidence of biological evolution are missing. Of the mutations present in human DNA, essentially all are neutral or deleterious, with beneficial mutations being nonexistent or extremely rare. Overwhelming evidence indicates the human genome is deteriorating rather than improving. Present knowledge strongly discredits the idea that beneficial evolution is now occurring or occurred in the past.

Mitochondrial DNA

The human body contains both mitochondrial DNA and nuclear DNA. Nuclear DNA, about 3 billion base pairs long, is contained within the nucleus of every cell in the human body. On the other hand, mitochondrial DNA (mtDNA) is located outside the cell nucleus. In humans there are about 100-10,000 copies of mtDNA present in each cell, normally consisting of 16,600 base pairs and 37 genes.[9] During reproduction, the father and mother each contribute nuclear DNA to their child, while the mother contributes 100% of the mitochondrial DNA. Because of this, changes from generation to generation can theoretically be traced through the female lineage. Mitochondrial DNA mutates faster than nuclear DNA, and changes among females of different ethnic groups can be studied to determine relationships.

Changes in mitochondrial DNA have been extrapolated into the past in an effort to date the origin of humans. While close human relationships can be determined, more distant ones are problematic. Mitochondrial DNA has been obtained from diverse ethnic groups in South America, North America, Europe, Asia, New Guinea and Australia in an effort to trace humanity to its roots. Computer programs have extrapolated known changes into the unknown. As discussed by Dr. Donald Johanson, attempts to develop a human phylogenetic tree based on mtDNA have been controversial.[10] Based on the data input and the logic of the computer program, radically different results can be obtained. Chimpanzee and human mtDNA has been used to extrapolate changes back in time to when the two allegedly diverged, based on belief in animal to human evolution.

As a counter to evolutionary thinking, creationists have begun studying mitochondrial DNA. The Institute for Institute for Creation Research in its GENE project analyzed the mitochondrial DNA from 827 high-quality sequences. They found that 83 percent of the mitochondrial genome has not changed in all of human history. In more than 99 percent of the locations where a change has occurred, only a small minority of people carried that specific change. On average, human mitochondrial DNA differs from the original Eve consensus sequence by only 21.6 nucleotides. Because mtDNA is subject to a high mutation rate, the lack of significant worldwide variation is evidence for a young mitochondrial genome.[11] Based on their analysis, anatomically modern humans are relatively recent.

Animal and Human Fossils

Evolutionists believe humans and chimpanzees evolved from a common ancestor more than six million years ago. A number of fossils has been discovered and arranged in a series to support their idea. If evolution theory is true, a series would be expected; however, design theory can also explain such a series. Although the Hominidae family (referred to as hominids) consisting of gorillas, chimpanzees, orangutans and humans[12] have physical similarities, this does not mean they are related by evolution to a common ancestor—they could be related by design.

Ian Tattersall, PhD, is Curator of the American Museum of Natural History in New York and wrote *The Human Odyssey: Four Million Years of Human Evolution* followed by *The Fossil Trail: How We Know What We Think We Know About Human Evolution.* Both books contain a compilation of important hominid skeletal remains, the basis for Table 10.1.

A number of hominid skulls and skeletal remains have been found which can be arranged in a hypothetical evolutionary series from ape-like hominids to anatomically modern humans. Brain cavities grow progressively larger, eyebrow ridges decrease, jaws shrink in size and foreheads change from sloping to upright. Other features not readily evident in skull photographs include changes in the thickness of the cranial cap and shape of the teeth in the upper jaw.

The arrangement of fossilized hominid skeletal remains in a series is a subjective process and not proof of animal to human evolution. For example, based on size alone, a plastic pearl, steel ball bearing, marble, golf ball, billiard ball, tennis ball, baseball, softball, soccer ball and basketball can easily be arranged in order. Some might incorrectly conclude these round objects represent a step-by-step evolutionary series. To show the fallacy of this reasoning, if we include the additional characteristics of material, density, hardness, color and bounce, any order is debatable and consensus becomes impossible. Based on all six physical characteristics, these balls cannot be arranged in a logical series. Based on historical information, these balls were developed independently of each other and each one is an original design to fulfill a specific consumer need. These balls can only be rationally explained as a product of design and placed in a correct chronological order based on historical information.

The arrangement of hominid fossils in a series only proves intelligent people can arrange things. Any group of objects can be placed in a hypothetical order based on physical characteristics. Arranging fossils in a series as shown in Table 10.1 and assigning dates does not prove animal hominid to human evolution. A critical analysis of this process involves evaluating the completeness of the series, the reliability of the dates assigned, changes between skulls, the evolutionary mechanisms available and associated archaeological information.

The first item to consider in evaluating the skeletal remains in Table 10.1 is the completeness of the series. According to Darwin, evolution took place in a series of small steps. The skulls and skeletons listed in Table 10.1 exhibit significant steps and jumps. Even though there are twenty-three skulls in this series, there is no small step-by-step series of complete skeletons showing animal hominid to human evolution. There is no 100-step series from a three foot high ape with a small brain, sloping forehead, big teeth, protruding jaw, long arms, short hind legs and grasping toes that progressively changes into an anatomically modern human. There is no series showing how the pelvis of an orangutan or spine of a chimpanzee changes to its human counterpart. There is no series documenting how the foot of a gorilla slowly changes into a human foot. Instead, there are sudden leaps in skeletal structure. Features for

Table 10.1 – Sequence of Hominid Skeletal Remains
(Identified in *The Human Odyssey*)

Name	Skeletal Remains	Year Found	Assigned Age (yrs)	Brain Size (cc)
Lucy	Partial skull, skeleton	1974	3.2 m	440
Taung Child	Face only	1924	2.0 m	
Mrs. Ples, STS 5	Skull, no lower jaw	1947	2.6 m	485
Black Skull KNM-WT 17000	Fragmented skull	1985	2.5 m	530
Handy Man (OH 24)	Fragmented skull, no lower jaw	1960	1.8 m	600
KNM-ER 1813	Skull, few teeth, no lower jaw	1973	1.9 m	510
KNM-ER 1470	Skull, no teeth, no lower jaw	1972	1.9 m	750
Java Man	Skullcap	1891	700,000	940
Sangiran 17	Skull, few teeth, no lower jaw	1969	800,000	1,000
*Atapuerca 5	Skull, some teeth, lower jaw	1992	300,000	1,125
Peking Man	Skull, teeth, lower jaw	1925	300,000	
Turkana Boy	Complete skull, skeleton	1984	1.6 m	880
KNM-ER 3733	Skull, few teeth, no lower jaw	1975	1.7 m	850
*Boxgroove Man	Tibia (leg bone)	1994	500,000	
Tautavel Man (Argo XXI)	Skull face, some teeth	1971	400,000	1,160
*Petralona 1	Skull, some teeth, no lower jaw	1960	250,000	1,220
Rhodesian Man	Skull, some teeth, no lower jaw	1921	125,000	1,280
Old Man, Neanderthal	Skull, upper teeth, no lower jaw	1908	50,000	1,620
Shanidar 1, Neanderthal	Skull, upper teeth, no lower jaw	1953	50,000	1,600
Neanderthal skull	Skull, upper teeth, lower jaw		50,000	1,740
*Hobbit man	Skull, skeletal remain	2003	18,000	380
Old Man of Cro-Magnon	Skull, some teeth, lower jaw	1868	25,000	
Qafzeh (modern)	Skull, lower jaw, skeleton		92,000	
La Brea Woman (modern)	Skull, few teeth, no lower jaw	1914	9,000	

Note: All listings are in the sequence as they occur in the book. Skulls identified with an asterisk are not mentioned by Tattersall.
References:
(1) Tattersall, Ian (1993). *The Human Odyssey: Four Million Years of Human Evolution.*
(2) Johanson, Donald and Blake Edgar (1996). *From Lucy to Language.*
(3) Foley, Jim (2005). Prominent Hominid Fossils. The TalkOrigins Archive.

many of the skulls in Table 10.1 have a considerable amount of variation and do not fit into a progressive step-by-step series. According to Darwinian evolution, a definite step-by-step series would be expected, but this has not been discovered. On the other hand, design theory can explain the large skeletal differences between Lucy, Turkana Boy, Boxgrove Man, Neanderthals, Hobbits and La Brea Woman.

Human evolution can be hypothesized to have many branches; however, a step-by-step progression from the trunk to each branch is still a basic requirement. The absence of a step-by-step progression does not necessarily disprove evolution, but an appeal must be made to missing evidence. Filling in the gaps with fanciful drawings or conjecture is not the same as actual fossils.

A second item to be addressed in evaluating the skeletal remains in Table 10.1 is the chronological sequence. Do these skulls and skeletons fall in a definite chronological order? With the exception of Hobbit man, La Brea Woman and Old Man of Cro-Magnon, the assigned ages for these skeletons are conjectural at best. Radiocarbon dating is only credible back to about 30,000 years for directly dating skeletal remains under good circumstances. The assigned dates beyond 30,000 years represent unreliable dating methods, indirect dating or conjecture. Indirect dating is often based on the radiometric dating of overlying volcanic ash. In order to use the age of overlying ash as the lower age for a skeleton, the assumption must be made that the skeleton and ash have never been disturbed from their original location. Ash that has been moved and deposited over a skeleton is meaningless. It is quite possible for volcanic ash, formed before the skeleton was laid down, to be deposited over it by wind or water. Thermoluminescence (TL) and electron spin resonance (ESR) and are two additional dating methods, but are unreliable. A precise method of dating most of the skulls in Table 10.1 is beyond present science. Their firm chronological order is not known.

A third item to be discussed in evaluating Table 10.1 is the change process. If we hypothesize these skeletal remains are due to evolution, does it have a plausible mechanism? As discussed in Chapter 9, biological evolution has not been demonstrated in extensive breeding experiments. In addition, close examination of the human genome does not show evidence of ongoing beneficial evolution (99.9% or more of all mutations are neutral or deleterious). Furthermore, random mutation and natural selection are not a plausible explanation for creating the changes observed in these skulls and skeletons based on the principle of irreducible complexity. In contrast, design is a viable mechanism to explain these different hominids.

The skulls and skeletons of Table 10.1 are firm evidence of hominids that lived on Earth. Other than anatomically modern humans (La Brea woman and Old Man of Cro-Magnon), all remaining hominid species are extinct. Based on current information, (1) no complete step-by-step series for hominid development exists, (2) dates for most of the fossils in Table 10.1 are conjectural and (3) design theory has the best mechanism to explain these fossils rather than evolution.

Discerning Between Animal and Human

Brain size or cranial capacity is one criterion used to distinguish animals from humans.[13] About 90% of humans have a cranial capacity in the range of 1040-1595 cubic centimeters with the extreme range being 900-2000 cc.[14] Of the 12,000 American skulls in the U.S. National Museum collection, the smallest skull measures 910 cc in volume.[15] Modern human cranial capacity averages approximately 1,350 cc.[16] In contrast, the cranial capacity is approximately 175-540 cc for orangutans, 275-500 cc for chimpanzees and 340-752 cc for gorillas.[17]

Brain size alone is not an indication of intelligence or humanness. The human brain is not just a large chimpanzee brain, but is significantly different at the molecular level. Mario Cáceres and his associates identified 169 gene expressions in the cortex of human brains that were different from chimpanzee brains.[18] Phillip Khaitovich and his associates identified combined gene expression differences of 9-10.5% between human and chimpanzee brains.[19] Gene expression is measured from brain tissue samples and reflects the cellular RNA levels and proteins found. The cerebellum region of the brain shows the most difference, and ongoing research will continue to identify the molecules involved. Because of their brains, humans can speak and understand complex language, write, use tools, create art, make music, develop sophisticated designs and interact on a much higher level than chimpanzees.

Some of the skulls in Table 10.1 are difficult to explain by looking only at their features and cranial capacity. For example, Rhodesian Man has a cranial capacity of 1,280 cc and eyebrow ridges.[20] His face is flat, similar to modern man. The skull of Petralona 1 has a cranial capacity of 1,220 cc[21] and appears similar to Rhodesian Man. Do these skulls represent humans with heavy eyebrow ridges and a small brain? Atapuerca 5 has a cranial capacity of 1,125 cc and eyebrow ridges.[22] Is he a human with a primitive looking skull? Sangiran 17 has an estimated brain cavity of 1,000 cc and heavy eyebrow ridges.[23] Is Sangiran 17 an animal hominid with a large brain or a small-brained human with a thick skull and heavy eyebrow ridges? It is impossible to firmly classify some skulls in Table 10.1 as either animal hominid or human based on the information available.

To precisely classify some of the skulls in Table 10.1, more information is needed than currently available. What makes us human is substantially more than the size or shape of our skulls. A skull detached from of its skeleton represents a fraction of the information about the total skeleton. A complete skeleton detached from its fleshy parts represents a small part of the information about the complete body. A complete body without life represents a tiny portion of the information about the living hominid. A living hominid without observing its abilities cannot be judged as animal or human. Some of the things which define humans are their tools, weapons, shelter, art, speech, writings, technology, culture and religion. For most of the skulls listed in Table 10.1, this information has been lost forever.

Skulls and skeletons firmly associated with artifacts tell us something about their makers. Animals do not use tools other than perhaps a stick for poking or rocks for pounding. Animals cannot fashion tools. It takes intelligence to chip rocks into a sharp blade for cutting. It takes skill to shape rocks into spear points, knives, arrow points and sharp scrapers. It takes intelligence and skill to build a fire. Most of the skulls listed in Table 10.1 have no artifact support. There is no way to determine if their owners were capable of making tools, weapons, shelter or art. Currently, it appears the only skulls and skeletons associated with all five evidences of controlled fire, fashioned tools, formed weapons, created art and burial rituals are anatomically modern humans and Neanderthals. Evidence of the artifacts, technology and rituals left by Neanderthals are the simplest while those left by anatomically modern humans are more advanced, complex and elaborate.

Famous Hominid Fossils

Lucy is probably the most famous of the skulls and skeletons listed in Table 10.1. In 1974 the partial skeleton of a three and one-half foot tall hominid (believed to be female), given the name Lucy, was found in Ethiopia. Lucy's skeleton is typically described as forty percent complete[24] even though only 47 out of 207 bones were found.[25] The skull is severely fragmented with the majority of its pieces missing. Lucy's skull was reconstructed and her brain capacity estimated to be 375 to 500 cc.[26] Lucy is typically portrayed as a small, upright, bipedal animal with an ape head. Lucy cannot be directly radiocarbon dated and has been assigned an age of 3.2 million years. Tattersall gives no details in support of her age in his book.

Handy Man (*Homo habilis*) was discovered by Dr. Richard Leakey in 1960 in the Olduvai Gorge of east Africa. The skull of Handy Man OH 24 is severely fragmented with many missing pieces. The ape-like skull has heavy eyebrow ridges and an estimated cranial capacity of 600 cc.[27] Handy Man is claimed by some as the first hominid to use tools based on fractured rocks found at the site. The fractured rocks may be natural objects or, at the most, pounding stones used by apes. Animal bones found at the site with marks on them were interpreted by Leakey as cut marks from defleshing during butchering.[28] However, scanning electron microscope examination of these marks compared with sample marks made by sharp flakes show them to be different. Bone cuts with a sharp flake are shallow and very fine, while marks associated with Handy Man are deep and wide similar to a bone being struck by a dull rock edge. Rob Blumenschine, PhD and archaeologist at Rutgers University, extensively evaluated the Handy Man bone cuts and concluded the bones were probably scavenged from animal kills and pounded with rocks to break them open and extract the marrow for eating.[29] A collection of rocks found at the site was interpreted by some as a shelter created by Handy Man. Tattersall discounts the shelter rocks as a built-up structure and attributes it to natural cause.[30] A date of 1.8 million years is assigned to Handy Man. These skeletal remains cannot be directly radiocarbon dated, and in his book Tattersall gives no details in support of the assigned age.

Java Man has gained much more fame than deserved. In the fall of 1891, along the banks of the Solo River in Java, a skullcap was found. It was thick-walled, long and low, with no forehead and large eyebrow ridges. Even though no other parts of the skull were found, cranial capacity was estimated at 940 cc.[31] A year later and some distance away, a femur (upper leg bone) was found. A total of three teeth were also found.[32] Eugene Dubois, the discoverer of these bones, believed they were all from the same individual, eventually named Java Man. As described in *The Wisdom of the Bones,* Dubois was determined to find the missing link and reap attendant fame.[33] Dubois vigorously promoted Java Man as the transitional link between apes and humans. The skullcap of Java Man is ape-like, although the femur is essentially human. Because of the distance separating the skullcap, leg bone and teeth, it is questionable they are from the same individual. The Java Man skullcap cannot be directly radiocarbon dated and has been assigned a conjectural age of 700,000 years.

Peking Man is one of the more controversial "fossils" in Table 10.1. Between 1929 and 1937, fragments of about 30 skulls were found at Choukoutien (now Zhoukoudian) about 40 km (25 miles) from Peking (now Beijing), China. Included in the fragments were 11 mandibles (lower jaws) and 147 teeth,[34] although descriptions of the remains vary among references. These bones were found among extensive ash remains, more than 2,000 fashioned stone tools and more than 100 different kinds of animal bones.[35] These artifacts were found in a shallow cave which had collapsed. Some believe there was no cave, but that the face of a cliff was undercut during limestone quarrying and later collapsed.

Two individuals are primarily responsible for Peking Man. Excavation was initially begun and overseen by Dr. Davidson Black, a Canadian physician, who died suddenly in 1934. His work was carried on by Dr. Franz Weidenreich, a paleontologist who had recently fled to America from Germany and was teaching at the University of Chicago.[36] Weidenreich continued the excavation of Peking man until forced to leave in 1941 because of Japan's attack on the United States.

Peking Man is a mixture of ambition, conjecture and missing information. As described in *The Wisdom of the Bones*, Dr. Black was intensely interested in fossils, the missing link and fame. As part of his plan to explore remote parts of the world, Black took a job teaching in 1919 at the Peking Union Medical College in China.[37] In March 1927, a single tooth was found at the Zhoukoudian, China site. Black immediately went on a tour of Canada, America and England, where he promoted the single tooth as evidence of a new species linking ape and man.[38] With funding from the Rockefeller Foundation, excavation began in earnest at Zhoukoudian in 1929. In December 1929, a skullcap was discovered and Black immediately left on a tour of Europe with casts, slides, a well-polished lecture (he did not take the actual skullcap) and an assertion that he had found the missing link.[39] In 1932 Black was elected a Fellow of the Royal Society for his efforts. Black returned to China where he died in 1934 from heart failure. His work was carried on by Weidenreich, who made photographs, drawings, measurements and took notes of the original fossils.[40] Weidenreich allegedly had his assistant make plaster casts of every hominid fossil. When Weidenreich fled China before the war with Japan, all of the fossils except two teeth were mysteriously lost.[41]

A critical look at the evidence raises several troubling questions about Peking Man. We can comprehend loss of the Peking fossils due to the turmoil of war, but it is difficult to understand why no skull fragments found their way out of China over an eleven year period when they were readily available. It is further difficult to understand why no original fossil photographs taken by Weidenreich appear to be available. In addition, there are some "disconcerting anomalies" about the number of boxes of fossils and how they were packed before they disappeared.[42] When considering other frauds perpetrated on the public such as Nebraska Man and Piltdown Man, we should be skeptical of this missing evidence. The Peking Man "fossils" cannot be directly dated and have been assigned a conjectural age of 300,000 years.

Turkana Boy (sometimes called Nariokotome Boy) is a unique specimen discovered by Kamoya Kimeu near Lake Turkana in Kenya. This is a fairly complete skeleton of an estimated 12 year old boy. The main missing parts are the hands, feet and upper left arm bone. The size of the brain cavity is 880 cc with an estimated adult size of 909 cc.[43] Turkana boy is 160 cm (5 ft, 3 in) tall with an estimated adult height of 185 cm (6 ft, 1 in) and weight of 68 kg (150 lbs).[44] The hole in his neck vertebrae, through which the spinal cord passes, is about half the cross-sectional area of modern humans.[45] His pelvis is narrower than modern humans.[46] His skull appears primitive, while the overall body structure, weight and proportions are similar to today's ethnic Masai who live in Kenya. Based on skeletal remains, it cannot be determined if Turkana Boy was a thin, tall, small-brained boy with a primitive skull or a thin, tall, large-brained animal hominid. There are no artifacts associated with this skeleton, it cannot be directly radiocarbon dated and has been assigned a conjectural age of 1.6 million years.

Neanderthal fossils are the most recent, most numerous and according to some the most intriguing. Neanderthal bones have been found in Europe and Asia. Germany and France have produced the best Neanderthal remains. Tattersall gives a map of 71 sites[47] where Neanderthals lived while another source lists 25 locations identified as "classic Neanderthal."[48] Some of these sites lack any Neanderthal bones (their presence is based on artifacts believed to be of Neanderthal origin). The October, 2008 issue of *National Geographic* probably gives the most accurate list and identifies 19 sites by name where Neanderthal skeletal remains have been found (see Table 11.3).

In 1908 a partial skeleton was found in a cave near the village of La Chapelle-aux-Saints, France. The find consisted of a skull, mandible, most of the vertebrae, several ribs, most of the long bones of the arms and legs, plus some of the smaller bones of the hands and feet. Now known as the "Old Man" of La Chapelle-aux-Saints, the skeleton was stoop shouldered and had an attached skull with a cranial volume of 1620 cc.[49] Initially, this skeleton was considered to be from an ape-like creature, but in 1957 was re-examined and skeletal deformation was attributed to rickets and arthritis. Neanderthals probably looked similar to modern humans.

Neanderthals are somewhat of an enigma. They appear to have come out of nowhere, lived for a time and then became extinct roughly 30,000 years ago. Based on archaeological remains, Neanderthals were capable of controlling fire, fashioning tools, forming weapons, creating art and burying their dead. Neanderthals were not as culturally advanced as anatomically modern humans.

The skeletal reconstruction of Neanderthal shows them to be slightly different from an anatomically modern humans. Although parts of at least 400 Neanderthal individuals have been found, no complete skeleton has ever been discovered.[50] A full skeleton was constructed by anthropologists Gary Sawyer and Blaine Maley[51] from skeletal remains found at seven different sites. They based their reconstruction on a skeleton known as La Ferrassie 1, found in La Ferrassie Cave in the Dordogne Valley near present-day Les Eyzies, France in 1909. Skeletal remains show Neanderthals had protruding eyebrow ridges, a sloped forehead, a broad projecting nose, a large lower jaw, a weak chin and a larger than average skull that extended backwards into an occipital bun. Their cranial cavity ranges from 1,200 to 1750 cc (10% greater than the modern human average). Average height for a male Neanderthal was about 5 feet 5 inches and 5 feet 2 inches for the female.[52]

Small Ancient Archaeological Sites

In addition to fossils, a number of small archaeological sites have been discovered which shed additional light on the past presence of anatomically modern man and other hominids. Archaeology is important information in the search for man's origin. Ancient villages, campsites, caves and burial sites can be examined to determine when and where hominids lived. Artifacts from these can tell us about their diet, weapons, tools, art, technology and burial practices.

Terra Amata (a beach campsite on the Mediterranean Sea) near Nice, France is the oldest human archaeological site mentioned by Tattersall.[53] No hominid bones were found at the campsite; however, broken animal bones discovered include elephant, wild boar, deer, wild cattle and rabbit. Worked flint stones were discovered at the site. Other stones uncovered were interpreted as forming the periphery of an oval structure of bent saplings embedded in the ground. An age of 400,000 years is assigned to this location; however, Tattersall gives no details to support this date. Other sources indicate this age was determined by thermoluminescence dating soil at the site (a non-credible dating method).

Table 10.2 and 10.3 give a list of some well-known small archaeological sites. A combination of village sites, campsites, skeletal remains and drawings have been discovered in Europe, South America, North America and Australia.

Some sites and dates older than those listed in Table 10.3 have been discovered, but are challenged by scientists skeptical of their accuracy. For example, the Meadowcroft Rockshelter site in Pennsylvania has been radiocarbon dated to 13,000 years before present (BP), with some dates going back to 19,000 BP. The Cactus Hill site in southeastern Virginia has been dated to about 15,000 BP. Tlapacoya, Mexico has been radiocarbon dated to 24,000 BP. The Boqueirao de Pedra Furada site in Brazil has been radiocarbon dated to 25,000 BP with some dates going back to 48,000 BP.

Table 10.2 – Significant Small Ancient Hominid Archaeological Sites
(Europe and Asia)

Name	Remains	Date	Dating Method
Vogelherd Cave, Germany	Skeletal remains	3,000 – 1,900 BC	Radiocarbon
Ötzi Ice Man, Austria	1 mummy	3,370 – 3,100 BC	Radiocarbon
Oslonki, Poland	Village site	4,300 – 4,000 BC	Radiocarbon
Timna Valley, Israel	Copper mining site	4,500 BC	Radiocarbon
Mount Sandel, Ireland	Campsite	7,000 BC	Radiocarbon
Koelbjerg Woman, Denmark	1 female head	8,000 BC	Radiocarbon
Mugharet El-Wad Cave, Israel	100 human burials	10,500 – 8,500 BC	Radiocarbon
Cave of Lascaux, France	Cave paintings	14,000 BC	Estimate
Wadi Kubbaniya, Egypt	Campsite	16,000 BC	Radiocarbon
Ohalo II, Sea of Galilee, Israel	Campsite	21,000 BC	Radiocarbon
Lagar Vehlo Cave, Portugal	1 child skeleton	23,000 BC	Radiocarbon
Dolní Věstonice, Czech Rep.	3 skeletons	25,000 BC	Radiocarbon
Sungir, Vladimir, Russia	5 buried skeletons	25,000 BC	Radiocarbon
Red Lady of Paviland, Wales	1 male skeleton	27,000 BC	Radiocarbon
Kents Cavern, England	Jaw fragment	29,000 BC ?	Radiocarbon
Cro-Magnon Cave, France	6 skeletons	-----	-----
Cosquer Cave, France	Cave paintings	27,000 BP ?	Radiocarbon
Gorham's Cave, Gibraltar	Skeletal remains	30,560 BP ?	Radiocarbon
Mladec Man, Czech Republic	Partial male skull	31,000 BP ?	Radiocarbon
Chauvet Cave, France	Cave paintings	31,000 BP ?	Radiocarbon
Pestera cu Oase, Romania	Mandible	34,000 BP ?	Radiocarbon
La Ferrassie Cave, France	8 partial skeletons	35,000 BP ?	-----
Skhul Cave, Israel	3 skeletons + remains	40,000 BP ??	Artifacts
Niah Cave, Borneo, Malysia	Human skull	42,000 BP ??	Radiocarbon
Shanidar Cave, Israel	Skeletal remains	50,000 BP ??	-----
Amud Cave, Israel	Skeletal remains	60,000 BP ??	TL
Kebara Cave, Israel	Skeletal remains	60,000 BP ??	-----
Qafzeh Cave, Israel	Skeletal remains	90,000 BP ???	-----
Tabun Cave, Israel	Skeletal remains	90,000 BP ???	-----

Notes:
(1) Dates are based on radiocarbon dating and are from a number of unrelated sources. For a given site, dates of multiple artifacts by different investigators may not agree.
(2) Dates followed by a question mark are suspect.
(3) BP following a date means "before present."
(4) Dating methods identified by "TL" indicate thermoluminescence dating.

There is significant controversy over the accuracy of these dates because they upset the theory of human migration from Asia into North American immediately following the last ice age around 12,000 BC. Challenging the validity of these radiocarbon dates is not a design versus evolution issue, but rather one set of radiocarbon dates and information clashing with another.

Table 10.3 – Significant Small Ancient Human Archaeological Sites (North America, South America and Australia)			
Name	Modern Location	Remains	Date
Windover Pond	Titusville, Florida, USA	167 mummies	6,100 – 5,000 BC
Wizard's Beach Man	Reno, Nevada, USA	1 skeleton	7,200 BC
Spirit Cave Nevada	Fallon, Nevada, USA	1 skeleton	8,650 BC
Wilson-Leonard Site	Georgetown, Texas, USA	1 skeleton	9,000 BC
Meadowcroft Shelter	Pennsylvania, USA	Campsite	13,000 BC ?
Chinchorro	Arica, Chile	96 mummies	5,800 – 1,800 BC
Caral	Lima, Peru	Village site	2,600 – 2,000 BC
Real Alto	Valdivia, Ecuador	Village site	4,400 – 3,800 BC
Quebrada Jaguay	Quebrada, Peru	Village site	11,000 BC
Monte Verde	Puerto Montt, Chile	Campsite	12,500 BC
Kow Swamp People	Victoria, Australia	22 + skeletons	11,000 – 7,000 BC
Mungo Lake People	Lake Mungo, Australia	Skeletal remains	24,700 BC ??
Notes: (1) Dates for all sites are based on radiocarbon dating and are from a number of unrelated sources. For a given site, dates of multiple artifacts by different investigators may not agree. (2) Dates followed by a question mark are suspect.			

Radiocarbon dates from material at a site can vary widely, and credible dates at older sites can be difficult if not impossible to obtain. For instance, the Red Lady of Paviland skeleton was directly radiocarbon dated by different personnel at 18,460 BP, 26,350 BP and 25,840 BP. Artifacts found with the Red Lady included a mammoth tusk dated at 24,000 BP, three bone spatulas dated at 23,000 BP and an ivory carving at 21,000 BP. The Dolní Věstonice skeleton was dated at 25,640 BP based on associated charcoal at the site. The skeleton was so deteriorated it could not be directly dated. Kent's Cavern Man, Mladec Man and Pestera cu Oase are all based on directly dating human bone material. Bones from Mladec Man were so deteriorated that only his teeth could be dated.[54] These examples represent the difficulty of securely radiocarbon dating small archaeological sites.

Archaeological Sites Before 10,000 BC

Several authors give a list of ancient hominid sites, allegedly showing that man's history goes back millions of years. The book *From Lucy to Language* by Donald

Johanson, PhD, lists 66 of the most significant ancient hominid archaeological sites in the world.[55] *The Atlas of Archaeology* by Professor Mick Aston and archaeologist Tim Taylor identifies 108 hominid sites older than 10,000 BC in Europe, Africa, Asia, North America, South America, Southeast Asia and Australia.[56] Approximately 60 sites are assigned a date older than 30,000 BC. Their list represents essentially all the significant archaeological sites upon which the origin of man is based. Physical remains found at these sites are not disputable; however, dates and explanations of these remains are quite debatable, especially those more than 30,000 years old.

Some dates assigned to sites in *The Atlas of Archaeology* go back five million years, and their credibility is suspect. Radiocarbon dating has the capability to date organic material up to 10,000 years old quite well. Dates between 10,000 and 30,000 years old become increasingly uncertain. The realistic limit for radiocarbon dating is about 30,000 years because of the small amount of carbon-14 remaining in samples. Deteriorated samples add to the uncertainty of any date. A few dates beyond 30,000 years are based on thermoluminescence dating or electron spin resonance dating—two unreliable techniques. Most dates beyond 30,000 years are conjecture based on the hypothetical evolutionary ascent of man.

Border Cave in Africa is typical of archaeological sites touted as proving the antiquity of anatomically modern man well beyond 30,000 years. The cave is situated high in the Lebombo Mountains of South Africa where it borders the eastern edge of Swaziland. The cave was discovered in 1934 and first excavated by W. E. Barton in 1940. The cave was again excavated by Adrian Boshier and Peter Beaumont in 1970.[57] Subsequent work was done by K. W. Butzer, P. B. Beaumont and J. C. Vogel.[58] Stratum in the cave from top to bottom is around 2.75 meters (9 ft) deep. The complete skeleton of a four to six month old infant decorated with shell ornaments and red ochre was found in the cave. The child's skeleton was covered with a layer of ash. Remains of several modern adult humans were discovered along with thousands of stone implements and bones from 43 mammal species. Border Cave is claimed by some to date man's presence back to 200,000 BP.

There is little question anatomically modern humans lived in Border Cave based on their skeletons, tools, weapons, art, use of fire and burial rituals; however, dating is another matter. Border Cave is typical of several sites that either give old radiocarbon dates or cannot be radiocarbon dated at all. We can rightfully be skeptical of the old dates estimated for Border Cave.

In reviewing dates in *The Atlas of Archaeology*, there is legitimate concern about the credibility of approximately 60 sites assigned an age older than 30,000 years. Many of these dates are conjectural or tenuous at best. Some dates are contradictory with other well-dated human archaeological sites. A critical analysis of these sites will show they lack firm evidence supporting a long age for anatomically modern humans beyond 30,000 years.

The Appearance of Civilization

Large permanent settlements mark the appearance of civilization and are an additional way to evaluate the past presence of man. Villages and cities are an indication of when humanity began to develop complex social structures, technology, farming and animal husbandry. The careful excavation of these large archaeological sites has revealed much about ancient man and his development. Artifacts and their stratigraphic sequence at a site can be used to create a chronological history. Advances in shelter construction, tools, weapons, art, pottery, food sources, burial rituals and technology are indicators of cultural development. Typically, a large settlement provides a wide variety of materials to permit secure radiocarbon dating.

The oldest known major permanent villages and cities in the world are listed in Table 10.4. These villages are documented in *Civilizations: Ten Thousand Years of Ancient History* by Jane McIntosh and Clint Twist. Dr. McIntosh, an archaeological author, editor, consultant and teacher, discusses the development of ancient civilization and identifies important ancient cities within Europe, the Middle East and Asia. Several ancient cities within the Fertile Crescent, Mesopotamia, Indus Valley, Nile River and Yellow River are identified.

Abu Hureyra in the Fertile Crescent is the largest and oldest known permanent village yet found on Earth. The book *Village on the Euphrates* discusses the archaeological analysis of this settlement and represents a major work in defining the development of early civilization.[59] The ancient village of Abu Hureyra (located on the west side of the Euphrates River in present-day Syria) was archaeologically excavated in 1972 and 1973. The mound of Abu Hureyra covers 28.5 acres and is approximately the same size as Catal Hüyük and three times the size of the ancient Jericho mound. Abu Hureyra gives a continuous record of man's development for approximately 4,250 years at one location.

At its beginning, Abu Hureyra was a primitive pre-pottery, Stone Age, hunter-gather village. The houses were simple pole and reed structures. The inhabitants' food consisted of wild animals and wild grains. As time progressed, their houses became more permanent structures made of mud brick. Pottery was developed, animals and grains became domesticated and plaster was discovered and used to cover their mud brick houses. No metals or clay writing tablets were found. All of the tools for food preparation, work and hunting were made of stone or bone. Village artifacts include querns, rubbing stones, beads, bones, shells, axes, adzes, sickle blades and arrowheads.

Archaeological evidence suggests the village progressed from a hunter-gather to a farmer-herder type of existence. Stratigraphic analysis of animal remains shows the villagers' diet changed from gazelle meat to sheep, goats, cattle and pigs. Food grains went from wild wheat, rye and legumes to domesticated wheat, rye and other crops. As pottery developed, food cooking improved. Boiling permitted the softening of grains and the thorough cooking of dried meat before eating. This change was evidenced by the reduction in tooth wear among the skeletons. Some older skeletons showed extreme tooth wear, hypothesized to have occurred from eating hard grains and dried meat. (Tooth wear can also be caused by grit found in flour.)

Table 10.4 – Major Ancient Human Archaeological Sites
(Identified in *Civilizations*)

Name	Modern Location	General Location	Date of Origin
Catal Hüyük	Central Turkey	Fertile Crescent	6,500 BC
Jericho	Jordan Valley, Israel	Fertile Crescent	8,500 BC
Abu Hureyra	Tabqa, Syria	Fertile Crescent	11,150 BC
Akkad or Agade	S. of Baghdad, Iraq	Mesopotamia	Before 2,300 BC
Uruk or Erech	S. of Baghdad, Iraq	Mesopotamia	4,000 – 4,300 BC
Ur	S. of Baghdad, Iraq	Mesopotamia	4,500 – 4,200 BC
Giza Pyramids	West of Cairo, Egypt	Nile, Egypt	2,575 – 2,465 BC
Step Pyramid	Sakkara, Egypt	Nile, Egypt	2,650 – 2,575 BC
Memphis	South of Cairo, Egypt	Nile, Egypt	2,925 BC
Hierakonopolis	N. of Aswan Dam, Egypt	Nile, Egypt	3,900 BC
Dholavira	Western India	Indus Valley	2,500 BC
Mohenjo-Daro	Sindh, Pakistan	Indus Valley	2,600 BC
Harappa	Sahiwal, Pakistan	Indus Valley	2,600 BC
Anyang	Henan Province, China	Yellow River	1,766 BC
Jiahu	Henan Province, China	Yellow River	7,000 BC

Note: Jiahu, China is the only city not given in McIntosh's book. Dates are from miscellaneous sources.
Reference: McIntosh, Jane and Clint Twist (2001). *Civilizations: Ten Thousand Years of Ancient History.*

The skeletal remains of 52 juveniles and 50 adults were found during the excavation of Abu Hureyra. Females averaged about 155 cm (5 ft, 1 in) in height and males 162 cm (5 ft, 4 in). The people of Abu Hureyra appear to have been small, muscular and wiry. Many were buried under the floors of their houses, especially the women. The village was divided into individual households. Family units appear to have existed and duties were segregated by sex based on skeletal damage owing to specific chores. The women showed bone damage hypothesized to have occurred by kneeling while grinding grain by hand on stone querns.

Abu Hureyra shows a progression in shelter construction over time, with Abu Hureyra I being the oldest. These early dwellings consisted of round pit huts, eventually abandoned when Abu Hureyra II was built. Abu Hureyra II was composed of rectangular mud brick houses with narrow alleys. Most of the walls were made of two or three rows of sun-dried, mud-straw bricks. Some of the houses had black or white clay floors. In some instances, gypsum plaster was used on the buildings to improve their looks and durability. The maximum population of Abu Hureyra at its peak is estimated between 7,000 and 10,000 people.

Organic remains found at Abu Hureyra were thoroughly radiocarbon dated. The six oldest dates from Abu Hureyra I listed in *Village on the Euphrates* give an average calibrated date of 11,150 BC. The six youngest dates for Abu Hureyra II give an average calibrated date of 6,900 BC,[60] for a total time span of 4,250 years.

Table 10.5 – Major Mesoamerican and South American Cultures			
Culture	Beginning Date	Thriving Date	Main Locations
Aztec	1250 AD	1325 – 1500 AD	Tenochtitlan, Mexico
Inca	1200 AD	1400 – 1500 AD	Machu Picchu, Llactapata, Peru
Teotihuacan	400 BC	150 – 450 AD	Teotihuacan, Mexico
Maya	1000 BC	250 – 900 AD	Tikal, Copan, Bonampak, Uaxactun, Palenque, Rio Bec
Olmec	1500 BC	1150 – 400 BC	La Venta, San Lorenzo Tenochtitlan, Tres Zapotes
Andean	2500 BC	2100 – 1400 BC	Casma Valley, Peru
Norte Chico	3500 BC	3000 – 2500 BC	Caral, Supe Valley, Peru. Also Huaura, Pativilca and Fortaleza Valleys.

The Giza pyramids are probably the most well-known structures mentioned in Table 10.4 and have been extensively dated. These are unique because archaeological and historical records permit building a chronology back to their origin. Many cities have no continuous archaeological record and must rely totally upon radiocarbon dating to define their time in history. The three major Giza pyramids (the great pyramids of Khufu, Khafre and Menkaure) are archaeologically dated to around 2575-2465 BC and were constructed within a time span of less than one hundred years.[61] With the development of radiocarbon dating, the Giza pyramids were extensively dated using charcoal found in mortar to determine their age. In 1984 the Edgar Cayce Foundation sponsored radiocarbon dating of the pyramids (in an attempt to prove the pyramids were built around 10,500 BC as claimed by followers of the psychic Edgar Cayce) and in 1994 the David H. Koch Foundation supported another round of dating.[62] The radiocarbon dates are all older than the historical-archaeological dates, but average to within less than ten percent.

As focus is shifted to the Americas, a different timetable of development is observed. Mesoamerica is an area roughly defined from present-day central Mexico to Costa Rica where the Aztec, Teotihuacán, Maya and Olmec cultures developed. Further south, the Inca civilization emerged along the western edge of South America in present-day Ecuador, Peru, Bolivia, Argentina and Chile. The Maya culture, along with the Aztec and Inca civilizations, existed independently until conquered by the Spaniards starting in 1519 AD. Table 10.5 lists seven significant Mesoamerica and South American cultures. Based on the excavation and radiocarbon dating of small archaeological sites mentioned in Table 10.3, such as Monte Verde, Chile, we can hypothesize that people first appeared in South America around 12,500 BC.

A Review of Radiocarbon Dating

Radiocarbon dating permits determining the age of archaeological sites whose time in history would otherwise be unknown. Many cities, villages, campsites and skeletons cannot be associated with any continuous history. In some instances, settlements existed before writing, which made written historical records impossible. Some sites flourished for a time and were then abandoned when their inhabitants died, migrated or were carried away by their enemies. These settlements slowly became covered with the sands of time. In these instances, accurately establishing a time in history is heavily reliant upon radiocarbon dating.

Table 10.6 – Radiocarbon Dates for Sungir Human Burials in Vladimir, Russia

Laboratory	Item Dated	Direct Radiocarbon Dating Results	Calibrated Equivalent Calendar Date
NSF Arizona AMS Facility University of Arizona Tucson, Arizona	Sungir 1	19,200 BP	21,045 BC
	Sungir 2 and 3	26,200 – 27,200 BP	28,964 – 29,457 BC
Oxford University AMS Laboratory Oxford, England	Sungir 1	22,900 BP	25,795 BC
	Sungir 2 and 3	23,800 – 24,100 BP	26,771 – 27,039 BC

Note: Calibrated equivalent calendar dates were obtained using the CalPal radiocarbon calibration program (Weninger, 2005).
Reference: Kuzmin, Y. V., et al (2004). *Nuclear Instruments and Methods in Physics Research Section B: Beam Interactions with Materials and Atoms*, vol. 223-224, pp731-734.

Archaeologists have come to rely on radiocarbon dating to determine the age of organic remains such as bones, seeds, wood and campfire charcoal. As discussed in Chapter 4, radiocarbon dating extrapolates the decay of carbon-14 backward in time to when a living object died. Since this method relies on extrapolation, the trustworthiness of the process depends on understanding all factors that can affect it.

Radiocarbon dating must be calibrated and corrected. As previously mentioned, radiocarbon dates must be corrected due to the variation of carbon-14 in the atmosphere over time. The IntCal04 radiocarbon calibration curve goes back 26,000 calendar years and is used to calibrate radiocarbon measurements. Even though we can have a high degree of confidence in the accuracy of calibration curves that have been developed, this confidence does not extend to all organic samples radiocarbon dated. The upper limit for radiocarbon dating is realistically around 30,000 years.

The amount of carbon-14 available for measurement in a sample decreases rapidly with time, as shown in Table 4.4. Every 5,730 years the amount of carbon-14 in a sample decreases by half. At 28,650 years only 3.12% of the original carbon-14 remains. Dating accuracy becomes increasingly uncertain between 10,000 and 30,000 years because of the rapidly decreasing percentage of carbon-14.

Other variables can also affect the accuracy of radiocarbon dating. Errors can occur due to deterioration, contamination, fractionation, cleaning and carbon isotope measurement. As the percentage of carbon-14 in a sample decreases from 10,000 years and upward, these effects are magnified and significant error can result. Knowledgeable archaeologists and anthropologists are aware of the problems of accurately dating archaeological objects. Dr. Donald Johanson, the discoverer of Lucy, in briefly discussing radiocarbon dating says that "no dating method comes without constraints or potential pitfalls."[63]

Table 10.7 – Radiocarbon Dates for Horton Quarries Wood from the Jurassic Period

Laboratory	Laboratory Code	Direct Radiocarbon Dating Results	Calibrated Equivalent Calendar Date	C13/C12 Ratio Compared to PDB Standard
Grochron Labs	GX-21666-AMS	24,005 ± 600 BP	27,009 BC	-22.9%
Grochron Labs	GX-21611-AMS	22,730 ± 170 BP	22,591 BC	-24.0%
Grochron Labs	GS-21612-AMS	28,820 ± 350 BP	31,810 BC	-25.3%
ANSTO Labs.	OZC201	20,700 ± 1200 BP	23,002 BC	-16.6%

Notes:
(1) Laboratory code GX-21666-AMS and OZC201 are from the same wood sample.
(2) Calibrated equivalent calendar dates were obtained using the CalPal radiocarbon calibration program (Weninger, 2005).
Reference: Snelling, Andrew (2000, Mar.). Geological conflict: Young radiocarbon date for ancient fossil wood challenges fossil dating. *Creation ex nihilo* 22(2):44–47.

The Sungir burial site located at Vladimir, Russia illustrates the difficulty in radiocarbon dating old material. Four bodies were found at the site and three were recently radiocarbon dated to more accurately determine their age. Two major laboratories located in Oxford, England and Tucson, Arizona were selected to date the skeletons using collagen extracted from bones. The results presented in Table 10.6 show that the calibrated calendar dates for Sungir 1 range from 21,045 BC to 25,795 BC. Disparity among the dates was attributed to the two different techniques used for extracting collagen from the bones.[64] Differing results by these two reputable laboratories highlight the problem of obtaining accurate dates on old material.

Mladec man illustrates the difficulty of dating some skeletal remains. Mladec man was discovered in the Mladec Caves of Moravia (Czech Republic). The skeletal remains of several individuals were found along with stone and bone tools, ornaments and other art artifacts. Radiocarbon dating of animal bones found in the caves gave ages of 8,500 to 42,000 years before present (BP). The May 2005 issue of *Nature* announced the "successful" direct radiocarbon dating of Mladec man. Teeth from an upper jaw and an ulna (arm bone) were radiocarbon dated. The dates range from 26,330 to 31,500 years before present.[65]

Radiocarbon dates can often conflict with other known information. Three separate samples of fossilized wood from the Horton Quarries at Edge Hill near the village of Ratley, England were radiocarbon dated. The fossilized wood was encased in limestone believed to be from the Jurassic period based on index fossils. One sample was adjacent to a belemnite fossil believed to be 189 million years old. The pieces of wood appeared to be fragments of a tree branch. When broken apart by hand, the fossilized wood splintered and had a woody structure. All three samples were sent to Geochron Laboratories in Cambridge, Massachusetts for radiocarbon dating. One additional sample was sent to the Australian Nuclear Science and Technology Organization (ANSTO) in Lucas Heights, Australia.[66] The results from AMS radiocarbon dating these samples are given in Table 10.7 and range from 22,591 to 31,810 BC. Obviously fractionation, carbon infiltration and contamination have combined to result in a wrong radiocarbon date since the wood is from the Jurassic period (200-145 millions years ago).

Several things can affect the accuracy of radiocarbon dating. R. E. Taylor, PhD, in his book *Chronometric Dating in Archaeology* discusses some of the issues. He points out five critical assumptions of radiocarbon dating, including the assumption that the carbon isotope ratios in samples have not been altered except by carbon-14 decay.[67] Taylor mentions that because of microcontamination during the AMS dating process, the maximum ages that can be resolved are between 40,000 and 50,000 years. For example, carbon dioxide entrapped in marble, which should have no carbon-14 because of its old geological age, yielded an apparent age of 47,960 BP due to microcontamination.[68] Taylor points out that when bones deteriorate and their collagen drops below 5% of the original amount, carbon-14 dating can give seriously inaccurate results.[69]

The true age for old samples that are radiocarbon dated can be far from certain. Inaccuracy can occur by carbon loss in the sample, possible carbon infiltration, laboratory cleaning processes, measurement contamination, carbon-13 fractionation, carbon-14 fractionation and measurement errors. As pointed out in Chapter 4, carbon-13 fractionation effects in warm evergreens and conifer trees can average 27% less than carbon-13 found in the atmosphere. Carbon-14 fractionation is considered twice that value or 54%. It is well known that in some samples the ratio of carbon-13 to carbon-12 (which should be constant) varies from known standards and thus indicates a disproportionate change (see Table 10.7). Several important factors can affect the accuracy of radiocarbon dating to cause increasing uncertainty from 10,000 to 30,000 years. Beyond 30,000 years, any radiocarbon date should be viewed with great skepticism. Unfortunately, once a date is attached to a sample, it becomes endlessly quoted.

Radiocarbon dates should be supported by evidence from archaeology. Ancient man did not live in a vacuum and his existence must be confirmed to an extent by his tools, weapons, food, shelter and trash. There should be evidence of the people who lived before and after him. If archaeology cannot build a credible chain of existence back to a specific human archaeological site, then unsupported old radiocarbon dates can be validly rejected. For example, if the 48,000 BP date for the Boqueirao de

Pedra Furada site in Brazil is accepted, it turns archaeology upside down. This site is roughly four times older than the Monte Verde site in South America and would lead one to believe man originated at Pedra Furada and then spread to the rest of the continent.

Speculative dates well beyond 30,000 years for human existence are commonplace in anthropology and archaeology. One enterprising individual has built a chronology detailing the existence of humanity back six million years.[70] Old dates assigned to human artifacts include musical instruments, art, cave paintings, rock engravings, bone tools, stone tools, worked wood and constructed shelters. These dates can all be legitimately challenged. For example, one must question the credibility of a 230,000 year old date assigned to human dung. It is relatively easy to find a conjectural date to quote in support of one's belief in human evolution.

Thermoluminescence (TL) dating is sometimes used to date archaeological sites that are too old to be radiocarbon dated. Thermoluminescence dating is based on the accumulated radiation dose within crystalline material because of natural environmental radiation. This technique measures the accumulated radiation since the crystalline material was last zeroed out by heating. Sample material to be dated is electrically heated in an inert atmosphere at a rapid rate to 500°C. The resultant optical glow is then filtered and measured by a photomultiplier during the heating process. The glow curve is hypothetically related to its age.[71] Pottery, crystalline materials (quartz, feldspar, calcite, zircon and volcanic glass) and flint heated above 300-400° C have been measured by this technique.[72]

Electron spin resonance (ESR) dating is another method used to measure the accumulated radiation dose in a crystalline material because of naturally occurring radiation (since the sample was last zeroed out by heating). A sample to be dated is placed in a microwave chamber located within a strong external magnetic field. The sample is then subjected to microwave energy from a microwave generator. An ESR spectrometer measures the amount of energy the sample absorbs. The resultant microwave absorption curve, with respect to the magnetic field, is related to its accumulated radiation dose and age.[73] Electron spin resonance dating has been used to date tooth enamel, flint, coral and egg shells.[74, 75]

Variables that affect the accuracy of thermoluminescence dating and ESR include (1) how well the material was zeroed out initially in the past, (2) the amount of exposure to natural radiation (capable of varying from site to site and over time), (3) the accumulated effects of radiation can vary among minerals contained within a single sample and (4) natural temperature variation may influence radiation effects.[76] Both thermoluminescence and ESR are experimental dating techniques and can give wildly inaccurate results. Currently no standard calibration curve exists for these techniques and they are unreliable extrapolations.

Any date that shows anatomically modern humans existed on Earth beyond 30,000 years is strongly suspect and should be seriously challenged. In addition, radiocarbon dates should be separated from unreliable thermoluminescence dates, ESR dates and conjecture. Anomalous radiocarbon dates due to sample deterioration, contamination and carbon-13 to carbon-12 ratio changes should be rejected. Unreliable dating techniques, misapplied dating technology and known questionable dates must not be permitted to dominate ideas about the origin of man.

Conclusion

When properly accessed, biology discredits the theory that chimpanzees and humans evolved from a common ancestor. The limitations of evolutionary mechanisms cannot rationally explain the differences in DNA between the two species. Biological evolution lacks a valid mechanism for the development of new DNA information, molecules, proteins, genes, functions and features—random mutation and natural selection are insufficient for the task. Based on a 4.23% difference between chimpanzee and human DNA, more than 125 million nucleotide substitutions, additions or deletions had to occur between the two species. Some evolutionists believe humans developed from a small population of around 4,000 hominid adults in Africa.[77] The required 125 million nucleotide changes absolutely would not occur in a small population of this size in six million years due to random mutation. The total or significant absence of beneficial mutations in the human genome shows that constructive evolution is not occurring and strongly discredits its occurrence in the past.

Design theory proposes that similarity between other hominids and humans is due to a common designer or common design philosophy. DNA is firm evidence of similarity between humans and chimpanzees. We can hypothesize that DNA from a previous hominid was the basis from which anatomically modern humans were designed.

Fossil evidence supports the appearance of increasingly complex and diverse life on Earth over time. New species appear suddenly. The fossil record reveals that a significant number species first appeared in the Cambrian explosion. Over time more complex marine life, fish, reptiles, dinosaurs, birds and mammals appeared. Hominids are among the most recent to abruptly appear in the fossil record.

Species expansion (i.e. product expansion) on Earth may be interpreted as due to intelligent cause and best explains the abrupt steps between species and their sudden appearance without intermediates. The extreme differences between Lucy, Turkana Boy, Boxgrove Man, Neanderthals, Hobbits and anatomically modern man are best explained by design rather than evolution.

Design theory can readily explain the skulls and skeletons of Table 10.1. Based on DNA, fossils and archaeological evidence, we can hypothesize that several hominid species lived on Earth during the past several million years. The appearance of each species was sudden rather than gradual and was due to intelligent cause. These hominid species had different cranial capacities, skull features, skeletal characteristics, arm lengths, leg lengths and capabilities. Many of these species became extinct, but some of their remains have survived. Anatomically modern man appears at the end of this species chronology. His superior intelligence permitted him to control fire for technological purposes, fashion tools, make advanced weapons, build complex shelters, develop sophisticated art, perform elaborate burial rituals, domesticate animals, build boats and grow crops. Man's abilities allowed him to spread over the Earth and dominate it.

The appearance of anatomically modern man on Earth is relatively recent, although a precise date is yet to be determined. At present, there is no firm evidence to date the appearance of anatomically modern humans on Earth beyond 30,000 BC. If additional discoveries push the date of man back to 40,000 years, it will not invalidate the material presented herein. Virtually one hundred percent of the archaeological evidence for human existence occurs from the settlement of Abu Hureyra and forward. From 10,000 BC and backward, approximately 108 small archaeological sites have been identified, and of that amount, approximately 60 hominid sites are assigned a date older than 30,000 years.[78] If anatomically modern man existed on Earth for the past 500,000 years, then the number of old human sites should be in the tens of thousands. The small number of known human sites represents a small population with a recent past and strongly discredits a long age for man's existence.

During the last two centuries our knowledge of Earth and its inhabitants has grown significantly. If we go back two hundred years in time, any discussion about the origin of man was conjecture. Two hundred years ago our understanding of the fossil record, radiocarbon dating, biology, archaeology and anthropology were either rudimentary or nonexistent. At that time, evidence was not available to develop any type of an empirically based explanation. Ideas of origin were primarily based on philosophical or religious motivations.

With our current scientific knowledge, we have enough evidence to form a plausible hypothesis about the origin of humanity. Philosophical naturalism, which originated more than twenty-five hundred years ago, is absolutely wrong. Traditional young-Earth creation theology which proposes a 4004 BC date for the appearance of man is wrong. Darwinian evolution, proposed one hundred fifty years ago in a time of scientific ignorance, is utterly wrong. These ideas were conceived without adequate empirical support. A hypothesis based on design theory and evidence from biology, paleontology, archaeology and radiocarbon dating offers the best explanation for the sudden appearance of hominids on Earth. The appearance of anatomically modern humans on Earth 20,000 to 30,000 years ago was due to intelligent cause.

Your faith should not stand in the wisdom of men, but in the power of God. (1 Corinthians 2:5) KJV

Chapter 11

The Bible and the Origin of Man

Genesis clearly teaches that God created the first human. In simple terms, He formed man out of dust from the ground. The only physical difference between a pile of dirt and the human body is the arrangement of the atoms. But what a difference! God took dust from the ground and manipulated it to form DNA, genes, chromosomes, proteins, enzymes, fats, muscles and bones to make Adam. He rapidly assembled these together and then breathed life into man. This was an enormous task requiring intelligence, knowledge and power substantially beyond man's ability. Yahweh is both our creator and the one true God of humankind.

We can readily accept that God created man, even though the Bible does not give a detailed explanation. No other creature has man's exact genetic makeup and capabilities. Humans are distinctly different from animal hominids such as gorillas, orangutans and chimpanzees. To understand that humans share some DNA with animals neither conflicts with nor invalidates the biblical account of man's creation.

Bible students and Christians should not put God in a theological box and demand that He create in a specific manner—especially in the absence of specific biblical teaching. If God chose to begin life on Earth with a microscopic organism and subsequently create all other animal and plant species over extensive time, that is within His purview. If God made several hominid species and culminated His work with the creation of anatomically modern man, that is His business. If God made animals and man to have similar DNA, genes, functions and features, that is fully within His power. Empirical evidence of this nature does not disprove God's ability to create and specifically to create man in His image.

An all-inclusive description of the development of human civilization is not given in the Bible. The traditional setting for the Garden of Eden is on the Arabian Peninsula based on the description given in Genesis 2:10-14; however, the precise location has been lost. Adam and his immediate descendants (anatomically modern humans) lived near the Cradle of Civilization and spread outward over time. After the flood, life began anew on the Arabian Peninsula as Noah and his descendants migrated south from the mountains of Ararat to the present region of Syria, Israel and Iraq. Most likely Noah's descendants founded the ancient cities of Jericho and Abu Hureyra. Early life and civilization flourished in this part of the world.

The Bible and empirical science should harmonize where they intersect. Chapter 10 gives an explanation for the origin of anatomically modern man based on biology, genetics, fossils, archaeology, radiocarbon dating and design theory. *Exemplar* creation can harmonize with design theory for the origin of man and his appearance 20,000 to 30,000 years ago. *Exemplar* creation can agree with the verified empirical data of science, yet respect the integrity of inspired scripture.

There should be no inherent conflict between the Bible and physical evidence about how man came about, when he first appeared and where he initially lived. Knowledge of biology strongly supports man as a product of intelligent cause. The Bible clearly states that God created man. Fossil remains, radiocarbon dating, permanent villages and small archaeological sites support the relatively recent appearance of man on Earth—as does the Bible. Archaeology supports the existence of early man within the region described for the Garden of Eden.

Explaining Genesis Genealogies

A hypothesis for the appearance of humankind between 20,000 to 30,000 years ago conflicts with traditional explanations of biblical genealogies. Any explanation of the Genesis creation story which proposes history beyond 4004 BC conflicts with a rigid explanation of the genealogies found in Genesis Chapters 5 and 11. Inflexible explanations do not allow for additional time beyond 4004 BC.

Anglican Archbishop James Ussher (1581-1656) used biblical genealogies to arrive at a creation date for the world of 4004 BC.[1] His date for the creation of Adam and Eve conflicts sharply with archaeological evidence for the appearance of man presented in Chapter 10. Ussher's date conflicts strongly with scientific evidence presented in Chapter 4 which supports an old Earth and universe.

Ussher used genealogies in the Bible to develop a chronology from Jesus to Adam. The birth of Jesus Christ is essentially zero. The end of Israel's Babylonian captivity is 537 BC. The reign of King David is dated at 1056-1016 BC. The exodus of the nation of Israel from Egypt is dated at 1491 BC. Abraham's journey out of Ur of the Chaldees occurred in 1921 BC. The flood of Noah is given a date of 2349 BC, and the creation of Adam and Eve in the Garden of Eden took place in 4004 BC.[2]

Most of the dates proposed by Ussher have been contested. Disputes about dates from the call of Abraham to the birth of Jesus involve shifting events a hundred years or so to align with different explanations of extra-biblical history (or with competing biblical chronologies). However, dates assigned to Noah's flood and the creation of Adam have caused major controversy and cannot be easily reconciled.

More time is needed than Ussher's 2,083 years between the creation of Adam and the call of Abraham. More time is needed to harmonize the Bible with archaeological evidence. The time between Adam and Abraham must be expanded by roughly 16,000 years to agree with an early date of 18,000 BC for man's appearance.

Ussher and other biblical scholars use the genealogies in Genesis Chapters 5 and 11 to determine the number of years from Adam to Abraham. Genesis 5:3-32 establishes the time from the creation of Adam to Noah's flood, while Genesis 11:10-29 establishes the time from Noah's flood to Abraham. Table 11.1 lists ten names in each of these two genealogies.

Table 11.1 – Genealogies from Adam to Abraham (Based on the KJV)					
Name	Age at Begetting Next in Line	Remaining Years of Life	Name	Age at Begetting Next in Line	Remaining Years of Life
Adam	130	800	Shem	100	500
Seth	105	807	Arphaxad	35	403
Enosh	90	815	Salah	30	403
Cainan	70	840	Eber	34	430
Mahalalel	65	830	Peleg	30	209
Jared	162	800	Reu	32	207
Enoch	65	300	Serug	30	200
Methuselah	187	782	Nahor	29	119
Lamech	182	595	Terah	70	135
Noah	500	450	Abraham	100	75
Total Adam to Noah	1,556 yrs	7,019 yrs	Total Shem to Abraham	490 yrs	2,681 yrs
Note: The spelling of all names is based on the NKJV.					

The genealogies in Genesis Chapters 5 and 11 vary somewhat among three different ancient manuscripts. The Masoretic Hebrew text (the basis for the KJV), the Septuagint and the Samaritan Pentateuch all differ slightly in the years given.[3] Regardless of the differences, a strict reading of the Genesis genealogies does not allow enough time for the appearance of humans approximately 20,000 to 30,000 years ago.

A study of other genealogies in the Bible helps us to understand Genesis Chapters 5 and 11. Most Bible students are aware that the genealogy of Jesus given in Matthew 1:1-17 is abbreviated. It is not wrong, but for purposes of memorization was intentionally shortened. Some of the insignificant and evil kings were purposely omitted, showing that in the Hebrew culture a genealogy (1) did not have to represent a father to direct son relationship, (2) could represent a father to descendant relationship and (3) could be abbreviated.

Matthew 1:17 records fourteen generations from Abraham to David, fourteen generations from David to Josiah and fourteen generations from Jeconiah to Jesus Christ. These three periods represent approximately 910, 436 and 513 years respectively.[4] It would be unusual for these three different time periods to have the same number of generations. We can be confident that all three periods have missing direct descendant names and are not complete genealogies. Matthew's genealogy represents a series of father to descendant relationships, where the descendant could be a son, grandson, great grandson or distant male relative.

Table 11.2 – A Comparison of Matthew and Chronicles Genealogies
(From David to the Babylonian captivity)

Matthew's Genealogy (Mat. 1:6-12)	Chronicles Genealogy (1 Chron. 3:5-17)	Relationship	Age at Begin of Reign	Length of Reign	Scripture
David	David	son of Jesse	30	40 yrs.	2 Sam. 5:4
Solomon	Solomon	son of David	---	40 yrs.	2 Chr. 9:30
Rehoboam	Rehoboam	son of Solomon	41	17 yrs.	2 Chr. 12:13
Abijah	Abijah (Abia)	son of Rehoboam	---	3 yrs.	2 Chr. 13:1-2
Asa	Asa	son of Abijah	---	41 yrs.	2 Chr. 16:13
Jehoshaphat	Jehoshaphat	son of Asa	35	25 yrs.	2 Chr. 20:31
Jehoram	Jehoram (Joram)	son of Jehoshaphat	32	8 yrs.	2 Chr. 21:5
---	Ahaziah	son of Jehoram	22	1 yr.	2 Kings 8:26
---	Joash (Johoash)	son of Ahaziah	7	40 yrs.	2 Kings 12:1
---	Amaziah	son of Joash	25	29 yrs.	2 Chr. 25:1
Uzziah	Azariah (Uzziah)	son of Amaziah	16	52 yrs.	2 Chr. 26:3
Jotham	Jotham	son of Azariah	25	16 yrs.	2 Chr. 27:1
Ahaz	Ahaz	son of Jotham	20	16 yrs.	2 Kings 16:2
Hezekiah	Hezekiah	son of Ahaz	25	29 yrs.	2 Chr. 29:1
Manasseh	Manasseh	son of Hezekiah	12	55 yrs.	2 Kings 21:1
Amon	Amon	son of Manasseh	22	2 yrs.	2 Kings 21:19
Josiah	Josiah	son of Amon	8	31 yrs.	2 Kings 22:1
---	Jehoahaz	son of Josiah	23	3 mo.	2 Chr. 36:2
---	Jehoiakim (Eliakim)	2nd son of Josiah	25	11 yrs.	2 Kings 23:36
Jeconiah	Jehoiachin (Jeconiah)	son of Jehoiakim	18	3 mo.	2 Kings 24:8
---	Zedekiah (Mattaniah)	uncle of Jehoiachin	21	11 yrs.	2 Kings 24:17
Shealtiel	Shealtiel	son of Jehoiachin	captive	captive	1 Chr. 3:17

Notes:
(1) The spelling of all names is based on the NKJV.
(2) The first king of Israel, Saul, is not listed in Matthew's genealogy since he was from the tribe of Benjamin rather than the tribe of Judah.
(3) Athaliah who ruled Israel for seven years as Queen between Ahaziah and Joash is not listed (2 Chronicles 22:10-12).

Some of the missing generations in Matthew's genealogy can be proven based on information in the Old Testament. Table 11.2 gives a portion of Matthew's genealogy (from David to Jeconiah) compared to a genealogy of Israel's kings found in 1 Chronicles. Notice that in Matthew's list, three father-to-direct-son generations are left out following Jehoram. One more father-to-direct-son relationship is left out following Josiah. These four names were probably omitted because of the evil nature of those kings and to aid in memorization. We can be certain Matthew was familiar with 1 Chronicles 3:5-17 and his failure to mention these kings was not an oversight, but an intentional action. Abbreviating a genealogy was a perfectly acceptable practice. In Hebrew culture, genealogies represented a father to *descendant relationship*, in contrast to Western culture where genealogies represent a father to *direct son relationship*.

The word "beget" is translated from the Hebrew word *yâlad* (Strong's 03205) in the Old Testament and is often used in chronologies. *Yâlad* occurs 498 times in the Old Testament and is translated "beget or begat" 201 times, "bare" 110 times, "born" 79 times, "bring forth" 25 times, in addition to other miscellaneous words in the KJV. The Brown-Driver-Briggs Hebrew lexicon gives the English words "bear," "bring forth" and "beget" as acceptable translations of the verb *yâlad.*[5] When "beget" is used within the context of a Hebrew genealogy, it does not necessarily mean a direct father-to-son relationship.

An understanding of Hebrew culture helps shed light on the genealogies of Genesis. The Hebrew people were proud of their heritage, and knowledge of family genealogy was important. A male Hebrew during the time of Christ took great pride in claiming Abraham as his father (Luke 1:73 and John 8:39) although around seventy generations removed. A man from the tribe of Judah could claim to be a son of David (Matthew 20:31 and Luke 18:38) while roughly forty generations away. In referring to relatives, Hebrews used the words "father, son and beget" significantly differently than Western culture.

An analysis of Matthew's genealogy and the Genesis genealogies reveals an important issue about completeness and detail. Matthew's genealogy states neither the age of the father when the son was born nor the age of the father at death. Likewise, information is lacking in the Old Testament for several key figures to build an age related genealogy of the kings of Israel. The age at which Saul began reigning and his age at death are missing from the Old Testament. The age at which Solomon, Abijah and Asa became king is missing. Key statistics about the most important people in Israel, their kings, is not found in scripture. A genealogy of Israel's kings, styled after Genesis Chapters 5 and 11, cannot be built based on information in the Old Testament. If one assumes the genealogies of Genesis have no gaps, this makes them superior to all other genealogies in the entire Bible in completeness and detail—a highly unlikely proposition.

The translation of the genealogies in Genesis Chapters 5 and 11 can be easily revised to reflect their true intent. Genesis 11:10-26 is given as an example of how the Hebrew text can be translated into English for additional clarification. The words in italics explain the descendant relationships.

> Genesis 11:10 These are the generations of Shem: Shem was an hundred years old, and begat Arphaxad two years after the flood:

11:11 And Shem lived after he begat Arphaxad five hundred years, and begat sons and daughters.
11:12 And Arphaxad lived five and thirty years, and begat *a son whose descendant was* Salah:
11:13 And Arphaxad lived after he begat *a son whose descendant was* Salah four hundred and three years, and begat sons and daughters.
11:14 And Salah lived thirty years, and begat *a son whose descendant was* Eber:
11:15 And Salah lived after he begat *a son whose descendant was* Eber four hundred and three years, and begat sons and daughters.
11:16 And Eber lived four and thirty years, and begat *a son whose descendant was* Peleg:
11:17 And Eber lived after he begat *a son whose descendant was* Peleg four hundred and thirty years, and begat sons and daughters.
11:18 And Peleg lived thirty years, and begat *a son whose descendant was* Reu:
11:19 And Peleg lived after he begat *a son whose descendant was* Reu two hundred and nine years, and begat sons and daughters.
11:20 And Reu lived two and thirty years, and begat *a son whose descendant was* Serug:
11:21 And Reu lived after he begat *a son whose descendant was* Serug two hundred and seven years, and begat sons and daughters.
11:22 And Serug lived thirty years, and begat *a son whose descendant was* Nahor:
11:23 And Serug lived after he begat *a son whose descendant was* Nahor two hundred years, and begat sons and daughters.
11:24 And Nahor lived nine and twenty years, and begat *a son whose descendant was* Terah:
11:25 And Nahor lived after he begat *a son whose descendant was* Terah an hundred and nineteen years, and begat sons and daughters.
11:26 And Terah lived seventy years, and begat Abram, Nahor, and Haran.

The above explanation of the genealogies in Genesis Chapters 5 and 11 is plausible in light of other genealogies in the Old Testament and our current knowledge of extra-biblical history. This allows these genealogies to harmonize with evidence for the appearance of humanity substantially earlier than Ussher's 4004 BC date. This does not imply the Bible is in error, but that Western explanations have been in error.

Traditional young-Earth creationists are mixed on their view of the Genesis genealogies and a date for the origin of man. Henry Morris, founder of the Institute for Creation Research, says the date for the creation of man can be extended back to 10,000 BC.[6] He allows missing generations in the Genesis chronologies. Ken Ham, founder of Answer in Genesis, rigidly holds to a date of around 4000 BC for the creation of man and claims no generations are missing.[7] Once missing generations are allowed, it becomes easy to extend the appearance of Adam back 20,000-30,000 years ago. In considering the long lives of these people, 16,000 years between Adam and Abraham do not represent that many life spans.

Cultural Considerations

Several verses of a cultural nature are mentioned in Genesis and often used in an attempt to date the biblical origin of humans. Some Bible commentators say that Adam and Eve could not have lived earlier than around 8000 BC based on the following cultural and technological references given in Genesis.[8]

- Genesis 2:15 mentions Adam as a cultivator of plants in the garden.
- Genesis 4:2 says Cain was a tiller of the ground while Abel was a keeper of sheep.
- Genesis 4:20 refers to Jabal as a tent dweller who owned cattle.
- Genesis 4:17 declares that Enoch built a city.
- Genesis 4:21 states Jubal was the father of all those who play the lyre and pipe.
- Genesis 4:22 makes the claim that Tubalcain forged tools of bronze and iron.

The growing of crops and the tending of cattle (goats, sheep and cows) in Genesis 2:15, 4:2 and 4:20 does not conflict with archaeological support for the appearance of humans 20,000 to 30,000 years ago. According to the Bible, agriculture and animal domestication began with Adam's sons. The earliest extra-biblical date for the harvesting and eating of grains is around 16,000 BC, supported by evidence found at Wadi Kubbaniya near Aswan, Egypt. In addition to edible grains, the bones of "wild cattle" were found at this site. Small-scale grain harvesting and cattle herding could have existed before this time without leaving lasting evidence.

The building of permanent living structures does not conflict with archaeology. Genesis 4:17 says that Enoch built a "city" in contrast to scattered groups living in the open, in tents or in caves. The Bible gives no description of what kind of city he built. The earliest archaeologically confirmed village is Abu Hureyra, dated back to 11,150 BC. This does not invalidate Enoch from having built a city at an earlier time. His city may have consisted of a collection of people living in wood and skin huts similar to that found in Abu Hureyra.[9] Whatever materials it was made of, Enoch's city disappeared with the passage of time and has not been found.

The use of musical instruments as a cultural indicator does not conflict with archaeology. Genesis 4:21 says Jubal was the father of all those who play the lyre and pipe. A rudimentary lyre can be considered any type of stringed instrument, while the pipe is any kind of wind instrument. Nicholas Conard from the University of Tübingen, Germany and his colleagues found an 18.7 cm (7.36 inch) long flute carved from mammoth ivory with three finger holes. The flute was found in

Geißenklösterle cave in the mountains near Ulm, in southern Germany among archaeological deposits dated 30,000 to 37,000 years old.[10] Even though the accuracy of these radiocarbon dates are suspect, this flute give credibility to the early appearance of musical instruments described in the Bible.

The mention of forged bronze and iron in Genesis 4:22 is a technology indicator that appears to conflict with archaeology. The significant mining and refining of copper did not occur until Timna Valley, Israel around 4500 BC (see Table 10.2), even though small copper artifacts dating back to about 9000 BC have been found. The development of bronze and iron followed copper.

Genesis 4:22 says, "Tubalcain, who forged all kinds of tools out of bronze and iron." (NIV) If we accept this translation as accurate, then he developed metals forging technology at a very early date. In support of an early date, we can hypothesize Tubalcain did not pass on his knowledge. Perhaps he died suddenly and his metallurgical expertise was lost. Archaeologists are currently unable to verify the early technical influence of Tubalcain; however, a lack of confirmation does not necessarily disprove the biblical account.

A more reasonable view of Tubalcain is that English translations of Genesis 4:22 are overstated. This verse may be translated, "And Zillah she also bore Tubalcain, a sharpener[03913] of all stones[02794] of brass and iron." The Hebrew word *lâtash* (Strong's 03913) also occurs in 1 Samuel 13:20; Job 16:9; Psalm 7:12 and Psalm 52:2, and is best translated "sharpen." The Hebrew word *chôrêsh* is identified by Strong's numbers 02796, 02795, 02794, 02793, 02792, 02791, 02790 and 02789, and is translated a number of ways in the Old Testament. *Chôrêsh* occurs in Job 2:8; Job 41:30; Isaiah 30:14 and Ezekiel 23:34, and is respectively translated "potsherd, stones, sherd and sherds" in the KJV. In these verses, *chôrêsh* applies to some type of sharp object. Genesis 4:22 may be referring to nothing more than a man who was skilled in sharpening pieces of copper or iron-rich ore by pounding or chipping them to make knives and scrapers.

When properly explained, the cultural and technological references given in Genesis Chapters 2 and 4 do not conflict with archaeology. The earliest artifacts representing human achievement have no doubt been lost. The biblical historical account can be accepted as an accurate record of man's earliest accomplishments.

Neanderthals

Neanderthals are an enigma and a problem for both evolutionists and creationists. Their exact relationship to anatomically modern humans is unknown and may very well remain a mystery due to a lack of historical information and physical evidence. Neanderthals appear to have come out of nowhere, lived for a short period of time and then became extinct roughly 30,000 years ago.

Humans and Neanderthals have been compared on several levels. Mitochondrial

DNA from at least four Neanderthals has been studied and shows some significant difference to modern human mtDNA.[11] On the other hand, the Neanderthal's cranial cavity averages about ten percent greater than that of the average modern human. Neanderthals are associated with controlled fire, fashioned tools, formed weapons, art and burial rituals, although not as advanced or as well documented as anatomically modern humans. Some Neanderthals lived in present-day Israel in the caves of Kebara and Tabun where their bones have been found.[12]

Table 11.3 – Neanderthal Sites with Skeletal Remains

Site Name[1]	Country	Artifacts and Comments[2]
Scladina Cave	Belgium	Child remains, about 10 years old
Spy Cave	Belgium	Skeletal remains, 2 adults
Vindija Cave	Croatia	Skeletal remains, radiocarbon dated 28,000 BP
Krapina	Croatia	Skeletal remains, 876 bones, possibly 80 people
Arcy-sur-Cure Cave	France	Inner ear temporal bone
La Chapelle-aux-Saints	France	One skull, partial skeleton
La Quina	France	Skeletal remains, 20 people
Pech de l'Azé	France	Skull and mandible, child
St. Césaire	France	Partial skull, skeletal fragments
La Ferrassie	France	Skulls, skeletal remains, deliberate burial
Le Moustier	France	Juvenile skull
Neander Valley	Germany	Cranium, skeletal remains
Gorham's Cave	Gibraltar	Skeletal remains, evidence of cannibalism
Shanidar	Iraq	Skeletal remains, deliberate burial
Tabun Cave	Israel	Skeletal remains, female
Kebara Cave	Israel	Vertebral column, ribs, pelvis, most complete found
Okladnikov Cave	Siberia	Skeletal fragments, child, mtDNA match
El Sidrón Cave	Spain	Skeletal remains
Teshik-Tash Cave	Uzbekistan	Skull, 9 year old child, mtDNA match

Notes:
(1) Neanderthal sites identified by name in *National Geographic* map of Neanderthal occupation. Sources are William Davies, University of Southhampton, Clive Finlayson, Gibraltar Museum and Chris Stringer, National History Museum, London.
(2) Artifacts and comments are from miscellaneous sources and not *National Geographic*.
(3) A date of 28,000 BP equates to 30,524 BC according to the CalPal radiocarbon calibration program.
Reference: Hall, Stephen (2008, Oct.). Last of the Neanderthals. *National Geographic*. p51.

One aspect of Neanderthal life that appears less than human is cannibalism. Neanderthals eating their own has been documented in at least one location. Paleoanthropologist Antonio Rosas of the National Museum of Natural Sciences in Madrid discovered evidence of cannibalism at El Sidron Cave in Spain.[13] This does not mean all Neanderthals were cannibals, but some definitely were.

As discussed in Chapter 10, the small number of Neanderthal sites suggests a small population and short span of existence. Some evolutionists hypothesize Neanderthals and humans began diverging 700,000 years ago and were fully diverged by

370,000 years BC.[14] If this is true, there should be thousands of Neanderthal sites rather than nineteen,[15] listed in Table 11.3. It is difficult to believe a population of 10,000 Neanderthals (or anatomically modern humans) lived for hundreds of thousands of years and left such scant evidence of their existence. Because of uncertainty when dating remains 30,000 years old and beyond, credible dates have not been established for Neanderthal's appearance and their length of time on Earth. The assumed divergence of humans and Neanderthals is completely speculative, as is the computer extrapolation of presumed genetic mutations to arrive at a divergence date of 370,000 years ago.

From a strict biblical perspective, Neanderthals may be explained at least three possible ways. They may have lived and become extinct before God created Adam. Based on artifacts left behind, they had abilities significantly above animal hominids although more primitive than anatomically modern humans. A second scenario is that Adam and Eve were the first Neanderthals. Through subsequent God directed change, the human race was altered genetically to its current anatomically modern human form. A third possibility is that Adam was an anatomically modern human and Neanderthals appeared after him. Perhaps Neanderthals were a result of the curse of Cain or associated with the Nephilim and the events of Genesis 6:2-4.

Based on the interpretation of available extra-biblical evidence, Neanderthals lived primarily before anatomically modern man. Clive Finlayson of the Gibraltar Museum dated charcoal found in Gorham's Cave, believed to be from a Neanderthal hearth. The charcoal was radiocarbon dated at 28,000 BP and hypothesized to be the last trace of Neanderthals on Earth.[16] A Neanderthal skeleton in Vindija Cave, Croatia was directly radiocarbon dated at 28,000 BP.[17] Neanderthals are generally considered to have gone extinct around 30,000 years ago. As previously mentioned, we can hypothesize that anatomically modern humans appeared between 20,000 and 30,000 years ago. Regardless of the exact dates, we can be confident that both were a product of purposeful design. The differences between the two, and the severe limitations of random mutation and natural selection, preclude their origin by evolution.

Hominids, Neanderthals and Modern Man

Animal hominids, Neanderthals and anatomically modern humans appear quite recently in geologic time. Donald Johanson and Maitland Edey in their book, *Lucy: The Beginnings of Humankind*, discuss the sudden appearance of hominids in the fossil record. They say, "There are, in fact, no ape fossils from anywhere after about eight million [years]."[18] They continue by saying, "At any rate, modern gorillas, orangutans and chimpanzees spring out of nowhere, as it were. They are here today; they have no yesterday, unless one is able to find faint foreshadowings of it in the dryopithecids [a small four-footed monkey-like creature]."[19] The fossil record clearly shows the sudden and relatively recent appearance of hominids, with anatomically modern humans being the last to materialize.

With evidence available today, no firm chronology of hominid existence can be established. Most likely, if we had a true age for the skulls and skeletons in Table 10.1 it would create additional problems for any type of an evolutionary series, similar to the discovery of Hobbits on the island of Flores in Indonesia. It is doubtful the skeletal remains listed in Table 10.1 truly occurred in the chronological order shown.

God may have created a dozen or more hominid species prior to anatomically modern man. Other than the obvious ape-like creatures, other species include Sangiran 17, Atapuerca 5, Turkana Boy, Tautavel Man, Petralona 1, Rhodesian Man, Neanderthals and Hobbits (*Homo floresiensis*). Some of these may have been created at the same time, while others were made in a sequential fashion. We can hypothesize that all of these hominids were created before anatomically modern man, and essentially all were extinct by the time of Adam. Neanderthals and Hobbits may have lived concurrently with anatomically modern man, but at different locations.

In considering the creation of life, God is ultimately the potter and we are the clay. Why, when and where God created a variety of hominids is strictly within His purview. We can only attempt to understand the details through a combination of scripture and empirical science. A true understanding of history does not devalue humankind as a creation of God nor disrespect Yahweh as a powerful creator. Current knowledge of hominids, Neanderthals, Hobbits and anatomically modern humans does not disprove the Genesis account of man's creation. On the other hand, the significant differences between hominids and the lack of a definite step-by-step series strongly discredit evolution. Bible students and Christians can fully attribute all hominids to the creative hand of God.

God as the Creator of Humankind

As mentioned, the Genesis story is an abbreviated, non-technical, historical account. The exact time, sequence and method God used to create mankind are not revealed in scripture, and we are left to our own resources to fill in the details. The fossil record shows the abrupt appearance of different animal species over extensive time. Most species were probably introduced during a series of major creative events, typified by the Cambrian explosion. At the end of this creative series, a number of different hominids were made, with anatomically modern humans at the very end.

Many different hominids were created by God and existed on Earth previous to Adam. Different features include brain size, jaw size, tooth construction, skull characteristics, eyebrow protrusion, arm length, leg length, hand structure, foot form, pelvic shape and spinal column construction. Their living and survival skills were rudimentary. Their lack of intelligence, combined with the absence of any strong offensive or defensive abilities made them vulnerable to bears, lions, cheetahs, leopards, hyenas, jackals and wild dogs when on the ground. The acquiring of food and shelter was a constant struggle. As an additional factor, their inability to make vitamin C ultimately resulted in sickness and death when they inadvertently became separated from natural sources for a significant length of time. As a result, most hominids became extinct.

Similarity between previous hominids and anatomically modern man is because of design and a common designer, rather than evolutionary descent. Biological evidence indicates humans were not a clean sheet design, confirmed by similar DNA, proteins, genes, pseudogenes, features and functions found in animals. We can hypothesize that God used DNA information from a previous hominid as the starting point for the design of anatomically modern humans.

Humankind was created in the image of God, unique above all other creatures on Earth. The present-day comparison of humans and chimpanzees (the closest living hominid) shows they are stunningly different. Genesis 1:26-28 records that God made man in His image to have dominion over all the Earth. God gave man a phenomenal brain, five senses, full color vision, exceptional language ability, a fully functional opposable thumb, dexterous hands and a distinctive upright stature. God made man to be a designer, builder and creator in his own right. Humans were formed as intelligent, self-aware, self-directed entities with free will. Adam and Eve were a direct product of God's handiwork, unique above all other living creatures.

In discussing hominids previous to the appearance of anatomically modern humans, their spiritual state is often discussed. Were most hominids simply animals? Were the more intelligent hominids capable of knowing right from wrong, believing in God and going to heaven? These are among a number of theological questions not specifically addressed in the Bible. Rather than speculate, we can believe that Yahweh as a sovereign, wise creator handled their ultimate fate in a just manner.

In summary, God formed Adam out of dust from the ground (Genesis 2:7) and built Eve of material from Adam's rib (Genesis 2:22) to start the human race. We can hypothesize they were created 20,000 to 30,000 years ago. Adam and Eve were the last of God's creation, other than a few animals formed for the garden. The Garden of Eden, located in a hot arid climate on the Arabian Peninsula, was a protective enclosure and allowed newly created man time to mature. After sinning, Adam and Eve were cast out of the Garden of Eden. Their children intermarried with each other and began the spread of humanity. Humankind survived by hunting, gathering, farming, herding and fishing. During this period of history, the cultures of widely separated small groups advanced and declined because of different circumstances. Technology and living skills were developed and lost. Just as today, the levels of civilization varied radically over the world. Until 10,000 BC the human population on Earth was quite small, based on the number of human archaeological sites identified previous to this date.

Origins Theology

Traditional young-Earth creation theology handed down through the early church tends to be Earth-centered and man-centered. The Earth was created first and then the rest of the universe was created around it on the fourth day. Man was created to rule over the animals and was totally disconnected from them. The making of Earth and man were God's only relevant creative acts.

Knowledge of astronomy has enhanced our appreciation of the heavens as God's creation. At one time people thought the other planets, Sun and stars revolved around the Earth. We have since realized the untruth of this belief. While Earth has lost its seat as center of the solar system, our expanded knowledge confirms God's creation to be much larger, grander, more fine-tuned and older than previously thought. The heavens show God as a timeless, ageless, intelligent, powerful designer who created over billions of years.

Knowledge of biology has improved our admiration of God as a planner, integrator, designer and creator of life. We now know anatomically modern man was not the only hominid God created. There were definitely others, and our biology is related through design. So while man has lost his seat as a special creation totally disconnected from all other creatures, our expanded knowledge shows God created life in a much greater, more complex, information rich, fine-tuned and integrated manner than we could have ever imagined.

Exemplar creation exposes the error of Earth-centered and man-centered theology. The primary purpose of Genesis is to show God as creator. He is the main focus. The Earth is insignificant compared to the rest of the universe, while man is only one of many living creatures in the animal kingdom—yet God's touch has made both unique. God's creative actions go significantly beyond man, Earth and six days.

Exemplar creation does not redefine traditional theology which views God as sovereign, eternal, powerful, intelligent, knowledgeable, holy, loving and just. Yahweh created the heavens and Earth, and all life on Earth. He is not time-bound. God's inspired word is truthful and trustworthy. Yahweh will not falsify evidence, lie, mislead or deceive. His word and work will harmonize.

A few related verses of scripture may help Bible students and Christians put the origin of man in a biblical perspective. One significant verse is 1 Corinthians 15:45 which says, "And so it is written, The first man Adam was made a living soul; the last Adam was made a quickening spirit." (KJV) According to this verse, Adam was the first man. The hominids before Adam were not fully human in God's sight.

Several verses such as Romans 9:21 and Jeremiah 18:6 metaphorically refer to God as a potter. Isaiah 64:8 says, "But now, O LORD, thou art our father; we are the clay, and thou our potter; and we all are the work of thy hand." (KJV) A potter has absolute control over the clay, and can mold, shape or destroy it at will. God as the creator of life can modify it at His pleasure. Hominids that God chose to make in the past were subject to His power and authority, and He determined their limitations, capabilities, intelligence and ultimate fate.

The way God thinks about His creation is beyond our understanding. Isaiah 55:8-9 says, "For my thoughts are not your thoughts, neither are your ways my ways, saith the LORD. For as the heavens are higher than the earth, so are my ways higher than your ways, and my thoughts than your thoughts." (KJV) Mankind cannot comprehend why God has done certain things no more than a bacterium can grasp human actions. We can only accept what God has accomplished and attempt to understand.

The historical record of Genesis is not all inclusive and only the basics are given. Many things about the heavens and Earth are not revealed in scripture. As 1 Corinthians 13:12 says, "For now we see through a glass, darkly; but then face to face: now I know in part; but then shall I know even as also I am known." (KJV) The history of human origins given in special revelation is incomplete. Perhaps some day we may know the rest of the story, but for the present that knowledge is imperfect.

Conclusion

A combination of the Bible, design theory, biology and archaeology provide the best explanation about man's past. Overcoming faulty ideas is the greatest hindrance to accepting a correct explanation of human origins. The actual record of our past does not agree with philosophical naturalism and Darwinian evolution, nor with traditional young-Earth creation theology and associated thought. If Bible students and Christians can move beyond these incorrect ideas, true history can be embraced which recognizes Yahweh as both creator and the one true God of humanity.

It is important that we attempt to harmonize scripture and empirical evidence in Christian science education. For Bible students and Christians unfamiliar with human origins, the material presented in Chapters 10 and 11 may be deeply unsettling. Addressing this issue may seem unnecessary and contentious. Some may even consider it heretical. However, this is the type of subject matter that biologists, archaeologists and anthropologists must deal with on a continual basis. Professors who teach science in Christian colleges are forced to come to grips with this information—they do not have the luxury of ignoring it. Fortunately, teaching about the origin of man can respect the integrity of the Bible, yet remain solidly based on physical evidence. These goals are not mutually exclusive. Christian schools and colleges that teach science can satisfactorily address the issue of human origins from a biblical perspective.

Has not God made foolish the wisdom of the world? For the foolishness of God is wiser than man's wisdom, and the weakness of God is stronger than man's strength. (1 Cor. 1:20, 25) NIV

Chapter 12

The Fantasy of Cosmological Evolution

On a dark, cloudless night the stars in the heavens are awe inspiring. They are even more so when one realizes that most objects visible to the naked eye outside our galaxy, typically considered stars, are actually other galaxies. Based on observations made by the Hubble Space Telescope in 1999, an estimated 125 billion galaxies exist in the visible universe.[1] Our galaxy, the Milky Way, considered to be of average size, contains an estimated 400 billion visible stars.[2, 3] As viewed from Earth, the universe is enormous and appears almost endless.

The universe is filled with stars and other unique phenomena. Stars range in color from blue to red and in size from small to large. For example, the diameter of the KY Cygni star in the Milky Way Galaxy is roughly 1,420 times that of the Sun. There are single stars and binary stars in addition to solar systems, globular star clusters, galaxies and galaxy clusters. There are supernovae, novae, novae remnants, quasars, pulsars, nebulae and interstellar dust. All of these are real objects within the observable universe.

Definition 12.1 – Ideas of Cosmological Evolution

Ideas which purport to explain the origin of the universe based on natural cause. The moons, planets, solar systems, stars, galaxies and other objects in the universe are attributed to undirected change. Known mechanisms available for evolution are gravitational attraction, explosion, nuclear fusion, cooling from gas expansion, heating from gas compression, increasing entropy, collisions and other undirected phenomena acting over time.

Cosmological evolution refers to ideas which purport to explain the origin of the universe (space, matter and energy) based on natural cause. The moons, planets, solar systems, stars, galaxies and other objects in the universe are attributed to undirected change. Known mechanisms available for evolution are gravitational attraction, explosion (e.g. novae and supernovae), nuclear fusion, cooling from gas expansion, heating from gas compression, increasing entropy, collisions and other undirected phenomena acting over time. Cosmological evolution includes ideas of the Big Bang, evolutionary star formation and the solar nebular hypothesis.

Ideas of cosmological evolution are driven by an absolute requirement for natural explanations. Although wrapped in a veneer of scientific terminology, mathematical concepts and geometric models, these ideas goes substantially beyond empirical science. The following material discusses the basics of evolutionary cosmology and a rationale for its rejection.

Big Bang Theory Basics

An understanding of Big Bang theory begins with the work of astronomer Henrietta Leavitt. While working at the Harvard College Observatory, Leavitt determined the cycles of twenty-five Cepheid variables (a type of star that varies in luminosity at a specific rate) and published the results in 1912.[4] Leavitt found that bright Cepheid stars pulse more slowly than dimmer ones, and that absolute brightness can be determined by their pulsation rate alone. Knowledge of a star's absolute brightness and apparent brightness permits use of the inverse square law to calculate its distance from Earth and is a well-understood method that astronomers use for measuring the distance to a star.

During the 1920s, astronomer Edwin Hubble used Cepheid variables to determine the distance to neighboring galaxies outside the Milky Way. He also measured the spectral redshift of those galaxies. Spectral redshift, based on the Doppler Effect, is a phenomenon that causes the light spectrum to shift toward the red end as the light producing object moves away from the observer. As the same object moves toward an observer, the light spectrum shifts toward the blue end. With Cepheid variable stars as a measuring stick, Hubble noticed that the farther away galaxies were, the greater their spectral redshift. He theorized that galaxies in the universe were moving away from each other.

Big Bang theory was developed to explain galaxy movement and the apparent expansion of the universe. If the universe is expanding similar to an inflating balloon, then its size can be extrapolated backwards to when it was much smaller, to the size of a planet, a moon, a beach ball, a marble and finally an infinitesimally small point (often called a quantum fluctuation or singularity). According to theory, this was the beginning of the universe as we know it.

Big Bang theory is the dominant idea for the natural origin of the universe, currently explained in a number of books. Georges Lemaître, a Belgian ordained Roman Catholic priest, in 1927, proposed an expanding universe as an explanation for spectral redshift. He later said the universe began as a "primal atom."[5] Before the Big Bang, time, space, matter and energy did not exist. Others have added to Lemaître's basic idea and say the Big Bang started as a singularity,[6] 100 million trillion times smaller than a proton.[7] This created an enormous amount of pure energy at an extremely high temperature of 1×10^{32} degrees Kelvin.[8] Within the first second, this primordial fireball rapidly expanded. Three minutes after the beginning, temperature dropped to 1 x 10^9 degrees Kelvin[9] and all the neutrons, protons and electrons in the universe were created.[10] At this point, the entire mass of the universe was contained within a sphere, six light minutes in diameter or more. This hot ball of matter and energy continued to expand and after 10,000 years cooled to 10,000° K.[11] Around this time the radiation era closed and the matter era began.[12] After 300,000 years of expansion, the temperature dropped to 3,000° K and atoms of hydrogen and helium began forming. Following this, miniscule density fluctuations occurred in the expanding universe which, under the influence of self-gravity, led to the formation of galaxies, galaxy clusters and superclusters.[13]

Some people incorrectly believe the Big Bang was an explosion. Instead, the Big Bang is theorized as a rapid expansion of the universe starting from an infinitesimally small point. Space expanded from this point, carrying all newly created matter and energy with it.

According to theory, natural cause can account for all elements and heavenly objects. The Big Bang yielded approximately 75 percent hydrogen and 25 percent helium atoms by weight. No carbon, oxygen, calcium, iron or other heavy elements were formed initially.[14] Hypothetically, the first stars were created when hydrogen molecules and helium atoms began collecting together by self-gravity 300,000 years following the Big Bang. Hypothetically, these initial stars created the remaining elements through nucleosynthesis. As the first stars progressed through their life cycles, some became novae, scattering their newly formed material throughout the universe. This material in turn became the basis for planets and other stars.

Big Bang theory goes beyond empirical science and makes a direct appeal to natural metaphysical cause. Before the Big Bang there was no universe—space, matter and energy did not exist. There was absolutely nothing, yet within a three minute period all energy and raw matter in the universe popped into existence.

Evolutionists often criticize biblical creation theology for proposing that a powerful, intelligent God created the heavens and Earth out of nothing (*creatio ex nihilo*). Evolutionists go one step further by believing all matter and energy in the universe were created out of nothing *by nothing.* Though accepted by a majority of astronomers, Big Bang theory makes a direct appeal to unknown phenomena and natural metaphysical cause.

Alternate Natural Hypotheses for the Universe

Ideas of cosmology are limited only by the human imagination, and at least three other naturalistic explanations can be given for the universe.

The oscillating universe hypothesis proposes that space, matter and energy were created by a Big Bang in the distant past. The universe is currently in an expansion phase. Ultimately gravity will bring a halt to expansion because of a greater than critical density. The universe will become static for a time and then collapse until it becomes a gigantic black hole or some other dense cosmological feature. At that point, expansion will again be triggered. The universe will oscillate between expansion and contraction forever.[15]

As a second hypothesis, we can propose that if one singularity occurred, then billions could have occurred. Each singularity created a mini-bang to form a galaxy or galaxy cluster. Expansion of each mini-bang has resulted in the separation movement among galaxies. Visualize billions of balloons packed tightly together. As each balloon expands the entire collection expands, representative of an expanding universe.

A third hypothesis is suggested by astronomer Halton Arp of the Max Planck Institute of Germany. He has proposed a continual creation model for a nonexpanding universe of indefinite size. According to Arp, matter and energy are eternal and the conservation of mass and energy applies to the whole universe, but may not apply to the visible universe. New objects form and grow because of complex unknown laws.[16] Newly created matter keeps the universe from running down, and new galaxies are created when they are ejected from older galaxies. As galaxies age, they grow in size and mass, and in turn eject new galaxies in a cascading process.[17]

All natural hypotheses for the origin of the universe have difficulty explaining energy and matter. All ideas based strictly on natural cause propose that energy and matter essentially popped into existence from nothing, based on a singularity, quantum fluctuation, primal atom, cosmic egg, quantum sea or zero-point field. All evolutionary theories make an appeal to unknown phenomena and natural metaphysical cause, and lack direct evidence in support of their claims.

The speculative ideas of cosmology are virtually unlimited. Hypotheses for the origin of the universe are constrained only by the human imagination and an appeal to natural cause. Empirical science based on observation, measurement and experimentation is conspicuously lacking among these inventive schemes.

Matter, Energy and the Big Bang

There are several troubling questions about the Big Bang. According to theory, the universe began as a singularity, yet how does a singularity occur when space, matter and energy do not exist? How can absolutely nothing result in essentially the spontaneous generation of all the energy and matter in the universe within three minutes? This is not a minor issue. Just about any physicist can come up with a speculative argument supporting singularities, but as far as empirical science is concerned, an event of this type has never been observed, measured or created experimentally. No physical evidence directly supports its occurrence. The singularity that started the Big Bang is a speculative idea developed to support a dogmatic requirement for natural cause.

One key issue about the Big Bang relates to the formation of matter from energy. Theoretically, a singularity created a vast amount of extremely hot energy at 1×10^{32} degrees Kelvin which rapidly expanded and cooled to form protons, neutrons and electrons. Although never observed, Big Bang theory makes the assumption that high temperature energy can spontaneously convert into matter. In truth, if the universe started out as ball of hot energy it would expand and nothing more. Gerald L. Schroeder, PhD, discusses this issue briefly in *The Hidden Face of God.* He says, "So in theory the universe should be an expanding ball of ever more dilute (cooler) radiation and no particles of matter."[18] It is a huge leap of faith to believe high temperature energy can instinctively change into matter.

Scientists have discovered that matter can naturally convert into energy by fission or fusion, but not the reverse. As far as empirical science is concerned, no natural process has been observed that converts *pure energy into matter*. Scientists using atomic accelerators have hit targets with fast moving protons, electrons and atomic nuclei to produce new kinds of matter, but have not created matter using only pure energy.

Many natural processes are not reversible or easily reversible. For example, suppose we take an egg, crack it open, put it in a pan and make scrambled eggs. It is humanly impossible to reverse the process and create an egg capable of hatching into a chicken. Just as the cooking of an egg cannot be reversed, it appears that pure energy does not naturally turn into matter.

Conversion of energy into structured matter is a problem yet to be resolved by proponents of the Big Bang. How does raw energy know how to make ordered matter? As Schroeder points out in discussing the Big Bang, "There'd be no simple logic that would predict that some of this energy would change into stable, lasting matter."[19] Without empirical experimental support, it is pure speculation to believe raw undirected energy knows how to form protons, neutrons and electrons that in turn combine together to form stable atoms. Such belief represents an appeal to unknown phenomena and natural metaphysical cause.

Big Bang theory, as a natural explanation for the origin of the universe, rests on a shaky beginning. The singularity that allegedly started it all is totally unsupported by empirical science and clearly illustrates the root of the problem. Any idea based strictly on natural cause will have major difficulty explaining the origin of energy and matter. Singularities, quantum fluctuations, zero-point fields and the quantum sea as an energy source are nothing but speculative musings. While thinking of this type may be within the bounds of scientific thought, accepting it as true represents a faith-based commitment to philosophical naturalism.

Design Theory

Design theory proposes that complex functional systems and many complex non-functional structures within the universe are best explained by intelligent cause.

For example, the gathering together of hydrogen and helium by self-gravity to form stars is in opposition to the natural diffusion of gases in a vacuum, and suggests a non-natural cause. Nuclear fusion inside stars to produce heat and light indicates a complex functional system. The grouping of the vast majority of stars into galaxies by natural means is improbable. Binary stars represent fine-tuning and the high percentage of binary star systems is statistically improbable. Spiral galaxies, ring galaxies and polar ring galaxies represent fine-tuning on a galactic scale.

The origin of the universe can be clearly framed as a choice between two competing ideas. Evolution theory appeals strictly to natural cause, while design theory allows an appeal to intelligent cause. The choice between these two ideas can be summarized as follows:

(1) Evolution theory – Space, basic matter and energy originated by natural cause due to an unobserved event that will probably never be repeated. After formation, undirected change used these raw materials to make all subsequent objects within the universe. Evolution theory ignores the severe limitations of natural cause and proposes that undirected change is a powerful constructive mechanism when given significant time.

(2) Design theory – The information, complex functional systems, mathematically improbable structures and fine-tuned arrangements observed within the universe are evidence of design. A design agent(s) originating from outside the limitations of the known universe is proposed to explain these phenomena. The known limitations of the laws of chemistry and physics preclude natural cause as a complete explanation for the universe.

To help clarify the difference between evolution theory and design theory, consider the existence of space, basic matter and energy as given. Somehow these raw materials came into existence and were the starting point for the universe. We can hypothesize whatever we like to explain these basic phenomena, and if properly worded, no one can prove us wrong (all ideas will be faith-based). With these starting materials established, which idea best explains how nature and the universe at large came about?

Ideas of evolution rely solely on natural cause, which is severely limited. To believe natural cause can create all objects found in the universe from the proposed starting materials requires a continuum of faith-based appeals. Virtually every different complex object requires an appeal to some unknown phenomena. Natural explanations are locked into appealing to missing information, undiscovered laws and mathematically improbable events. Evolution as an alleged powerful constructive mechanism relies on a shameless appeal to natural metaphysical cause.

The superiority of design theory lies in its mechanism. Humans can create information, complex functional systems, mathematically improbable structures and fine-tuned arrangements. Humans can create things not found in nature and substantially beyond the ability of natural cause. Human intelligence and power can be extrapolated to account for a designer on a cosmic scale. Science has yet to place a limit on what intelligence and power can accomplish. A design agent of sufficient ability could take the proposed starting raw materials and create everything observed in nature and the universe at large. Intelligent cause is a more rational explanation than natural cause—it makes the least assumptions.

The Aquarium Universe

Man's attempt to explain the origin of the universe is analogous to fish inside an aquarium trying to explain their surroundings. In our imaginary aquarium, all six sides are painted black to prevent its occupants from observing the outside world. As fish in the aquarium develop from eggs into minnows and finally adults, they initially accept their surroundings as natural—simply the way things are. As the fish mature, they begin to seek answers and question about where they came from. Why are they the only self-aware, intelligent species? Why is the water temperature always exactly 26° C and the pH at 7? They debate the origin and operation of the light within the aquarium that shines without interruption. They wonder why the water is free of pollutants and always well oxygenated. They speculate about the endless supply of food that magically appears. They question the origin of a multitude of plants, corals, worms, snails and lesser living things in the aquarium.

As the fish become more knowledgeable, they begin to question the aquarium's origin based on the constraints they observe. With natural cause as a preliminary assumption, the fish attempt to develop an explanation for their universe. They begin by examining the functional complexity of their own bodies and of other living things. They study the fine-tuned environment of the aquarium. As they examine every item available, no plausible answer exists for its origin based on the limitations of the aquarium. The constraints of their surroundings preclude any logical answer for the objects they observe. After exhaustive research, the fish are forced to conclude the answer lies outside the aquarium. An aquarium can neither create itself nor create the objects found within. The preliminary assumption of natural cause is abandoned in favor of an intelligent cause as an explanation.

Both an aquarium universe and our universe compel an appeal to intelligent cause. From a human perspective, the origin of the aquarium is apparent and the source of its design is clear—human intelligence and power. We marvel at the struggle the fish have in recognizing the inadequacy of natural cause. How can fish be so blind to the obvious limitations of their universe? An aquarium cannot create itself. And then we look at the constraints of our universe. We examine the paradox between the limitations of natural cause and the information and functional complexity observed about us. We marvel at life, Earth and the vast expanse of the cosmos. Objects from DNA to galaxies defy a natural explanation. In the final analysis, we must conclude that a powerful intelligence originating from outside the limitations of our known universe is required. Based upon science, design theory becomes a rational explanation.

Design Theory versus Biblical Belief

Exemplar creation and design theory easily complement each other. However, the Bible is primarily a religious book, while design theory is mainly a scientific idea. Design theory, patterned after human design, looks for evidence of intelligent cause as a rational alternative to natural cause. On the other hand, the Bible teaches that Yahweh is both our creator and God in addition to a number of other specific historical, theological and moral issues.

It is important to understand the fundamental differences between design theory and *exemplar* creation (discussed in Chapter 7), especially in regard to public education. Additional important distinctions between the two applicable to cosmology are as follows:

(1) Explanatory scope – Design theory is a means of explaining the information and functional complexity found in the universe as a rational alternative to natural cause. Design theory focuses on evidence of design and is specifically limited. The Bible presents God as an all-encompassing explanation of origins. He created the heavens and Earth.

(2) Material origin – Design theory has no specific explanation for the origin of space, basic matter and energy. The Bible in Genesis 1:1; John 1:1-3 and Colossians 1:16-17 attributes God as creator of the heavens and Earth. We can infer from these statements that God created space, matter and energy, although no details are given.

(3) Belief basis – Design theory is based on (1) knowledge of phenomena in the physical universe about us, (2) acknowledging the limitations of undirected change and (3) human design. The Bible is God's message to humankind and is the basis for creation belief.

(4) Agent identity – Design theory has no presupposition about the origin, identity, character, moral nature or competency of a design agent or agents. The Bible identifies God as the creator (sole design agent) of the universe. He is sovereign, eternal, powerful, intelligent, knowledgeable, holy, loving and just.

(5) Agent origin – Design theory proposes that a design agent or agents originated from outside the limitations of the known universe, and no other explanation is given. The Bible refers to the eternal nature of God in Gen. 21:33; Deut. 33:27; Psalm 90:2, 93:2; Isaiah 48:12, 51:57; 1 Tim. 1:17, 6:16 and Rev. 1:8. God existed before the heavens and Earth.

Big Bang Theology

Some Christian scientists and theologians explain verses in scripture as supportive of Big Bang theory and an expanding universe. It is a mistake to take poetical or metaphorical verses in the Bible which describe God "stretching out the heavens" and use them as support for an expanding universe. These verses could just as readily explain the heavens stretched out in a static universe. The following two verses (and more than nine similar ones) are often misused:

- Psalm 104:2 Who cover Yourself with light as with a garment. Who stretch out the heavens like a curtain. (NKJV)
- Isaiah 51:13 And you forget the LORD your Maker, who stretched out the heavens and laid the foundations of the earth . . . (NKJV)

To promote the Bible as supportive of Big Bang theory is to ignore the mistakes of past biblical leaders and theologians. Theologically, there is no compelling need to explain the state of the universe in regard to whether it is expanding, static or contracting.

Interpretations of spectral redshift from distant galaxies support an expanding universe, but not necessarily a naturalistic Big Bang. The rate at which the universe is expanding, the Hubble constant, has been determined to be around 71 km/s/Mpc.[20] Extrapolating the current expansion backwards is a logical process, but the extrapolation eventually breaks down. To extrapolate the expansion back to nothing and say it came from nothing is not logical. Empirical science does not support the naturalistic origin of the universe from a singularity which rapidly created all energy and matter in the universe within three minutes from absolutely nothing (*evolutio ex nihilo*).

From a theological perspective, in the beginning God created all matter and energy that make up our universe. Based on interpretations of spectral redshift and other information, we can hypothesize that God caused raw matter to expand as a simple method of spreading it throughout space. Over a period of time, God gathered hydrogen, helium and other materials together to form stars and galaxies. The separation movement of galaxies observed today reflects movement given to the initial raw materials. *Keep in mind the Genesis creation story and the remainder of the Bible do not specifically address this hypothesis.*

Biblical theology has gotten into trouble when it has tried to explain too much rather than too little. The church has gotten into trouble by basing theology on popular scientific theory, which subsequently changed. This is pointed out in *Universe by*

Design by Danny Faulkner, professor of Astronomy at the University of South Carolina. Dr. Faulkner says, "If and when the big bang falls out of favor and we have made it a central theme of our apologetics, then what will happen to our apologetics?"[21]

Geocentric theory is a prime example of an incorrect scientific idea. Eudoxus of Cnidus (400-350 BC), a Greek mathematician and astronomer, developed a theory of 27 concentric spheres to explain the movement of the Sun, Moon and five known planets around the Earth. In his scheme, Earth was located at the center of the inner sphere while the outermost sphere contained the fixed stars.[22] The Greek philosopher, Aristotle (384-322 BC), adopted Eudoxus' ideas with some modification. The Aristotelian universe consisted of 49 concentric transparent spheres nested about the Earth fixed at its center.[23] Claudius Ptolemy (100-170 AD) of Alexandria, an Egyptian astronomer, created a geocentric model of the solar system known as the Ptolemaic system.[24] His model consisted of eight concentric spheres around the Earth. The Moon, Sun and five planets occupied the first seven spheres while the fixed stars occupied the eighth sphere. Each planet moved in an epicycle as it circled the Earth within its concentric sphere.[25]

Geocentric theory originated outside the church and was then adopted by the church. Geocentric theory was not developed by the Hebrews, early New Testament Christians or the Roman Catholic Church. As originally developed, it was not based on the Bible. Over time the Roman Catholic Church accepted the "scientific" geocentric model of the universe and developed theology based on poetical and metaphorical verses such as Psalm 93:1, 96:10, 104:5; 1 Chronicles 16:30 and Ecclesiastes 1:5. The church attempted to ride on the coattails of theoretical science.

The merging of biblical theology with scientific theory went well until scientists began to expose error in the geocentric model. In 1543 AD, Nicholas Copernicus, a Polish astronomer, challenged the geocentric theory with *De Revolutionibus Orbium Celestium* (On the Revolutions of the Celestial Spheres). Copernicus proposed the Sun as center of the solar system with the planets revolving around it.[26] Copernicus released his book as he lay dying, apparently convinced the church would not persecute a man on his death bed. After his passing, astronomers Galileo Galilei and Johannes Kepler became major figures in promoting the heliocentric theory.

Galileo began studying the heavens with his hand crafted telescopes, which led to his argument for a Sun centered solar system in the *Dialogue Concerning the Two Chief World Systems,* completed in 1630. Because of his book, he was called to Rome and tried for heresy by the Roman Catholic Church. Galileo argued that metaphorical verses should not be used as the basis for belief about the solar system and proposed that special revelation and general revelation should agree.[27] Galileo ended up denouncing his book and thus avoided possible torture and martyrdom. Over succeeding years, his ideas slowly gained favor and the geocentric theory was eventually shown to be false.

In retrospect, the geocentric theory and theology based upon it were horribly wrong. This errant theology and the actions it spawned have resulted in serious injury to biblical belief and the church. We should learn from past mistakes and not repeat them. Biblical theology about origins must be sound to account for any discoveries by empirical science within the next millennium. Speculative theology has the possibility of being proven wrong and doing additional harm to biblical belief in the future—this must be avoided.

The beginning described in the Bible is a historical statement strongly supported by empirical science. Genesis 1:1 says, "In the beginning God created the heavens and the Earth," and beyond this we should be cautious because of the immense diversity and complexity of the universe. The Bible does not state how God accomplished this feat. No starting materials are mentioned and the timing, sequence and methods are not specified. However, the universe did have a beginning, and God initially created all space, matter and energy. God is a timeless, ageless, powerful intelligence who created the universe over a period of undefined time. The lack of a plausible natural explanation for the moons, planets, solar systems, stars, galaxies and other objects in the cosmos strongly support the biblical position.

Conclusion

Bible believers and Christians can easily accept God as creator, evidenced by the smallest microscopic life to the largest galaxy. At the atomic level, He assembled molecules into complex arrangements to make living things. He created plants and animals that appear to operate autonomously based on the intricate interaction of unseen molecules. At the other extreme, God created stars which also appear to operate autonomously. He gathered stars together to form billions of galaxies as highly visible beacons of light throughout the heavens.

Our most basic questions about origins and the universe can only be answered by accepting the Bible as God's word. We will never know where God came from, except that He existed from before the beginning. We will never know how space, matter and energy came into existence, other than they were created by God. We will never know in detail how the heavens and Earth came about, other than they were made by God. These are simple answers for the present, but the only answer given to satisfy the insatiable curiosity of humankind. God asks us to have faith in His word. There is no harm in using science to shed light on these basic questions, but speculative ideas should not replace our faith in the Bible. We must refuse to replace our belief in God with philosophical naturalism.

An in depth study about the origin of the universe is intellectually and philosophically challenging in addition to being emotionally overwhelming. A weighty subject indeed! However, a careful analysis gives credibility to the biblical creation account. Based on factual knowledge about the universe, intelligent cause and design are strongly evident. Based on *exemplar* creation, the Bible easily harmonizes with empirical science in regard to the age of stars and galaxies. When the severe limitations of natural cause are acknowledged, we can firmly reject philosophical naturalism and its ancillary ideas. Bible students and Christians can accept the Genesis creation story as an abbreviated, non-technical, historical account that is truthful and trustworthy.

In his book *Seeing Red: Redshifts, Cosmology and Academic Science*, Dr. Halton Arp makes some rather provocative statements about science and origins. He states, "One of the crusades of academic science is against religious creationism. Periodically there arises a messianic need to save the general public from the ignorant beliefs that humans were created in their present form some short time ago, say 8,000 years or so." In reference to the Big Bang he says, "Establishment science is the most blatant possible form of creationism. The claim is that not just humans, but the whole universe was created instantaneously out of nothing."[28] Many scientists, influenced by philosophical naturalism, worship at the altar of natural metaphysical cause.

Bible students and Christians must understand that the goal of philosophical naturalism is to enthrone natural cause as king, with a necessary side effect of dethroning God as creator. Since believers in philosophical naturalism view the creation account as myth, this job is long overdue. Cosmological evolution based on a consensus of scientific opinion is equivalent to fact. Lack of supporting empirical evidence and the limitations of the known laws of chemistry and physics are irrelevant. The ultimate goal is the validation of philosophical naturalism as the only acceptable faith-based belief. Regardless of how shallow or illogical, any answer supportive of natural cause is considered scientific. Philosophical naturalism has an answer for the origin of the universe, and the answer does not involve God.

I devoted myself to study and to explore by wisdom all that is done under heaven. What a heavy burden God has laid on men! (Ecclesiastes 1:13) NIV

Chapter 13

A Comparative Analysis

Regardless of their basis, ideas of origin promoted in the public arena should be open to scrutiny, examination and debate. Credible theories, philosophies and theologies about beginnings should reasonably explain a range of phenomena observed in nature and the universe at large. To that end, a comparative analysis of (1) philosophical naturalism, (2) traditional young-Earth creation theology and (3) *exemplar* creation are analyzed with empirical science as the measuring stick.

As applicable to origins, philosophical naturalism proposes that random change, undirected phenomena and extensive time have resulted in nature and the universe at large. God, theistic figures and intelligent powers are rejected as creative causes. By inference, philosophical naturalism requires that natural cause plausibly explain all things considered natural in the universe. Philosophical naturalism is supported by ideas of the Big Bang, evolutionary star formation, the solar nebular hypothesis, chemical evolution and biological evolution.

Traditional young-Earth creation theology proposes that God created the heavens and Earth in 4004 BC, though some extend this time back to 10,000 BC.[1] God created all plants, animals, man and the universe during six 24-hour days. Noah's flood, around 2349 BC, covered the entire Earth and destroyed all birds and air-breathing land animals outside the ark. The flood is responsible for the vast majority of coal deposits, sedimentary deposits and the fossil record. God is sovereign, eternal, powerful, intelligent, knowledgeable, holy, loving and just. Yahweh will not falsify evidence, lie, mislead or deceive.

Exemplar creation, defined primarily in Chapters 1, 6 and 11 of this book, proposes God as the primary creative cause for the universe, but allows time for limited undirected change. Genesis 1:1, 1:2 and 1:3 occur in chronological order, although the time between each verse is undefined by the Bible. The heavens and Earth were created by God in Genesis 1:1 and the "beginning" refers to an undefined period of time. In Genesis 1:2, the Earth is described as dark, covered with deep water and devoid of life at some point following creation. Significant change occurred between Genesis 1:2 and 1:3 when God created plant life, marine life, flying creatures and land animals. During this time the fossil record was laid down, sedimentary deposits occurred, fossil fuel reserves developed, dry land was raised and the ocean basins formed, in addition to other geological change. General revelation shows that extensive time and change occurred between Genesis 1:1, 1:2 and 1:3.

Exemplar creation interprets Genesis 1:3-31 as six consecutive 24-hour days in which God describes the Earth as having abundant plant and animal life, dry land and seas, in addition to a visible Sun, Moon and stars. On the sixth day as recorded in Genesis 1:26, God created man and woman, made the Garden of Eden and formed a small number of animals for the garden. The time for the six-day Genesis chronicle (when man was created) may be extended into the past beyond 4004 BC based on archaeological evidence and missing generations in the genealogies of Genesis Chapters 5 and 11. Although the majority of creation occurred prior to Genesis 1:3, God is not restricted from continuing to create in the universe. God is sovereign, eternal, powerful, intelligent, knowledgeable, holy, loving and just. Yahweh will not falsify evidence, lie, mislead or deceive.

Theories are developed to explain phenomena. Theory attempts to explain questions of origin, ultimate cause, operation and structure by taking an associated collection of data and inferring an explanation.

All theories of science in the public arena should be critically analyzed based on their strengths and weaknesses. Strong theories have large amounts of supporting evidence and little or no discrediting evidence. They have explanatory power and are able to rationally explain applicable physical data. On the other hand, weak theories must often develop tenuous explanations to discredit conflicting evidence and are often unable to logically explain relevant information.

As discussed in Chapter 2, the correctness of theory can be evaluated based on (1) the amount of specifically supportive evidence, (2) the amount of discrediting evidence, (3) failed or successful experiments, (4) the ability or inability to plausibly explain important phenomena within the realm of applicable theory and (5) failed or successful predictions.

Science is a search for dependable knowledge about the world based on physical evidence and its logical inferences. A dependence on verified empirical data is essential to effectively evaluate competing theories, philosophies and theologies claiming to explain our beginnings. Physical evidence must be a significant measuring stick.

In this chapter, comparative analysis is used to examine ideas of origin proposed by philosophical naturalism, traditional young-Earth creation theology and *exemplar* creation. Nineteen criteria are used to evaluate the explanatory power of these three ideas, and each is rated from zero to ten on how well it integrates applicable evidence. A rating of zero shows little explanatory power while a rating of ten demonstrates excellent explanatory power.

Comparative analysis is a powerful tool used by technology and business to evaluate competing products and ideas. It is easy to be attracted to one particular feature of a product and ignore obvious deficiencies. Analysis, narrowly confined to one topic, can overlook serious flaws in reasoning and evidence. Carefully structured debates often fail to discuss the full range of conflict among competing ideas. Comparative analysis is a method of rating multiple relevant criteria which define the performance of a product—in this case ideas about origins. A summary of this analysis is given in Table 13.3.

1. The Origin of Life on Earth

Chemical evolution proposes that the random combinations of amino acids, simple molecules and elemental materials over extensive time produced the first microscopic life. (Chapter 8 discusses this issue in more detail.) Proponents of chemical evolution directly appeal to missing information, unknown phenomena and undiscovered laws. Chemical evolution is strongly discredited by empirical science and clearly exemplifies the complete inadequacy of natural cause to create information and functional complexity. Chemical evolution cannot bridge the vast chasm between life and non-life. Philosophical naturalism and its ancillary ideas do not have a remotely logical mechanism for the first life and are rated zero.

The evidence against chemical evolution is overwhelming, and unless the scientific community is willing to accept this evidence any discussion of more complex issues is essentially futile. Without a willingness to recognize the limitations of natural cause and the legitimacy of design theory, it is pointless to discuss issues with less clarity. Chemical evolution by itself shows the bankruptcy of natural cause as an all-encompassing explanation of origins. As mentioned, scientists in general avoid use of the word impossible as it denotes finality and a strong sense of decisiveness. It almost sounds dogmatic. Regardless, we can say that based on the observed constraints of the universe, the known laws of chemistry and physics, mathematical probability, the severe limitations of natural cause, exhaustive scientific investigation and our knowledge of microscopic life, chemical evolution is impossible.

Traditional young-Earth creation theology proposes that God instantly created all life on the third, fifth and sixth days of creation. Although traditional young-Earth creation theology has a viable mechanism to explain life, this idea is downgraded to a rating of nine because of the time proposed.

Design theory and intelligent cause are necessary to rationally explain the information, complex functional systems, mathematically improbable structures and fine-tuned arrangements observed in single-celled life. A design agent of sufficient intelligence and power can create microscopic life. Intelligent cause is vastly superior to natural cause as a creative mechanism as evidenced by the accomplishments of human engineers, chemists and biologists in contrast to undirected phenomena.

Design theory has no time frame for the appearance of life on Earth and can accept evidence that life has existed for some time. Clearly identified microfossils dated around 2.9 billion years old have been found in a number of locations.[2] No fossil evidence of life forms older than 3.5 billion years has been found.[3] The carbon isotope analysis of 3.85 billion year old rocks (ratio of C12/C13 and fractionation effects) can be interpreted to support the existence of single-celled life at that time.[4]

Exemplar creation accepts the arguments and explanations of design theory given above. Proponents of this theology readily acknowledge the severe limitations of natural cause observed by empirical science. God created the first life on Earth in undefined time between Genesis 1:2 and 1:3. *Exemplar* creation does not conflict with empirical science in regard to time or the kind of life that first appeared on Earth, and is rated ten.

2. Minor Species Change

Minor species change in plants and animals is normally called microevolution and primarily consists of (1) random mutation, (2) natural selection, (3) species divergence, (4) hereditary variation, (5) genetic transfer and (6) adaptive change. (Chapter 9 discusses microevolution in more detail.)

Darwinian evolution attempts to use microevolution to support its claims. Minor species change has been confirmed by empirical science; however, small changes cannot be legitimately extrapolated to support Darwinian macroevolution. Slight changes cannot be extended to support the creation of new information, proteins, functions, features and dissimilar species. Because microevolution does not support Darwinian evolution, philosophical naturalism is rated zero.

Traditional young-Earth creation theology requires rapid microevolution to explain the variation found in dogs, cats, cows, horses and other animals. Each animal that went into Noah's ark represented one of its kind. After the flood, microevolution turned these kinds into the similar species, subspecies, breeds and varieties observed today over a period of approximately 4,350 years. Traditional young-Earth creation theology is downgraded to nine because of its problem with time and the rapid microevolution required.

Empirical science recognizes that undirected change (microevolution) could develop the many similar species, subspecies, breeds and varieties of plants and animals evident today. However, microevolution is limited in the amount of change it can produce and cannot be extrapolated to support Darwinian macroevolution. Extensive breeding experiments confirm the limitations of microevolution.

Exemplar creation can accept empirical science in regard to minor species change. This theology allows adequate time for minor species change to occur in undefined time between Genesis 1:2 and 1:3. *Exemplar* creation fully harmonizes with empirical science and is rated ten.

3. The Origin of Earth's Species

Biological evolution proposes that natural cause is responsible for all of Earth's species starting from microscopic life. Development of new species occurred by the hereditary accumulation of genetic material within populations from generation to generation over time. The primary mechanisms of biological evolution are random beneficial mutations within heritable DNA and natural selection. (Chapter 9 discusses biological evolution in more detail.)

Biological evolution is discredited by (1) the principle of irreducible complexity, (2) absent beneficial mutations, (3) a lack of beneficial protein isoforms in the human genome, (4) a lack of improved alleles, (5) the absence of neutral mutations and recessive deleterious mutations, (6) limited change in breeding experiments and (7) genetic entropy. In light of known evidence, random mutation and natural selection are not rational mechanisms to explain Earth's species. Since some fossils and evidence of similarity among species can be interpreted to support biological evolution, philosophical naturalism is upgraded to a rating of one.

Traditional young-Earth creation theology proposes God created plant life on the third day, birds and marine life on the fifth day and land animals on the sixth day. This theology proposes that all living and extinct species were instantaneously spoken into existence by God from nothing during three days in 4004 BC. Currently there are approximately 59,000 different vertebrates which include 3,802 mammal[5] species, 9,955 bird[6] species, 6,317 amphibian[7] species, 8,734 reptile[8] species and 30,200 fish species.[9] Insects are roughly estimated to exceed one million species.[10] The number of single-celled microscopic organisms such as eukaryotes (protozoa) and prokaryotes (bacteria, cyanobacteria and archaea) are roughly estimated to exceed one million species.[11] In addition, plant species are estimated at 400,000.[12] Theology that demands the creation of all these species from nothing in three days pushes the envelope of credibility—at least for most people. Although traditional young-Earth creation theology has a viable mechanism to explain species, this idea is downgraded to a rating of seven because of the time proposed.

Design theory and intelligent cause are necessary to logically explain the fossil record, similarity and the multitude of Earth's species. The fossil record contains many gaps and jumps, best explained by design theory. Many of Earth's species have similar DNA, molecules, proteins, genes, pseudogenes, functions and features. In living things, similarity is evidence of a common designer or a common design philosophy. Earth's species represent massive product expansion over extensive time with evidence of original design, design by menu selection and design iteration improvement. Most species were introduced during a series of expansion events, typified by the Cambrian explosion. The Cambrian Period marks the beginning of nearly all modern animal phyla.

Exemplar creation accepts the arguments and explanations of design theory given above. Proponents of this theology readily accept the severe limitations of random mutation and natural selection observed by empirical science. God is an ageless, intelligent, powerful designer who planned, integrated and created all of Earth's original species over an undefined period of time between Genesis 1:2 and 1:3. *Exemplar* creation allows time during this period for species extinction to occur, species to be preserved in the fossil record and new species to be created. Similarity among species is due to a common designer. *Exemplar* creation harmonizes with empirical science, has a plausible mechanism and is rated ten.

4. The Origin of Galaxies

The majority of visible stars reside in galaxies rather than being evenly distributed throughout the universe. As part of Hubble Deep Field, in 1999 NASA estimated the number of galaxies in the visible universe at 125 billion.[13] The grouping of most stars into galaxies does not appear to be a natural phenomenon. Based on the known laws of chemistry and physics, no plausible natural mechanism exists to cause a homogeneous universe of stars to self-form galaxies. As stated by one astronomy reference book, "The specifics of [natural] galaxy formation are unknown though there is no shortage of theories."[14]

Evolutionary explanations propose galaxies began forming as a consequence of density fluctuations and self-gravity in expanding primordial hydrogen and helium gases around 300,000 years following the Big Bang. Hypothetically, density fluctuations grew in strength under the influence of self-gravity. Clouds developed that subsequently fragmented into stars to make up galaxies.[15] However, the density fluctuations proposed are strictly speculative. We can just as readily hypothesize that any significant density fluctuations diffused and disappeared after 300,000 years. It is more logical to believe randomness would increase rather than decrease.

According to theory, galaxies were primarily formed by self-gravity within gigantic gaseous clouds. One major problem with this idea is that gravity has no "off switch." Molecules and atoms attracted together by self-gravity into a gaseous cloud tens of thousands of light years in diameter would continue to collapse. All matter would accelerate and converge at increasing velocities toward the center of mass based on the inverse square law. One enormous ball of matter would be the end result. No known natural phenomena exist to turn off gravity and stop inward radial movement during this alleged galaxy formation process.

Hypothetically, irregularities in gaseous cloud shape caused angular rotation to increase (due the conservation of angular momentum) as the cloud collapsed. While some angular rotation is possible, radial velocity toward the center of mass would be the major component of particle movement and would ensure cloud collapse. There are many elliptical galaxies that are almost round (e.g. M87, M32 and NGC 4881) without significant coordinated rotation, yet they exist. Formation by self-gravity would have caused these structures to completely collapse.

Spiral galaxies such as NGC 5194, NGC 4603, NGC 1288 and Messier 101 look like slowly rotating giant pinwheels. Their rotation represents an enormous amount of angular momentum which undoubtedly did not originate from a collapsing gaseous cloud. Star movement in a spiral galaxy is a fine-tuned balance between gravitational attraction and centrifugal force. Gravity attempts to collapse the galaxy, while centrifugal force attempts to scatter the stars apart. Massive black holes are hypothesized to exist in the center of spiral galaxies to provide enough gravity to keep them together. Other unusual features of spiral galaxies include the arrangement of stars to form distinct spiral arms and their flattened shape. The natural shape of an object formed by gravitational collapse is spherical (e.g. stars, globular star clusters and slightly elliptical galaxies) rather than a disk. When all factors are considered, there is no plausible natural explanation for the angular momentum, balance of forces, distinct spiral arms and flattened shape of spiral galaxies.

Ring galaxies and polar ring galaxies are truly unique and pose a severe challenge to natural cause. The Sombrero Galaxy (shown on the cover) is about 50,000 light years in diameter and has a large, bright spherical core within a disk of stars surrounded by a uniform dust ring. Hoag's Object is a ring galaxy with a bright yellow core surrounded by a ring of hot blue stars having an outer diameter of around 121,000 light years. Polar-ring galaxies are represented by NGC 4650A and contain an elongated central core surrounded by a ring of stars. (Photographs of these galaxies can be found on the Internet.) The precise circularity, concentricity, uniform thickness and homogeneous distribution of material in these ring structures cannot be rationally explained by natural cause.

In summary, cosmological evolution proposes the formation of all galaxies by natural cause. Hypothetical density fluctuations and self-gravity among hydrogen and helium gases cannot explain the many different galaxy types and their unique features. Proponents of natural galaxy formation directly appeal to unknown phenomena and indirectly to natural metaphysical cause. Time is not a magic wand capable of curing the severe limitation of natural cause. Philosophical naturalism and supporting cosmological evolution do not have a plausible mechanism for galaxy formation and are rated zero.

Traditional young-Earth creation theology proposes God made all galaxies on the fourth day of creation. Theology that demands the creation of 125 billion galaxies in one day pushes the envelope of believability. If all galaxies were created in one day, then all stars and all galaxies should be of the same age—clearly not the case. The argument that God made the stars and galaxies with different apparent ages is not a credible explanation. Apparent age makes God a great deceiver and puts Bible believers in an indefensible position. Although traditional young-Earth creation theology has a viable mechanism to explain the origin of galaxies, this idea is downgraded to a rating of seven because of its conflict with time.

Design theory and intelligent cause are necessary to plausibly explain the statistically improbable collection of most stars into galaxies. A design agent is necessary to explain spiral galaxies and (1) their angular momentum, (2) the fine balance between gravity and centrifugal force, (3) their segregated spiral arms and (4) their flattened shape. A design agent is necessary to explain ring and polar ring galaxies. Self-gravity cannot plausibly explain galaxy formation except as a temporary state prior to the total collapse of all stars and matter into the galactic core.

Exemplar creation accepts the arguments and explanations of design theory given above. Proponents of this theology can accept empirical science and the severe limitations of self-gravity and undirected phenomena. God is an ageless, timeless, intelligent, powerful designer who created the initial galaxies during the undefined past of Genesis 1:1 and may have created additional ones prior to Genesis 1:3. There is no requirement for all galaxies to have been created at once, and time is allowed for limited undirected change to occur. *Exemplar* creation does not conflict with empirical science in regard to time or the age of galaxies, has a logical mechanism and is rated ten.

5. Origin of the First Stars

Philosophical naturalism requires the formation of all stars by natural cause. Big Bang theory proposes the universe started as a quantum fluctuation that created an enormous amount of high temperature energy, in turn spontaneously creating all basic matter. Atoms of hydrogen and helium were the first elements to form as the universe expanded outward and cooled. Hypothetically, the first stars nucleated out from hot hydrogen and helium gases.

Stars hypothetically form in the present by the self-gravitational collapse of interstellar dust and gases observed in giant clouds[16] such as the Orion Nebula (M42), Lagoon Nebula (M8) or M33. Hypothetically, during collapse a disk forms around the accreting star as a consequence of angular momentum in the initial cloud.[17]

Astronomers have observed dense globules, protostars, protoplanetary disks, stars surrounded by gas and dust, nebulae, stellar dust and alleged stellar nurseries. Yet despite a consensus of scientific opinion that new stars are currently evolving, no birth of a star has been observed in detail to positively conclude its origin is strictly due to natural cause (similar to maggots appearing in raw meat, leading to the theory of spontaneous generation). Stars appear to be forming, albeit at a very slow rate, but the details of their formation are not known to absolutely determine ultimate cause.

Hydrogen and helium are the two primary gaseous elements in stars; however, their natural ability to diffuse contradicts speculative ideas that self-gravity can form a star. Our Sun is estimated to be composed of 71% hydrogen and 27% helium by weight based on spectral analysis of its surface gases.[18] As gases, hydrogen and helium readily diffuse. For example, a solid is inherently constrained to its given shape. A liquid inherently settles to the bottom of its container under the influence of gravity. A gas such as hydrogen will inherently expand to completely fill an empty container in *defiance* of strong Earth gravity.[19] Rather than accumulate, gas innately disperses. As the temperature of a gas increases, so does the movement of its molecules and atoms, in turn increasing its ability to diffuse.

Self-gravity within a gas is a real force, but essentially insignificant. Self-gravity between hydrogen molecules and helium atoms is extremely weak and particle motion (the random movement of molecules and atoms as they collide elastically with each other) is sufficient to prevent agglomeration. Particle motion rather than self-gravity defines the behavior of gases. Even when hydrogen and helium are compressed to high pressures (i.e. 10,000 psi), they do not agglomerate because of self-gravity.

The temperature effect on gas diffusion strongly contradicts evolutionary ideas of star formation. Even when cooled, a gas will still diffuse. This is especially true of hydrogen and helium, which remain a gas at very low temperatures. The normal boiling point of hydrogen under one atmosphere of pressure is 20.28° K and of helium 4.22° K,[20] temperatures near absolute zero. As liquids, hydrogen and helium will agglomerate, while as a gas they readily diffuse. Big Bang theory proposes that the temperature of the universe dropped to around 3,000° K after 300,000 years of expansion,[21] at which time stars began forming. Hydrogen and helium at this high temperature would never agglomerate by self-gravity. These hot conditions are the exact opposite of what is needed for natural star formation. Extreme cold approaching absolute zero is more favorable. As confirmed by empirical science, the natural characteristics of gases strongly discredit speculative ideas of star formation by self-gravity.

A study of planets and moons in our solar system further illustrates the difficulty of star formation by natural cause. A planet must have a mass equivalent to Mars or larger to have any appreciable hydrogen atmosphere (Mars has essentially no atmosphere). Hydrogen and helium have a strong tendency to rise to the upper atmosphere of a planet where they are lost in space. Weak gravity, heavier gases, solar heating, solar wind and diffusion all contribute to hydrogen loss from a planet's surface. Liquid hydrogen has a density of 0.071 g/ml and liquid helium has a density of 0.125 g/ml under one atmosphere of pressure at their boiling point.[22] If a planet the diameter of Mars were formed out of 70% liquid hydrogen and 30% liquid helium by weight (the approximate composition of the Sun), it would have a density of approximately 0.087 g/cm^3. This contrasts with the current density of Mars at 3.940 g/cm^3. The resultant planet would be 43 times lighter than Mars, with a proportional drop in gravitational attraction. Because of its low gravity, this planet would disappear from the loss of hydrogen and helium rather than growing by attracting additional gases. A collection of hydrogen and helium the size of Mars within our solar system would slowly diffuse into the vacuum of space, rather than grow to become a star.

In support of evolution, it can be hypothesized that star formation begins when interstellar dust and debris collect together to form a dense planetoid capable of attracting hydrogen and helium. Over time enough hydrogen and helium collect to form a star. Unfortunately, this puts the cart before the horse, because in the early universe no dust or debris existed. The first stars were formed out of pure hydrogen and helium—heavier elements were allegedly created later through nucleosynthesis. The first stars and subsequent generations went through their life cycles and exploded to create the dust and debris now observed in space.

Binary stars pose an added difficulty for stellar evolution. A binary star system consists of a pair of stars that orbit about their common center of mass and are held closely together by mutual gravitational attraction. The typical average separation between two binary stars is 10 to 20 astronomical units.[23] (As a comparison, the diameter of Jupiter's orbit is approximately 10 astronomical units.) More than half of all visible stars are actually binary star systems. Where an accurate census of a particular portion of space has been made, up to 90% of stars can be in a binary system.[24] Astronomers have identified more than 100,000 binary star systems.[25] The natural formation of two stars in a binary system from a single collapsing gaseous disk is even more implausible than a single star forming by this method. In short, the large percentage of binary stars observed today is highly improbable.

Nuclear fusion in stars is alleged to be a natural phenomenon. Fusion is the apparent process that creates heat and light within a star, even though the details are not fully understood. For the last fifty years scientists have strongly pursued the development of sustained nuclear fusion, a highly desirable energy source. They have used a variety of approaches in their attempt to attain sustained fusion, including conditions of high pressure and temperature typical in stars. Thus far, the brightest minds, best technology and billions of dollars have been unsuccessful. If their efforts have proven anything, it is that sustained nuclear fusion which generates excess energy appears to be unnatural and a highly elusive process.

In summary, theory for the evolutionary formation of the first stars is strongly contradicted by empirical science. Stars evolve in the sense they burn up their fuel, go through various stages, grow old and eventually die. Stars cannot evolve in the sense they naturally form from hydrogen and helium in the vacuum of space. The phenomena of diffusion, random particle motion, elastic collision and the Ideal Gas Law define the behavior of hydrogen and helium gases. Self-gravity among gas particles is an insignificant force and cannot cause agglomeration, regardless of gaseous cloud size. Philosophical naturalism and its ancillary ideas do not have a plausible mechanism to explain the first stars and are rated zero.

Traditional young-Earth creation theology proposes all stars were created by God on the fourth day of the Genesis account. As mentioned, the Milky Way, considered an average sized galaxy, contains an estimated 400 billion visible stars.[26, 27] The number of visible galaxies is estimated at 125 billion[28] which equates to 5×10^{22} stars in the universe. Dr. Simon Driver quotes an estimate of 7×10^{22} visible stars in the universe obtained by astronomers at the Australian National University.[29] Regardless of the exact number, theology demanding the creation of all stars in one day pushes the envelope of believability. If created in one day, then all stars should be of the same age. Most stars are of different ages and clearly older than 6,000 years. Although traditional young-Earth creation theology has a viable mechanism to explain the origin of stars, this idea is downgraded to a rating of seven because of the time proposed.

Design theory and intelligent cause are necessary to plausibly explain the formation and complex function of stars. Stars may be viewed as nuclear fusion devices for converting matter into energy designed to operate unattended for millions of years. Self-gravity is not a rational explanation for the formation of the first stars out of helium and hydrogen gases at a temperature of 3,000° K. No natural mechanism has been identified capable of compressing unrestrained gases into a dense mass in space. The observed collection of gases, dust and debris into a circumstellar disk in space, believed to be a part of the star formation process, can be hypothesized as due to a non-natural, shaped, high-gravity force. The large percentage of binary stars in the universe is highly improbable. Given time, a design agent of sufficient intelligence and power could build the stars we observe.

Exemplar creation readily accepts the arguments and explanations of design theory given above. Proponents of this theology can accept empirical science and the severe limitations of self-gravity and undirected phenomena. God is an ageless, timeless, intelligent, powerful designer who created the first stars during the undefined past of Genesis 1:1 and additional ones prior to Genesis 1:3. God may be presently creating stars. There is no requirement for all stars to have been created at once. Time is allowed for stars to age and undergo undirected change. *Exemplar* creation does not conflict with empirical science in regard to time or the age of stars, has a plausible mechanism and is rated ten.

6. The Origin of our Solar System

Philosophical naturalism requires the formation of our solar system by natural cause. The solar nebular hypothesis proposes that a large irregular cloud of debris, dust and gases in space collapsed into a disk-shaped cloud, rotating around a central point. Hypothetically, a cloud of this nature provided the reservoir of mass and momentum, and a likely environment for planet formation.[30] Because of the irregular shape of the initial cloud, it gained rotational speed as it collapsed (due to the conservation of angular momentum) and allegedly flattened. The majority of mass collected at the center, forming the Sun, while a small amount of material agglomerated within the plane of the rotating disk to form the planets in their present-day orbits.

The solar nebular hypothesis proposes that an irregular cloud collapsed by self-gravity into a disk. If a cloud collapsed by self-gravity as alleged, it would result in a smaller, faster rotating, *spherical* cloud. Collapse would not cause flattening. No known natural phenomena exist capable of causing a cloud in the vacuum of interstellar space to form a disk. While two simultaneous explosions on either side of the cloud might cause flattening, a natural event of this type is highly unlikely. The solar nebular hypothesis cannot credibly explain a circumstellar disk.

Table 13.1 – Mass and Momentum in the Solar System
(All data are in Earth units)

Body	Orbit Period	Orbit Radius	Mass	Angular Momentum
Mercury	0.241	0.387	0.055	0.034
Venus	0.615	0.723	0.815	0.693
Earth	1.000	1.000	1.000	1.000
Mars	1.881	1.524	0.108	0.133
Jupiter	11.862	5.203	318.098	725.877
Saturn	29.457	9.555	95.095	294.717
Uranus	84.010	19.191	14.542	63.753
Neptune	164.789	30.109	17.156	94.378
Pluto	248.593	39.529	0.002	0.013
Sun			332998.493	35.001

Notes:
(1) The planetary orbit radius is an average.
(2) The Sun is considered to be a homogeneous solid mass that rotates once every 30 days.
(3) In these calculations, the Sun represents 99.87% of the mass and 3% of the angular momentum in the solar system. This is the maximum momentum possible.
(4) Other sources typically give less than 2% for the Sun's angular momentum. This is based on a model of the Sun having a light outer shell with a dense core that rotates at a slower rate.
(5) All values are in Earth units which is 1.000. Values for orbit period, orbit radius, mass and angular momentum are given in relation to Earth as a standard.

The solar nebular hypothesis cannot plausibly explain how the planets ended up with most of the angular momentum and the Sun ended up with most of the mass. Missing angular momentum in the Sun is a major problem for this hypothesis.[31] Rotation of the Sun about its axis and the orbits of the planets around the Sun represent angular momentum. Since the Sun and all the planets move in the same direction, their momentum is additive. When considering only the planets and the Sun (excluding the Kuiper and asteroid belts), the planets have approximately 97% of the angular momentum in the solar system and the Sun has 3% or less. On the other hand, planets have 0.13% of the mass and the Sun has 99.87% as shown in Table 13.1. The Sun has the great majority of mass and should have the great majority of momentum in the solar system. The Sun should also be rotating significantly faster than its current speed. Missing angular momentum in the Sun strongly discredits the solar nebular hypothesis.

To understand the momentum problem further, assume a rotating disk formed as proposed by the solar nebular hypothesis. The diameter of this disk is hypothesized to have been substantially larger than our present solar system. All mass in the disk rotated about a central point. *Somehow* a small amount (0.13%) of debris, dust and gases collected together to form the planets in their present orbits. *Somehow* the remaining debris, dust and gases (99.87%) separated from the planets and lost essentially all their angular momentum. *Somehow* self-gravity then caused them to collapse and form the Sun. According to the known laws of physics, this cannot naturally occur. The law of the conservation of momentum by itself discredits the solar nebular hypothesis.

Self-gravity is not a plausible mechanism for planetary formation. Interstellar debris, dust and gases have been observed in the cosmos and their presence is well documented. However, self-gravity among rocks is extremely weak and when they accidentally bump against each other do not clump together by gravitational attraction. Any attraction is easily overcome by the resultant velocity following collision. For example, billions of years and adequate material have not created a new planet in the asteroid belt.

In summary, the solar nebular hypothesis proposes that our solar system was formed from a collapsing cloud of interstellar debris, dust and gases. This hypothesis is not supported by our knowledge of chemistry and physics. Proponents of natural solar system formation indirectly appeal to unknown phenomena and natural metaphysical cause. Philosophical naturalism and its supporting ideas do not have a remotely plausible mechanism for solar system formation and are rated zero.

Traditional young-Earth creation theology proposes our solar system was created during the first four days of the Genesis account around 4004 BC. A special light existed for the first three days until the Sun and Moon were created on the fourth day. At that point, the special light was removed. The age of the Earth in this explanation strongly conflicts with radiometric dating, radiocarbon dating, ice core dating and tree-ring chronologies. Although traditional young-Earth creation theology has a viable mechanism to explain the origin of our solar system, this idea is downgraded to a rating of seven because of its conflict with time.

Design theory and intelligent cause are necessary to explain the complex function of the Sun, the orbiting planets and the retrograde (reverse) rotation of Venus. Whatever the starting materials, a design agent formed them into the observed planets and Sun. Most momentum was given to the planets, yet most mass was concentrated in the Sun. We can hypothesize a design agent gave Venus its retrograde rotation as a reminder of his presence. Given time, a design agent of sufficient intelligence and power could build the solar system we observe.

Exemplar creation accepts the arguments and explanations of design theory given above. Proponents of this theology can readily accept empirical science and the severe limitations of self-gravity and undirected phenomena. God created our solar system in the undefined past of Genesis 1:1. Time is allowed prior to Genesis 1:3 for undirected change to occur. *Exemplar* creation does not conflict with empirical science in regard to the age of the Moon and Earth, has a plausible mechanism and is rated ten.

7. A Unique Earth for Life and Humanity

The favorable conditions found on Earth, agreeable to life, are unique in our solar system. Earth orbits a Sun that is relatively stable and at just the right distance away. If too far from the Sun, the Earth would be freezing cold while if too close it would be insufferably hot. Earth's twenty-four hour rotation period is just right for plants and animals. Tilt of our planet's axis at 23.5° provides four seasons, more variable weather and enlarges its habitable zone. Earth's strong magnetic field protects its inhabitants from harmful solar storms, and field orientation in relation to the Sun maximizes this protective effect. Gravity on Earth is strong enough to hold its atmosphere to prevent loss into space. Ozone in the upper atmosphere filters out essentially all harmful radiation with a wavelength of 300 nanometers or less while being transparent to the majority of beneficial radiation. The Moon is just the right size and distance to provide good tidal action.

Earth's water and atmosphere are also favorable to life. Earth has an abundance of surface water not found on any other planet in the solar system. Water is critical to our climate and weather systems. Earth's unique atmosphere is composed of 78 percent nitrogen and 21 percent oxygen. If the percentage of oxygen were increased substantially, fires would burn uncontrollably and oxidation would be accelerated. If the atmosphere had less oxygen, breathing would be more difficult and higher elevations might be uninhabitable. Unlike other planets, there are no significant harmful gases in Earth's atmosphere.

One key component of Earth's unique biosphere is the nitrogen cycle and the nitrogenase enzyme. Nitrogen exists primarily as a diatomic molecule (N_2) in the atmosphere and is virtually inert because of the strong triple bond between the two atoms. Nitrogenase enzymes, whose primary job is to break apart the nitrogen molecule, are found in certain bacteria on the roots of plants. These very large, complex enzymes are capable of breaking the strong nitrogen bond and combining one nitrogen atom with three hydrogen atoms to form ammonia (NH_3). This process is called nitrogen fixation[32] and is responsible for roughly half of the global input of newly fixed nitrogen.[33] As an added complicating factor, nitrogenase is irreversibly destroyed by exposure to oxygen. To prevent its destruction, plants and bacteria fully compartmentalize the enzyme to protect it from damaging oxygen in the atmosphere, while at the same time giving it access to atmospheric nitrogen.[34]

Earth's unique conditions include mineral and fossil fuel reserves favorable to human technology. Metals such as iron, copper, aluminum, silver, lead, gold, platinum, cobalt, mercury, molybdenum, tungsten, nickel and uranium are conveniently located in concentrated deposits near the surface of the Earth's crust. Sufficient concentrations of these metals exist to make extraction and refining possible, forming the basis of human technology. In addition, Earth's abundant fossil fuel reserves in the form of oil, coal and gas are the primary energy sources for industry and manufacturing.

There are literally hundreds of things—high temperatures, low pressures, poison gases, atmospheric hydrocarbons, contaminating elements, an unstable Sun, a missing ozone layer, low levels of oxygen, no nitrogen, a weak magnetic field and scarce surface water—that could make Earth uninhabitable. Based on mathematical analysis, the favorable conditions that exist on Earth are improbable. The vast majority of features that make Earth uniquely favorable to life are not found on other planets and moons in our solar system.

Philosophical naturalism requires the formation of Earth by natural cause, even though the wide range and number of attributes that interact together to create our planet's habitat are highly implausible. When all criteria are considered, proponents of natural cause directly appeal to a long list of unlikely events to explain Earth's unique conditions. Philosophical naturalism and its ancillary ideas lack a credible explanation and are rated zero.

Traditional young-Earth creation theology proposes God created the Earth in six days. Although this theology has a viable mechanism to explain Earth's habitat, this idea is downgraded to a rating of nine because of the time proposed for the process.

Design theory and intelligent cause are necessary to plausibly explain the fine-tuning and statistically improbable conditions found on Earth. Based on the number of unique characteristics favorable to life, Earth strongly exhibits the attributes of a designed system. Hostile conditions verified on the other seven planets, eighteen spherical moons and one planetoid (Pluto) in our solar system illustrate Earth's uniqueness. Given time, a design agent of sufficient intelligence and power could build the favorable conditions for life and human technology observed on Earth.

Exemplar creation accepts the arguments and explanations of design theory given above. Proponents of this theology can accept empirical science and the severe limitations of random change and undirected phenomena. God created the Earth and fine-tuned it as a habitat for life in undefined time prior to Genesis 1:3. *Exemplar* creation does not conflict with empirical science in regard to time and the age of the Earth, has a completely adequate mechanism and is rated ten.

8. Distant Starlight

Starlight that has been traveling through space for millions of years is just now reaching the Earth. This light contains specific time related events which include supernova, Cepheid star pulsations and normal star intensity fluctuations. Starlight is a major problem for those who believe the Earth and stars were created in 4004 BC—more time is needed. As discussed in Chapter 4, there is no rational explanation for distant starlight in the context of a six thousand year old universe. God would not falsify the events recorded in starlight. Falsification makes God a great deceiver and places Bible believers in an indefensible position in trying to reconcile His word and physical evidence. Traditional young-Earth creation theology is rated zero because of its severe conflict with empirical science and distant starlight.

Based on distant starlight, the case against a young age for the Earth and universe is overwhelming. None of the books—*The Revised & Expanded Answers Book,*[35] *Refuting Compromise*[36] and *The New Answers Book*[37]— in support of traditional young-Earth creation theology have a reasonable explanation for the starlight and time problem. The evidence of empirical science shows that a young age for the stars is a scientifically bankrupt idea. Unless proponents of young-Earth creation theology are willing to accept the legitimacy of starlight data, it is essentially futile to discuss other evidence indicative of significant age.

Evidence from astronomy permits the logical explanation of events recorded in starlight. Some phenomena recorded in starlight millions of years old are just now reaching Earth. Astronomers use the inverse square law and other means to measure the distance to stars and galaxies. When distance is known, the speed of light is a dependable measuring stick to date past events. Philosophical naturalism and its ancillary ideas harmonize with this evidence and are rated ten.

Exemplar creation accepts the empirical evidence of astronomy showing the universe is incredibly large and very old. Proponents of this theology can readily accept distant starlight that is millions of years old and a universe that is significantly older than 4004 BC. *Exemplar* creation easily harmonizes with empirical science in regard to starlight and time, and is rated ten.

9. Moon and Earth Meteor Craters

The Moon and Earth have been subjected to extensive meteor impacts in the past. Today, the surface of the Moon is still heavily marked with impact craters. Scientists have counted 8,596 craters more than 1 km (0.62 miles) in diameter on the its surface.[38] Based on the diameter of the Moon in relation to the Earth, a proportionally greater number, estimated at 115,500 impact craters one kilometer in diameter or larger, have occurred on Earth in the past. Weathering effects have removed evidence all craters except for 172 known sites. (Chapter 4 discusses meteor impact craters in more detail.)

Traditional young-Earth creation theology does not allow enough time for the occurrence of Moon and Earth impact craters. A recent Earth has inadequate time for 115,500 meteors to strike it without destroying most life. In addition, there is not enough time for natural processes to remove evidence of these craters. God would not falsify impact craters on the Moon nor intentionally remove evidence of them on Earth. Falsification makes God a great deceiver and places Bible believers in an indefensible position in trying to reconcile the Bible and physical evidence. Traditional young-Earth creation theology is rated zero because of its inability to plausibly explain meteor impact evidence within the time frame proposed.

Evidence from astronomy and geology show that the Moon and Earth have been subjected to extensive meteor impacts in the distant past. The absence of Earth impact craters is attributed to weathering and other natural processes over significant time. Theories of natural cause easily explain Moon craters and the absence of Earth impact craters when sufficient time is allowed. Philosophical naturalism and its ancillary ideas agree with this information and are rated ten.

Exemplar creation proposes that the Moon and Earth were created by God in the undefined past of Genesis 1:1. Proponents of this theology can accept impact craters on the Moon and that a proportional number occurred on Earth in the distant past. They can accept the removal of essentially all of Earth's impact craters by weathering over extensive time prior to Genesis 1:3. Noah's flood did not remove evidence of these craters. *Exemplar* creation easily harmonizes with the evidence of empirical science and is rated ten.

10. Radiometric Dating of Igneous Rocks

The age of igneous rocks, formed by the cooling of once molten material, can be determined using radiometric dating techniques. Dating techniques extrapolate backward in time the radiometric decay of specific isotopes trapped in a rock upon solidification. Some of the oldest rocks on Earth have been radiometric dated to be 3.8 billion years old. (Chapter 4 discusses radiometric dating in more detail.)

While radiometric decay and dating is a well-understood process, it is not always precise as clearly pointed out by young-Earth creationists.

- The isochron radiometric dating of rock samples from the Cardenas basalt of the Grand Canyon gave radiometric ages of 1.07 to 1.34 billion years.[39]
- Dated rock samples using four different radiometric techniques from Beartooth Mountain gave an age range of 1,520 to 2,886 million years.[40]
- Dated rock samples using four different radiometric techniques from Bass Rapids in the Grand Canyon gave an age range of 665 to 1,379 million years.[41]

These results show significant variation in the dating process, which critics use to attack the validity of radiometric dating. Regardless of the data spread, this evidence shows the Earth is substantially older than six thousand years.

God would not falsify unstable isotope decay in rocks. Falsification makes God a great deceiver and places Bible believers in an indefensible position in trying to reconcile His word and physical evidence. Traditional young-Earth creation theology strongly conflicts with empirical science in regard to unstable isotope decay and radiometric dating evidence, and is rated zero.

The evidence of radiometric dating supports an old Earth and Moon. Isotope decay is a consistent, natural phenomenon. The precise measurement of isotope decay rates and their extrapolation into the past can be used to measure the age of igneous rocks, and indicate the Earth is approximately 4.6 billion years old. Philosophical naturalism and its ancillary ideas readily harmonize with evidence and are rated ten.

Exemplar creation proposes that the Earth and our solar system were created by God in the undefined past of Genesis 1:1. Proponents of this theology can readily accept Earth rocks, Moon rocks and meteor rocks as being significantly older than 4004 BC. They can accept a solar system that is approximately 4.6 billion years old. Time is allowed previous to Genesis 1:3 for unstable isotope decay to occur. *Exemplar* creation can harmonize with the evidence of empirical science regarding unstable isotope decay and the radiometric dating of igneous rocks, and is rated ten.

11. Carbon-14 Dating of Organic Material

Radiocarbon dating is a useful scientific tool, and although is has its shortcomings and limitations, can be used to date once living animals and plants. It is possible to accurately date organic material up to 10,000 years old. Between 10,000 and 30,000 years, radiocarbon dating becomes increasingly uncertain because of the rapidly decreasing percentage of carbon-14 in organic remains and accompanying sample deterioration. We can be quite skeptical of any radiocarbon date beyond 30,000 years. (Chapters 4 and 10 discuss carbon-14 dating in more detail.)

Scientists who support traditional young-Earth creation theology attack radiocarbon dating by pointing out incorrect dates. For example, fossilized wood samples from the Jurassic period (200-145 million years ago) gave a radiocarbon date of 20,700 to 28,820 BP.[42] This wood age is much too young. Ten samples of coal from major coalfields across the United States, believed to have been formed millions of years ago were radiocarbon dated at approximately 50,000 years.[43] Five diamonds, theorized to have been formed millions of years ago, suggested a radiocarbon date of about 58,000 years.[44] The incorrect fossilized wood and coal dates are probably due to sample contamination, and the diamond dates can be attributed to inherent error in the AMS dating process when measuring old samples that are small and contain little residual carbon-14.

Traditional young-Earth creation theology has no good explanation for radiocarbon dates cross checked for accuracy against annual physical events. For example, Dr. C. W. Ferguson built a radiocarbon calibration curve based on the growth of bristlecone pine trees in the White Mountains of California going back almost 8,000 years. The IntCal04 radiocarbon calibration curve is based on a tree ring series that goes back 12,410 calendar years. A correlation curve spanning 29,100 calendar years was developed by measuring the radiocarbon age of organic material in alternating layers of white blooms and sediments in Lake Suigetsu, Japan. This data is in direct conflict with belief that Earth's biosphere was created in 4004 BC. God would not falsify this evidence. Traditional young-Earth creation theology strongly conflicts with empirical science in regard to radiocarbon dating and is rated zero.

Our knowledge about the chemistry and physics of carbon-14 permits the radiocarbon dating of organic material. The creation of carbon-14 in the atmosphere and its accumulation in plants and animals is a known phenomenon. The normal decay of carbon-14 in plant and animal tissues after death is understood. Radiocarbon dating shows that life has existed on Earth for 30,000 years and more. Philosophical naturalism and its ancillary ideas readily accept this evidence and are rated ten.

Exemplar creation proposes that Earth's biosphere was created by God in undefined time before Genesis 1:3. Proponents of this theology can accept the existence of plants and animals on Earth significantly prior to the creation of Adam and Eve. They can accept carbon-14 measurements and dendrochronology which permit building a calibration curve going back 12,410 years. They can accept the Lake Suigetsu, Japan sediment layers which allow making a carbon-14 measurement and sediment layer correlation curve spanning 29,100 calendar years. However, they can be skeptical of the accuracy of radiocarbon dated samples from 10,000 to 30,000 years old and beyond that cannot be checked against annual physical events. *Exemplar* creation has no inherent conflict with radiocarbon dating and is rated ten.

12. Ice Core Dating

Numerous ice cores drilled in Greenland and Antarctica represent physical evidence of more than 100,000 years of Earth history. In some instances, the ice is more than 3,000 meters (9,842 ft) thick. Annual dust layers and oxygen isotope annual temperature cycles easily correlate and show that each layer represents one year. (Chapter 4 discusses ice cores in more detail.)

Traditional young-Earth creation theology has no rational explanation for ice cores, and the number of years they represent directly conflicts with belief in a 4004 BC creation date. God would not falsify this evidence to intentionally mislead Bible believers. Traditional young-Earth creation theology is rated zero because of its disagreement with the empirical evidence of Greenland and Antarctica ice cores.

Ice cores permit the dating of annual snow layers, accumulated and compacted over thousands of years to great thicknesses. The counting of these layers by different methods indicates they represent more than 100,000 years of time. Philosophical naturalism and its ancillary ideas easily agree with this evidence and are rated ten.

Exemplar creation proposes that the Earth was created by God in Genesis 1:1. Proponents of this theology can accept ice cores as irrefutable physical evidence the Earth is more than 100,000 years old. The undefined period between Genesis 1:2 and 1:3 allows time for snow to accumulate in Greenland and Antarctica to great thicknesses. *Exemplar* creation can harmonize with the evidence of ice cores and is rated ten.

13. Antarctic Coal and Fossils

Coal and fossils have been found in Antarctica representative of plants and animals that lived before the present cold conditions. Due to the high altitude location of some coal deposits and animal fossils, they could not have been transported to these locations by flood waters without leaving sediment evidence within the ice. Antarctic ice cores show no evidence of a flood during the past 100,000 years and confirm the continent's current cold state during those years. (Chapter 4 discusses coal and fossils in more detail.)

Traditional young-Earth creation theology has no good explanation for Antarctic coal and fossils. Theology that proposes an Earth covering global flood, accompanied by catastrophic plate tectonics (which includes the rapid movement of Antarctica from a warmer region to its present location), is contradicted by this evidence. Antarctica has been in its present location with a cold climate for 100,000 years and more. Traditional young-Earth creation theology is rated zero because of its conflict with physical evidence and lack of a plausible explanation.

Geological theories explain the formation of Antarctic coal and fossils by natural cause. Deposits of coal and fossilized animal remains require that a much warmer climate existed on the continent in the distant past. Movement due to continental drift is theorized to have relocated Antarctica from near the equator to its present position over millions of years. Philosophical naturalism and its ancillary ideas harmonize with the evidence of coal and fossils in Antarctica, and are rated ten.

Exemplar creation proposes that Earth's continents and its biosphere were created by God in undefined time prior to Genesis 1:3. Time is allowed for Antarctic coal and fossils to form under much warmer climatic conditions than exist today. Time is allowed for the drifting of Antarctica to its present location. *Exemplar* creation has no conflict with related empirical evidence and is rated ten.

14. Solar System Age Phenomena

Planetary impact craters and planetary cooling are two time-related phenomena. Some time-related phenomena in our solar system point to an old age for its planets and moons while others point to recent events. In addition, some phenomena do not have a plausible natural explanation.

An age of 4.6 billion years for the solar system suggests the occurrence of several long term natural phenomena. Evidence indicates the eight planets and their eighteen spherical moons were once giant balls of molten material. In their molten state, self-gravity formed them into smooth spheres. They then began to cool and harden, and have been doing so for 4.6 billion years. Small planets and moons should be completely solidified, essentially have no global magnetic field, no volcanic activity and no significant geological activity (earthquakes or crustal movement). During this time, all planets and moons were hit by thousands of meteors over millions of years. Planets and moons without an atmosphere should show evidence of many impact craters. Satellite exploration of our solar system has confirmed these expectations to a large extent; however, there are some notable exceptions.

Neptune's single spherical moon, Triton, has several features that are extraordinary. Triton appears to be covered with frozen material, not unusual since several moons including Europa, Enceladus, Tethys, Dione, Rhea, Titania and Oberon also appear to contain frozen material on their surfaces. What is unusual about Triton is its relatively young looking surface, evidenced by the small number of impact craters.[45] Triton is very cold with an estimated surface temperature of -237° C (36° K), the coldest of any planet or moon in our solar system. Triton's surface is primarily frozen nitrogen with some frozen carbon dioxide, water, carbon monoxide and methane. Of the 162 confirmed moons in our solar system, 18 are spherical moons (the remaining 144 are irregular moons considered captured asteroids). Triton is the only spherical moon with a retrograde orbit (opposite the planet's rotation). In addition, its orbit is extremely circular with less eccentricity than any other spherical moon or planet.

Triton cannot be plausibly explained due to natural cause. Evolutionary ideas propose that Triton somehow formed as a giant molten ball and was somehow captured by Neptune. One hypothesis of natural cause proposes Triton's initial orbit was highly elliptical and much larger. Over time its orbit decayed and somehow became highly circular.[46] In the recent past, tidal heating (because of its closeness to Neptune) melted Triton's surface to remove all meteor impacts. The surface somehow refroze and attained its current low temperature.

There are several problems with the Triton capture, orbital decay and tidal heating hypothesis. If the initial orbit of Triton were larger, the tidal heating effect would increase with any decrease in orbit diameter. Triton should currently be at its *hottest* condition. Explanations of tidal heating and orbital decay are the opposite of current conditions. Natural explanations do not plausibly explain the relatively new surface of Triton, its current extremely cold condition or its highly circular orbit.

Saturn's elaborate ring system cannot be plausibly explained due to natural cause. Satellite examination of the rings indicates they are made primarily of ice crystals that are slowly evaporating. These rings are considered a short-term phenomenon, and eventually Saturn will have no rings. Their existence points to a recent origin. If Saturn's rings formed at the start of the solar system 4.6 billion years ago, they would have disappeared by now. The size, uniform width, color, definition, alignment, empty spaces and circularity of the rings are unique. In dramatic photos taken by the Cassini spacecraft, the giant rings appear to be an intricate series of curved pastel colored laser beams.[47] The innermost D ring has an inner radius of 67,000 km (41,600 miles) and the outermost E ring has an outer radius of 483,000 km (300,000 miles). The inner rings are approximately 5 meters (16 ft) in thickness.[48]

The most popular natural hypothesis proposes Saturn's rings resulted from one or more captured comets made primarily of ice. The problem is that any captured comet would have a highly elliptical orbit rather than a circular orbit. In addition, a comet would not disintegrate to create the five meter thick inner rings. A single disintegrating comet would not form the distinct, uniform, empty spaces observed among the rings, but would form one continuous band. A single comet would most likely consist of one color rather than the variety of distinct colors seen in Saturn's rings. It is highly unlikely multiple comets could align themselves to form concentric circular rings within a common plane. The *Encyclopedia of the Solar System* candidly admits the difficulty of a plausible natural explanation in saying, "The origin of Saturn's massive ring system remains an unsolved mystery."[49]

In summary, our solar system exhibits both old and new phenomena. Old features represent the majority of the evidence and include the cool, geologically stable moons and planets with their extinct volcanoes and many impact craters. They represent Mars, Venus and Mercury and the loss of their planetary magnetic fields. The two most prominent recent features are the relatively crater-free surface of Triton and the rings of Saturn.

Traditional young-Earth creation theology proposes God created the solar system in 4004 BC. Apparent age is often given as an explanation for meteor impact craters and cooled planets within the solar system. God would not falsify this evidence. Falsification makes God a great deceiver and places Bible believers in an indefensible position in trying to reconcile scripture and age related phenomena. Traditional young-Earth creation theology can explain the new features, but not the old. This theology strongly conflicts with evidence related to extensive time and is rated two.

Scientific exploration has revealed the existence of new and old phenomena within our solar system. Natural change and extensive time explain old phenomena quite well. However, natural cause cannot plausibly explain the new features of the smooth surface of Triton and the rings of Saturn based on the undirected laws of chemistry and physics. Philosophical naturalism and its ancillary ideas lack a mechanism for the recent phenomena identified and are rated eight.

Exemplar creation proposes that God created the solar system in the undefined past of Genesis 1:1. Proponents of this theology can hypothesize that, except for Earth, after the planets and moons were created in our solar system they were left to the influence of natural processes. This includes cooling of these bodies from a molten state, loss of any magnetic field and meteor impacts from debris within the solar system. Proponents can hypothesize God created phenomena during the recent past in the heavens as a sign to modern man. Two of these signs include the relatively impact free surface of Triton and the rings of Saturn. Genesis 2:4 implies God did something to the heavens during the six-day Genesis chronicle—perhaps in reference to these phenomena. *Exemplar* creation can harmonize with old and new phenomena in the solar system and is rated ten.

15. Sedimentary Deposits

Sedimentary rock is formed by the accumulation and compaction of fine material from a variety of sources. Clay, shale and sandstone may be formed when particles, eroded by water, are carried to an area and deposited. Limestone and chalk may be formed by the slow accumulation of small marine organisms on the ocean floor over extensive time. Dolomite, salt, and gypsum may be formed when they are evaporated or precipitated out of mineral rich water. Volcanic ash is formed when a volcano erupts and spews out small bits of pulverized rock and glass. Conglomerate is a mixture of stones that have been cemented together by smaller particles over time. Sedimentary rock, as opposed to igneous or metamorphic rock, makes up the majority of the Earth's surface.

Sedimentary deposits on the Earth's surface vary from thin to extremely thick. For example, exposed Precambrian rock with a light covering of sediment can be found in Maine. The other extreme is represented by the Williston Basin in North Dakota with sediments thousands of feet deep resting on Precambrian rock.

Earth's crust contains a number of basins (deep depressions) filled with sedimentary deposits. Table 13.2 lists 25 large geologic basins identified by the oil exploration and drilling industry containing sedimentary deposits more than 10,000 feet deep with 20 distinct sediment layers or more.[50] These basins represent one percent of the Earth's surface and encompass thousands of square miles. As an example of their structure, the Williston Basin in North Dakota contains 27 distinct formations (deep sediment layers) that extend to a depth of 4,555.2 meters (14,945 feet) where they rest on solid igneous rock. Formations in the basin consist of shale, limestone, dolomite, sandstone, salt (more than 100 meters thick), gypsum, anhydrite, potash, conglomerate, volcanic ash, chalk and organic materials.

In addition to the basins in Table 13.2, more than 700 additional ones have been identified by the petroleum industry.[51] According to geologists Gerard C. Bond and Michelle A. Kominz, geologic basins "contain most of the world's hydrocarbon reserves."[52]

Most sedimentary deposits are developed by the slow processes of erosion, settling, accumulation, evaporation and precipitation. Thousands to millions of years are a fundamental requirement. A long age for the Earth allows time for the natural formation of soil from rock by weathering. Extensive time permits shale and clay deposits to form by the settling out of fine particles from deep stationary water. Extensive time permits the formation of thick sediment deposits consisting of alternating layers of shale, limestone, dolomite, sandstone, gypsum, salt, volcanic ash, chalk, fossils and organic materials.

Catastrophic events can readily account for sedimentary deposits containing fossils. Local and regional catastrophic events include mudslides, avalanches, earthquakes, tsunamis, hurricanes, flash floods and volcanoes. Because of the delicate nature of some animals and plants found fossilized in sediments, they were rapidly buried within a matter of minutes. Animals and plants that die and remain on the surface of the ground do not form fossils, but are eaten by scavengers or decay and leave no trace. Polystrate fossils (fossils that extend through two or more layers of different strata) are best explained by closely occurring catastrophic events. Earth's fossil record is best explained as being built over a long period of time by many local and regional catastrophic events.

Table 13.2 – Twenty-five Deep Sedimentary Basins on Earth			
No.	Name and Location	No.	Name and Location
1	Beni Mellal Basin, Morocco	14	Yazd-Kerman-Tabas Basin, Iran
2	Tunisian Basin, Tunisia	15	Farah Basin, Afghanistan
3	Oman Interior Basin, Oman	16	Helmand Basin, Afghanistan
4	Western Desert Basin, Egypt	17	Ghadames Basin, Libya
5	Adana Basin, Turkey	18	Manhai-Subei Basin, China
6	Iskenderun Basin, Turkey	19	Jiuxi Basin, China
7	Moesian Platform, Bulgaria	20	Tung t'in - Yuan Shui Basin, China
8	Carpathian Basin, Poland	21	Tarim Basin, China
9	Baltic Basin, USSR	22	Szechwan Basin, China
10	Yenisey-Khatanga Basin, USSR	23	Tampico Embayment, Mexico
11	Yukon-Porcupine Province, Alaska, USA	24	Bogata Basin, Colombia
12	Williston Basin, North Dakota, USA	25	Bonaparte Basin, Australia
13	Beaufort Sea Basin/McKenzie River Delta		
Reference: Morton, Glenn (2001). The Geologic Column and its Implications for the Flood, Talk of the Origins Archive.			

Scientists who support traditional young-Earth creation theology propose that most of Earth's sediments and fossils can be explained by a global Noah's flood. However, within the time constraints of the flood, it cannot explain sediments in the Williston Basin including (1) the formation of 27 thick layers of different sedimentary materials, (2) the formation of thick clay and shale deposits from the settling of fine particles in tranquil water, (3) the eruption of volcanoes and the deposit of thick ash on top of water deposited sediments, in turn covered by more water laid sediments, (4) the slow formation of thick evaporites and precipitates such as salt, dolomite and gypsum, (5) the death and slow accumulation of small marine creatures to form deep limestone and chalk deposits and (6) the formation of 12.15 meters (39.9 ft) of sediment per day for 375 days Noah was in the ark. This basin and others like it are compelling evidence against formation due to a single flood and are evidence for multiple events over extensive time. Traditional young-Earth creation theology cannot plausibly explain Earth's sedimentary deposits within the time constraints of Noah's flood; however, because catastrophic flooding and related fossil formation can be interpreted to support this theology, it is upgraded to a rating of one.

Geological theories can plausibly explain the formation of Earth's sedimentary strata by natural cause over thousands and millions of years. Repetitive local and regional catastrophes over time best explain fossil deposits. Unfortunately, most school science textbooks, encyclopedias, reference books and science articles that discuss sediments and fossils never mention flash flooding or local catastrophic mechanisms. There appears to be an aversion to any mention of catastrophic events because of their association with Noah's flood and catastrophists such as Immanual Velikovsky. Philosophical naturalism and its ancillary ideas can harmonize with sedimentary evidence and are rated ten.

Exemplar creation allows time in the undefined past prior to Genesis 1:3 for extensive natural sediment deposits to occur. Proponents of this theology can readily accept fossil formation due to repetitive local and regional catastrophic events over extensive time. *Exemplar* creation easily harmonizes with the evidence of empirical science and is rated ten.

16. Earth's Primordial Atmosphere

Philosophical naturalism requires that life originated from non-life by natural means. One popular idea in support of chemical evolution proposes that the first life on Earth developed in a primordial soup. This is based on the 1953 experiments by scientists Stanley Miller and Harold Urey at the University of Chicago. As previously discussed, they conducted experiments to simulate conditions for the natural formation of amino acids necessary for life. Miller and Urey mixed methane, ammonia, water vapor and hydrogen gases together, and then subjected the mix to electric sparks in a sealed apparatus that excluded oxygen. After several days, some of the twenty essential amino acids found in living things had accumulated in a trap.

Chemists know that oxygen must be excluded in any test similar to the Miller-Urey experiment. If methane, hydrogen and oxygen are mixed together and subjected to an electric spark, the mixture is most likely to explode. Biochemists know that if oxygen is present during a Miller-Urey type experiment, compounds of oxygen rather than amino acids are the result. Oxygen is inherently a chemically reactive element. To produce the amino acids and molecules found in life in this type of experiment, oxygen must be specifically excluded.

For the Miller-Urey experiment to plausibly represent conditions for the formation of amino acids, Earth had to have an oxygen-free (anaerobic) atmosphere composed of methane, ammonia, hydrogen and water vapor. Scientific research into Earth's past atmosphere supports the absence of any significant methane or hydrogen. The atmosphere was primarily composed of nitrogen, water vapor and carbon dioxide,[53] which by themselves are not sufficient to synthesize many useful biomolecules necessary for life.[54]

Some scientists say Earth's atmosphere has always contained oxygen.[55, 56] Sedimentary layers deep into the Earth's crust contain oxygen compounds and evidence of oxidation at every level. The most common elements in our planet's crust by rank are 47.4% oxygen, 27.7% silicon, 8.2% aluminum, 4.1% iron, 4.1% calcium, 2.3% sodium, 2.3% magnesium and 2.1% potassium.[57] Virtually all elements in the Earth's crust are combined in some form or another with oxygen.

In summary, ideas of chemical evolution propose that at one time Earth had conditions favorable to the natural development of life. One leading hypothesis proposes that deep into our planet's past (at least 3.8 billion years ago or more) it had an atmosphere rich in methane, ammonia, hydrogen and water vapor, without any oxygen. This gaseous mix, plus lightning, allegedly filled Earth's vast ocean with amino acids and biomolecules to result in a rich chemical soup. Yet scientists have failed to find evidence supporting this hypothesized atmosphere. They have failed to find evidence in sedimentary strata of amino acids and other molecules from alleged oceans of primordial soup that existed for hundreds of millions of years. Physical evidence does not support the requirements and claims of this hypothesis. Philosophical naturalism and supporting ideas of chemical evolution are rated zero for their explanation of Earth's primordial atmosphere.

Traditional young-Earth creation theology proposes that Earth's atmosphere has always contained oxygen and nitrogen as major gases, though the ratio may have undergone limited change over time. By the third day of creation, Earth had a completely livable atmosphere. Young-Earth creation theology has no major conflict with empirical science and is rated ten.

Exemplar creation proposes the Earth was created by God in the undefined past of Genesis 1:1. Proponents of this theology can accept that Earth began as a molten ball of material and later acquired an atmosphere. There are no rigid requirements about our planet's early atmosphere, although it undoubtedly changed over time both in composition and pressure. Time is allowed prior to Genesis 1:3 for change to occur. *Exemplar* creation can accept the evidence of empirical science, has no conflict with its findings and is rated ten.

17. Coal Formation

According to geological theory, most coal is of *autochthonous* origin and was formed by the on-site accumulation of organic material in peat bogs, marches and swamps over millions of years. This process is similar to present-day peat swamps identified in Borneo where coal formation is occurring.[58] Root penetration into strata under coal beds by trees and other vegetation is evidence that organic material grew, collected and transformed into coal at the same location. Leaves, grains, twigs, nuts, seeds, cones, pollen, bark and woody structures from plants native to a common area indicate on-site accumulation. Some continuous coal seams of relatively uniform thickness and structure extend for hundreds of miles[59] and strongly support coal formation by the growth and accumulation of organic material in vast swamps.

Coal may also be of *allochthonous* origin,[60] formed when vegetation from one area was transported to another area. Local or regional floods stripped vegetation from the land, collected it at settling points and then covered it with sediment. Over time, accumulated organic material transformed into coal due to the action of heat, pressure and chemical action. There is strong evidence for the water transport formation of coal. Upright fossilized trees extending through multiple coal seams support the rapid accumulation of organic material by water transport. Marine animals and plants not native to swamps are often found in coal beds and imply water transport. Boulders, sand, clay and silt sometimes found in coal indicate flooding. The absence of root penetration into soil below a coal bed implies a transport process. Coal often appears directly over granite, limestone or other rock with a clean separation between the two.[61] Some coal beds more than 90 meters (300 ft) thick[62] may indicate coal formation by water transport. In Vermillion County, Illinois, a coal forest with upright trees covering more than 2,400 acres was discovered. This forest is hypothesized to have been buried within two months time.[63] Coal formation by water transport, accumulation and burial is strongly supported.

Coal and sediments often occur in multiple layers. In Illinois and West Virginia, coal deposits with more than fifty alternating layers of coal and sediment have been found.[64] These are best explained by a series of swamps or floods over time that deposited alternating layers of organic material and sediments. Multi-layered coal seams discredit Noah's flood as an explanation. A single catastrophic flood cannot plausibly result in fifty alternating layers of coal and sedimentary rock.

Some coal deposits may have resulted from a combination of on-site accumulation and accumulation by transport. It is possible for a swamp with extensive peat deposits and root penetration into underlying strata to be overlaid by transported sediments. This can result in alternating layers of coal and sedimentary rock with grown in-place trees extending through several layers. All coal formations and their unique features can be plausibly explained by a combination of on-site accumulation and accumulation by transport.

In summary, theory in support of traditional young-Earth creation theology proposes that essentially all coal was formed due to Noah's flood through accumulation by water transport. According to Ussher's chronology,[65] there are approximately 1,650 years between the creation of the Earth and Noah's flood. This is the time allowed for organic material responsible for coal deposits to grow. Earth contains enormous deposits of coal, with current estimates for all known coal reserves at 12.7×10^{12} metric tons.[66] Earth has a current land surface area of approximately 1.489×10^{8} square kilometers (57.5×10^{6} sq. miles) which equates to 71.2 kg (157 lbs) of coal for every square yard. It is extremely doubtful Earth could have contained the necessary amount of biomass prior to the flood to create this amount of coal. In addition, fifty or more alternating layers of coal and sediments cannot be formed by a single flood of any type. The formation of Earth's coal reserves at a single point in time because of a single event is not plausible. Traditional young-Earth creation theology cannot explain Earth's coal deposits; however, because catastrophic flooding and coal of allochthonous origin can be interpreted to support this theology, it is upgraded to a rating of one.

Natural processes can explain coal deposits by on-site accumulation and accumulation by water transport. Repetitive local and regional flood events over extensive time best explain some coal deposits. Unfortunately, most public school science textbooks, encyclopedias, reference books and science articles about coal never mention coal formation due to accumulation by water transport. As mentioned, there appears to be an aversion to any discussion of catastrophic events because of the association with Noah's flood and the ideas of Immanual Velikovsky. Philosophical naturalism and its ancillary ideas can harmonize with geological science in regard to coal and are rated ten.

Exemplar creation allows time in the undefined past between Genesis 1:2 and 1:3 for the formation of Earth's coal reserves by natural processes and God directed quasi-natural processes. Proponents of this theology can readily accept coal formation by on-site accumulation or accumulation by water transport. *Exemplar* creation easily harmonizes with the evidence of empirical science and is rated ten.

18. Origin of Humankind

The vast majority of archaeological evidence supports the recent appearance of humankind on Earth. It appears that Neanderthals and anatomically modern humans are the only skeletal remains associated with all five activities of controlled fire, fashioned tools, formed weapons, created art and burial rituals. The small number (about 108) of hominid archaeological sites more than 10,000 years old and the rise of civilization within the last 10,000 years support a recent appearance for humanity. Archaeological evidence does not show a continuous chain of human existence going back 500,000 years and more. Legitimate radiocarbon dates in support of anatomically modern man taper off around 30,000 years ago. (Chapter 10 and 11 discuss the origin of man in more detail.)

The difference in DNA between humans and chimpanzees is 4.23% and represents approximately 125 million nucleotide substitutions, additions and deletions. For this amount of change to occur, roughly 20 mutations per year had to become fixed and then spread throughout a small population of hominids over a six million year period—a highly unlikely scenario especially when many of these mutations are neutral.

Random beneficial point mutations are necessary for step-by-step evolution based on natural selection. More than 99.9% of all known protein isoforms (mutations) in humans are neutral or deleterious, and prevent a step-by-step series from occurring. This evidence strongly discredits animal hominid to human evolution.

Some proteins and molecules are significantly different between humans and chimpanzees. A comparison of proteins between humans and chimpanzees show 71 percent are different by at least one amino acid. The HAR1 gene has a difference of 18 nucleotides between the two species with no intermediates. Dissimilarities such as these discredit evolution as a mechanism.

Hominid remains identified in Table 10.1 do not show a step-by-step evolutionary development of humans from animal hominids. There are many gaps, steps and excursions in skeletal remains. There is no 100-step series showing a three foot high ape with a small brain, sloping forehead, big teeth, protruding jaw, long arms, short hind legs and grasping toes that progressively changes into an anatomically modern human.

In summary, the theory of human evolution is an extension of biological evolution. Allegedly, humankind and the chimpanzee originated from a common ancestor roughly six million years ago. This theory is discredited by an analysis of information from biology, archaeology and paleontology. Since some hominid fossils and DNA evidence can be interpreted to support human evolution, philosophical naturalism is upgraded to a rating of one.

Traditional young-Earth creation theology proposes God created man on the sixth day of creation. Evidence from radiocarbon dating and archaeological sites strongly conflict with a date of 4004 BC for the origin of man. Humans have been on Earth substantially longer than six thousand years. Although traditional young-Earth creation theology has a viable mechanism for the origin of man, because of its strong conflict with archaeological evidence in regard to time, it is downgraded to a rating of six.

Design theory proposes that intelligent cause best explains man's phenomenal brain, five senses, full color vision, exceptional language ability, fully functional opposable thumb, dexterous hands and distinctive upright stature. Random mutation and natural selection cannot plausibly account for the improvement between humans and the nearest hominid. The 4.23% difference in DNA between humans and chimpanzees is primarily due to design. Human and chimpanzee DNA are each roughly 99.9% homogeneous and discredit the idea that significant random change has occurred in either species since inception. The recent rise of civilization, radiocarbon dating and archaeological evidence support a hypothesis for the sudden appearance of anatomically modern man 20,000 to 30,000 years ago.

Exemplar creation accepts the arguments and explanations of design theory given above. God is an omnipotent creator who made man according to Genesis 1:26-27. Proponents of this theology can accept similar DNA, molecules, proteins, genes, pseudogenes, features and functions found in both animals and humans (similarity is evidence of a common designer). They can accept an expansion of the genealogies of Genesis Chapters 5 and 11 to allow time for the spread of humanity over the Earth and the development of agriculture, metals technology and civilization evidenced by the oldest towns and cities. *Exemplar* creation has no known conflict with verified empirical data regarding the origin of man and is rated ten.

19. Language Families

A study of the origin of human language is complicated, if for no other reason than the sheer number of languages. The book *Ethnologue: Languages of the World* lists 6,912 known living languages in the world,[67] grouped into families based on the similar pronunciation or spelling of common words. According to theory, a single language can evolve into several languages over time because of separation and isolation to create a family of associated languages. *Ethnologue* lists one hundred spoken language families.[68]

Languages within a language family can be linked together, while between families no linkage exists. Linguists who accept evolution would like to link language families together. From an evolutionary perspective, all language families should converge back to a single language similar to following the branches of a tree back to its trunk. Currently, there appears to be no hope of achieving this goal. Radically different language families cannot be united based on similarity because none exists. When explanations are limited to natural cause, the origin of language families is a complete mystery.

Dr. David Premack, emeritus professor of psychology at the University of Pennsylvania and author of several books and articles on language development, succinctly summarized the relationship of evolution and human language when he said, “Human language is an embarrassment for evolutionary theory.”[69]

Natural cause cannot satisfactorily explain the existence of one hundred spoken language families. We can understand how the first humans could have developed a rudimentary language. We readily comprehend how the vocabulary of that language could have increased over time and the ability to express ideas could have become more sophisticated. We have documented how word pronunciation can drift over time. We have identified how one language can diverge into two or more related languages because of geographical separation. However, we cannot plausibly explain how a primitive human who spoke one language (e.g. English) could develop a radically different language (e.g. Chinese).

In summary, philosophical naturalism requires a natural explanation for language. Biological evolution cannot plausibly explain development of the human voice box and associated functions. Evolution cannot explain development of the human brain to permit talking and understanding speech. An unbridgeable gap exists between chimpanzees and humans in regard to speech ability. There is no good natural explanation for the origin of one hundred spoken language families and 6,912 languages within the short history of anatomically modern humans. Philosophical naturalism and its ancillary ideas cannot plausibly explain the development of human language families and are rated zero.

According to the Bible, God confounded human language at Babel sometime after the flood of Noah. Up until Babel, the entire human race spoke one language. At Babel God erased each person’s language and implanted a new language in their mind. People were grouped by their languages and spread over the Earth. How God accomplished this feat is currently beyond our scientific understanding. We can no more understand what happened at Babel than an ignorant savage can understand how a DVD player can record a movie, erase it and record another. We can hypothesize that the number of original languages created at Babel equate roughly to the language families identified today. In other words, one hundred original languages at Babel evolved into more than 6,912 living languages. The biblical historical record is the only account that makes sense in the light of what is known about genetics, history, languages and human behavior.

Traditional young-Earth creation accepts the biblical story of the tower of Babel as an explanation for today's languages; however, the traditional date of around 2250 BC set by Ussher for the tower of Babel is too recent. Radiocarbon dating supports the spread of humanity throughout the Earth sometime between 8,000 and 12,000 BC. The confusion of languages had to occur before humans were scattered over the Earth, rather than after. Because of its strong conflict with radiocarbon dating, archaeology and the spread of humanity, traditional young-Earth creation theology is downgraded to a rating of five.

Exemplar creation accepts the biblical account of the tower of Babel in Genesis 11:1-9 as an explanation for the origin of human language families. Proponents of this theology can accept an archaeological hypothesis for the appearance of man 20,000 to 30,000 years ago. They can accept the tower of Babel, the confusion of human languages and the spread of humanity over the Earth sometime after Noah's flood. They can accept a date prior to 12,000 BC for Babel, giving time for the one hundred or so original languages to be scattered over the Earth and develop into the many languages today. *Exemplar* creation has a plausible explanation, does not conflict with empirical science and is rated ten.

Conclusion

The philosophical, scientific and biblical communities must realize that their ideas of origin are not sacrosanct. Philosophical ideas, scientific theories and biblical theologies about origins should be held accountable to physical evidence gained by observation, measurement and experimentation. To have a meaningful search for truth about origins, empirical evidence must be used as a measuring stick to objectively evaluate competing ideas.

Table 13.3 lists the results from comparing philosophical naturalism, traditional young-Earth creation theology and *exemplar* creation. A comparative analysis clearly shows their strengths and weaknesses.

Table 13.3 shows that philosophical naturalism, based on natural cause, has poor explanatory power in regard to *complex function.* As a destructive mechanism, random change can wear away, deteriorate and randomize things. As a weak constructive mechanism, undirected phenomena are capable of producing simple effects, elementary arrangements and mathematically probable structures. Some people believe evolution is a powerful constructive mechanism when given significant time; however, this has not been observed, demonstrated or verified. The severe limitations of natural cause strongly discredit philosophical naturalism and its ancillary ideas as a total explanation of origins.

Many evolutionists overlook the severe limitations of natural cause while flippantly rejecting any appeal to intelligent cause. Richard Dawkins, a leading proponent of evolution says, "The complexity of living organisms is matched by the elegant efficiency of their apparent design."[70] He goes on to say, "Biology is the study of complicated things that give the appearance of having been designed for a purpose."[71] Dawkins dogmatically denies evidence of design due to God, even though life appears to be designed. He believes somehow natural processes created the first life and then climbed "Mount Improbable" to form all of Earth's species.

Table 13.3 shows that traditional young-Earth creation theology has poor explanatory power in regard to evidence that indicates *extensive time*. The appeal of traditional young-Earth creationists to an omnipotent, supernatural power does solve their theology problems. A holy, righteous God would not falsify physical evidence to intentionally mislead or deceive. Traditional young-Earth creation theology strongly conflicts with empirical science in regard to time.

Young-Earth creationists reject evidence supporting a long age for the universe. Ken Ham, a leading proponent of recent creation says, "And when He [God] created the universe, He created it fully functional, with the appearance of age—even though it wasn't old."[72] Ham firmly believes the Earth, Moon, stars and universe were created around 4004 BC.

It is ironic that both evolutionists and young-Earth creationists use almost the same argument. Evolutionists say that even though life gives the *appearance* of being designed, it is not. Young-Earth creationists say that even though the universe has the *appearance* of being old, it is not.

Both evolutionists and young-Earth creationists appeal to the unknown and undiscovered to salvage their respective beliefs. For example, evolutionists believe unknown laws will be discovered capable of self-organization to salvage chemical evolution. A natural means of creating information, producing beneficial mutations and forming complex functional structures will be found to save biological evolution. Complex natural laws with the ability to create basic energy and matter will be discovered to rescue cosmological evolution. On the other hand, young-Earth creationists say that perhaps new information will be discovered to overturn current data about starlight, ice cores, dendrochronology, radiometric dating and radiocarbon dating. "All dating methods are based on fallible assumptions and can't be trusted to give absolute dates about the past."[73] Strangely enough, both camps find common ground in an attempt to prop up their respective ideas.

As shown in Table 13.3, *exemplar* creation has the best total score of the three ideas evaluated. The Bible is not a scientific book, but where the historical record of the Bible and physical evidence intersect they should not conflict. Special revelation and general revelation should not clash. *Exemplar* creation can accept the arguments and explanations of design theory presented in this book and can harmonize with empirical science.

Table 13.3 – Comparative Analysis of Three Major Ideas of Origin
(Ratings are based on the power of each idea to explain empirical data)

Index	Criteria	Philosophical Naturalism	Young-Earth Creation Theology	*Exemplar* Creation
	Complex function indicators			
1	Origin of life on Earth	0	9	10
2	Minor species change	0	9	10
3	Origin of Earth's species	1	7	10
4	Origin of galaxies	0	7	10
5	Origin of the first stars	0	7	10
6	Origin of our solar system	0	7	10
7	A unique Earth for life and humanity	0	9	10
	Time indicators			
8	Distant starlight	10	0	10
9	Moon and Earth meteor craters	10	0	10
10	Radiometric dating igneous rocks	10	0	10
11	Carbon-14 dating organic material	10	0	10
12	Ice core dating	10	0	10
13	Antarctic coal and fossils	10	0	10
14	Solar system age phenomena	8	2	10
	Miscellaneous phenomena			
15	Sedimentary deposits	10	1	10
16	Earth's primordial atmosphere	0	10	10
17	Coal formation	10	1	10
18	Origin of humankind	1	6	10
19	Language families	0	5	10
	Total Score	90	80	190

Evolution and traditional young-Earth creation theology are in a stalemate as clearly pointed out in *The Evolution Controversy in America* by George E. Webb.[74] Comparative analysis in Table 13.3 identifies the fundamental issues in this struggle, with both positions having their strengths and weaknesses. Choosing between evolution and traditional young-Earth creation theology is a choice between the lesser of two bad ideas. Adherents of both ideas refuse to objectively look at the complete evidence. Evolutionists (driven by philosophical naturalism) refuse to accept the limitations of natural cause to create complex function and information. Traditional young-Earth creationists refuse to accept the conflict their theology has with time.

From a historical perspective, the Bible as interpreted by *exemplar* creation best explains our origins in an abbreviated, non-technical, narrative format. The Bible is the inspired word of God and is truthful and trustworthy.

From a scientific perspective, design theory best explains the origin of information, complex functional systems, mathematically improbable structures and fine-tuned arrangements observed in nature and the universe at large. Design theory best explains the DNA molecule, the DNA code, DNA information, the DNA copying system, complex biomolecules, functional proteins, unique enzymes, genes, pseudogenes, the origin of life and Earth's many species. Design theory best explains the fine-tuning of Earth as a habitat for life and the initial formation of our solar system, stars and galaxies.

On the other hand, natural cause as a weak constructive mechanism can explain simple effects, elementary arrangements and mathematically probable structures. Undirected change best explains snowflakes, mineral crystals, concretions, freeze-thaw rock circles, sand dune formations, dried mud crack patterns, clouds, sand particles and salt grains. Natural cause as a destructive mechanism can explain wearing away, deteriorating and randomizing things. Undirected change best explains point mutations, defective genes, protein isoforms, chromosome abnormalities, genetic disease, cancer, aging, limited species divergence, new viral strains and the spread of antibiotic-resistant bacterial populations. Undirected change best explains stellar ageing, stellar novae, meteor impacts, planetary cooling, weathering effects, erosion, sedimentation, volcanoes, earthquakes, hurricanes and continental drift.

Any objective rating of the nineteen criteria in Table 13.3 should obtain the same total relative rankings. Traditional young-Earth creation theology and philosophical naturalism should be comparable, while *exemplar* creation should have the highest total score. With our present scientific knowledge base, it is highly doubtful any new discoveries will overturn these ratings. Additional research will only confirm the analysis herein.

Do not be overcome by evil, but overcome evil with good. (Romans 12:21) NKJV

Chapter 14

Paradigm Shift

Philosophical naturalism and biblical belief are engaged in a war that has spanned centuries. The first skirmish occurred more than twenty-five hundred years ago and was initially a dispute between philosophical naturalism and the religions of the day. One of the first disagreements can be traced back to Anaximander (610-546 BC), a Greek philosopher, who proposed that life had a naturalistic origin, in contrast to the religious beliefs of the Hebrews, Sumerians, Egyptians, Babylonians, Assyrians, Phoenicians, Canaanites and Greeks. Within Europe, as time passed, the pagan religions lost credibility and were mostly abandoned, while Christianity gained influence. By 1700, biblical based theology about origins was the leading belief in the Western world. Philosophical naturalism had little influence and biblical theology dominated public thinking within and without the church. Thomas Aquinas (1225-1274) best summarized the prevailing attitude during this period when he said, "Theology was the queen of sciences." However, change was on the horizon.

Scientific understanding of the world accelerated during the 1700s and forced scientists, philosophers and thinkers of the day to reconcile their new knowledge with the Bible. Resolution focused primarily on the first eleven chapters of Genesis. Major topics of discussion included the age of the Earth, Noah's flood, geological features, fossils, our planet's many species and the origin of life. During the 1700s Christian thinking held high ground, but science had begun to expose weaknesses in young-Earth creation theology. At the onset, there was little concern because the majority of influential scientists, philosophers, scholars, thinkers and politicians within the Western world claimed to be Christian or were at least friendly to biblical thought. They in turn were supported by a large base of common people who accepted the Bible as true.

Three major books soon challenged Christian thinking about biblical origins. A three-volume book entitled *Principles of Geology* by Charles Lyell was published in 1830-1833, and explained Earth's geology as having developed over enormously long periods of time. The present was the key to the past. Slow change over vast time best explained geologic features and was a direct challenge to theology teaching that

God created the Earth in six days around 4004 BC.

Charles Darwin published *On the Origin of Species by Means of Natural Selection* in 1859, followed by *The Descent of Man* in 1871. Darwin's theory gave an evolutionary explanation for Earth's species and the origin of man, in direct opposition to the Genesis creation story. Darwin's books gave intellectual credibility to philosophical naturalism and represented the tipping point in this struggle. The Earth and all therein could be scientifically explained by natural cause. From that point onward, Christianity steadily lost influence and the ideas of philosophical naturalism gained ground. The writings of Friedrich Nietzsche (1844-1900) about philosophy, Karl Marx (1818-1883) about government and Adolph Hitler (1889-1945) about racial superiority added impetus to Darwin's beliefs. These men promoted atheism, class warfare and survival of the fittest.

Another change that occurred was the acceptance and popularity of theory. Rather than strictly appealing to empirical evidence, scientists began to theorize. Science became a search for dependable knowledge about nature based on physical evidence and *its logical inferences*. Theory allowed scientists to explain questions about origin, ultimate cause, operation and structure beyond the immediate reach of empirical science. Theory allowed scientists to address some of the important philosophical and religious issues of humankind—origins.

Science at one time was viewed as a means of exploring and understanding God's creation; however, this changed due to the influence of philosophical naturalism. Science came to be viewed as naturalistic, requiring natural theories and natural mechanisms based on natural cause. Charles Darwin was a major influence in this shift in thinking. In the beginning, conflict raged over Darwin's theory of evolution, but eventually skeptical scientists died and were replaced by eager young men who embraced his ideas with a passion. Today, science organizations, scientific journals, science museums and science education are dominated by people who believe in natural cause as a total explanation of origins. This includes essentially all science departments within biblical based colleges and universities. With the passage of time, philosophical naturalism has come to dominant scientific thought about origins.

A House Divided

The Septuagint translation of the Pentateuch is adjudged to present the best and most faithful rendering of the original Hebrew text[1] and is the basis for traditional errant interpretations of the Genesis creation account. As mentioned, around 275 BC the first five books of the Old Testament were translated from Hebrew into Greek and were the beginning of the Septuagint. During this process, Hebrew scribes embedded traditional young-Earth creation theology (a prevalent belief before, during and after the time of Christ) into their translation. Starting around 500 AD, Masoretic scribes further established this belief through the addition of vowel points and accent marks to the Hebrew Bible. Today, biblical scholars rely on the Septuagint translation (and Masoretic vocal marks to a lesser extent) for their interpretation of the Genesis creation account, and believe it is correct. They reason that Abraham was God's chosen man, his descendants God's chosen people and Israel God's chosen nation. The Hebrews were the copiers and keepers of the Old Testament writings. Hebrew and Greek were the native languages of the scribes who translated the Septuagint and they had the ideal linguistic background to make this translation.

Table 14.1 – American Adult Opinion about Origins

	Traditional Young Earth Creation	Theistic Evolution	Naturalistic Evolution Only
Total population	44–50%	30–38%	12–14%

Note: The above data represents a compilation of information from the five polls listed as references. Questions may vary slightly among polls. Generally 4-10% of the population in any specific poll has no opinion.
References:
(1) *Newsweek* Poll (2007). Conducted by Princeton Survey Research Associates International.
(2) Newport, Frank (2007). Majority of Republicans Doubt Theory of Evolution, Gallup News Services.
(3) CBS News Polls (2005). Poll: Majority Reject Evolution, CBS News.
(4) Mitchell, Susan (1996). *The Official Guide to American Attitudes.*
(5) Niemi, Richard and John Mueller, Tom Smith (1989). *Trends in Public Opinion: A Compendium of Survey Data.*

Currently, Hebrew scholars and theologians believe the traditional Genesis creation account in English is an accurate translation of the original text; however, they are completely puzzled, if not befuddled, by its conflict with empirical science. They cannot explain the irreconcilable conflict between special and general revelation. They cannot understand how a powerful, knowledgeable, intelligent God could inspire a story that is so obviously in conflict with physical evidence in regard to time. They find it difficult to believe God would lie, deceive or err. Thinking laymen are also deeply troubled by this conflict.

Our current theological and scientific problems with biblical origins stem from the Septuagint. This document is filled with translation errors and strongly reflects the origins theology of its translators. When the Septuagint is subjugated to the original God inspired Hebrew text of the Old Testament, a true story of biblical origins is allowed to emerge.

Within the biblical community, interpretation of the Genesis creation story can be divided into two extreme opposing positions, with a range of belief in between. Traditional young-Earth creation theology proposes that the Earth, Moon, stars, plants, animals and man were created approximately six thousand years ago—God spoke and they appeared. Adherents of this theology claim it is based on a straightforward, literal reading of the Genesis account. There will never be agreement with science and the Bible regarding origins—science is wrong.

The opposite position held by some of the biblical community is that the Bible is simply a religious book, relevant only in matters of faith and practice. Supporters say the Genesis creation story is in irreconcilable conflict with science. The first eleven chapters of Genesis are nothing but myth and metaphor, and represent a compilation of pagan beliefs. The Earth and universe are billions of years old. Adherents believe evolution is a fact, and God used natural change as His method of creating. There will never be agreement with science and the Bible—the Bible is wrong.

Both camps need to realize their respective positions are seriously in error. Both beliefs are based on a misunderstanding of specific words and phrases in the Genesis creation story. In addition, proponents of each perspective must either ignore or deny scientific evidence that strongly conflicts with their position.

Opinion polls have repeatedly verified the division within the biblical community over the Genesis creation account. Table 14.1 represents a compilation of data from several polls, including those conducted by Newsweek,[2] Gallup[3] and CBS[4] (in addition to two books[5, 6]) of American adults more than 18 years of age about creation and evolution. Approximately 44-50% say they believe in traditional young-Earth creation while the rest accept theistic evolution or naturalistic evolution. Even though polls show the creationist viewpoint is dominant, the biblical community is deeply divided over this issue. Polls show that 82-86% of American adults identify themselves as being Protestant, Roman Catholic or Christian. About 2% of adults identify themselves as Jewish, while 9-10% of adults say they are agnostics, atheists or have no religion.

Based on the above statistics, the primary battle over the evolution-creation issue is within the biblical community rather than the secular. Polls clearly point out that roughly 85% of American adults say they believe in God to some degree or another and respect the authority of the Bible. Therefore, the main battle is among those who say they believe in God, profess acceptance of the Bible or claim they are Christian.

Paradigm Shift

During the past three thousand years, religious belief changed as scientific knowledge and human culture advanced. Within Europe, man-made idols of wood, stone, bronze, silver and gold declined in appeal. Worship of the gods of nature in the form of the Sun, Moon, earth, sky, mountains, rivers, forests and animals diminished in status. The worship of emperors, kings and human deities decreased in popularity. Ancestor worship, spiritualism, mysticism, animism, pantheism and other forms of pagan worship lost favor. In Western culture, Christianity and acceptance of the Bible remained strong as scientific knowledge increased; however, this began to change in the 1700s. Today in many Western countries, Christianity is almost inconsequential. While a significant percentage of the population may still espouse Christianity, it has lost much of its relevance, especially in regard to the cultural war of values.

As pointed out in *Total Truth* by Nancy Pearcey, Christians have adopted a two-tier method of thinking and living. Their values have become compartmentalized. The way Christians view biblical teaching and how they act in their professional and public lives are completely separate. People can believe in God on a personal level, yet not have the conviction to express their beliefs in a tangible way. Christianity seems to have lost its power to affect people's lives in the public arena.

Few pastors and biblical leaders truly understand why Christianity is losing the culture war. They do not understand why people appear to be more selfish, materialistic, worldly, sinful, uncaring and unfaithful. The root problem is not a hostile news media, degenerate movie industry, material affluence, a busy lifestyle, a decline in stable families, working mothers, divorce, premarital sex, adultery, pornography, abortion, drugs, alcohol, gambling, corrupt politicians, liberal judges and the homosexual agenda. While these are all important influences and issues, they are secondary rather than primary.

The primary influences in the culture war include the Bible, science and philosophical naturalism. The first two chapters of the Bible are a key issue, because as traditionally explained they do not make sense. On a deep visceral level, people do not believe the Genesis creation story is credible, giving growth to skepticism affecting the entire Christian biblical perspective. People accept science as a dependable source of knowledge about the world. Science is based on what can be observed, measured and tested. Science has given us technology and a comfortable life style. Philosophical naturalism has hijacked science in regard to theories of origins and is using science to destroy belief in God as our creator.

How can we reverse this worldview? It begins with Christians acknowledging the historical impact of faulty theology. Around 1859 (the release of Darwin's first book) a paradigm shift occurred, and we are still reaping the results. No significant event has occurred to reverse this trend. A resurgence of traditional young-Earth creation theology within some circles has had little impact. Proposed alternate theologies, in an attempt to satisfactorily explain Genesis, are fraught with biblical and scientific problems. The rise of intelligent design has begun to challenge certain aspects of philosophical naturalism and evolutionary thinking. Although powerful, intelligent design is primarily a scientific idea rather than a theological enterprise and cannot help Christianity with its errant theology about origins.

The Christian community must realize that it is at total war with philosophical naturalism and must fight its own battles. Christianity and philosophical naturalism are antithetical beliefs. They are opposed over meaning, ultimate purpose and moral values. They are in violent disagreement about origins. All the chips are on the table and any strategy for winning must have real substance. To reverse the losing trend of the past 150 years, a strong Christian response is needed. We must go beyond superficial improvements in worship style, appearance and presentation. Even though methodology is important, it is not the answer. The war against philosophical naturalism must be waged with truth.

We tend to think that humanity has arrived at its present state of truth in a slow, methodical fashion with little difficulty, while in reality change has always been hard. For more than two thousand years, humankind accepted the four element theory of matter proposed by Empedocles until slowly disproved by science. For more than two thousand years, men accepted geocentric theory (Earth was the center of the solar system) as a fact until Copernicus, Galileo and Kepler discredited the idea. For more than two thousand years, people believed spontaneous generation was true until discredited by Louis Pasteur. For more than one thousand years, salvation through personal works was believed and taught essentially unopposed by the Roman Catholic Church. Martin Luther's teaching of salvation by God's grace through faith in Jesus Christ challenged this belief (he also objected to church materialism, sanctioned immorality and extra-biblical sacramental practices) and started the Protestant Reformation. All four of these historic changes were marked by years of research, study, debate, conflict and in some cases violence.

In the present, two faulty ideas dominate origins belief. Traditional young-Earth creation theology has been the reigning belief of biblical origins for more than twenty-two hundred years. Philosophical naturalism has existed for more than twenty-five hundred years and a dominant influence in science for the past 150 years, promoting evolution as a powerful constructive mechanism. It is now our turn to experience the overturning of these two ideas.

A Winning Strategy

A winning strategy begins by accepting the fact that our interpretation of the Hebrew text of Genesis Chapters 1 and 2 has been wrong for more than twenty-two hundred years. The Septuagint translation of the Genesis account is seriously in error, heavily influenced by the incorrect theology of its translators. The Septuagint has been the primary influence in steering origins theology in the wrong direction.

Christians must realize that traditional interpretations and translations of the creation story are not inspired, and theology developed by humans is not sacred. The Bible must be our ultimate religious authority. There is no inherent righteousness in clinging to creation theology that conflicts with a true understanding of scripture and physical evidence.

As Christians we must lay aside our parochial attitudes, private ambitions and personal pride. Our goal is the pursuit of truth based on special and general revelation. We have been wrong, we acknowledge our error, we accept the truth and will change.

Biblical based creation theology is important. While it is true one can be a Christian without a correct view of biblical creation (or knowing anything about it), many people are not open to Christianity because they believe the Genesis creation story is false. In their reasoning, this makes the rest of the Bible untrue. In addition, it is not uncommon for Christians to lose their faith based on the perception that Genesis conflicts with science and is therefore in error.

Educated, intelligent, talented, motivated Christians of sound judgment require solid answers. Christian leadership within our churches, Christian schools, Christian colleges, evangelical ministries and Bible based charitable organizations deserve credible explanations. A correct understanding of biblical origins provides a firm base for belief in God, the moral teachings of the New Testament, Christian ethics and acceptance of the Bible as true. This in turn provides Christians with a basis for making the tough judgment decisions often required in leadership positions.

A winning strategy requires retaking science education in Christian schools and colleges. Virtually all science programs in Christian colleges are dominated by evolutionary thought—a condition that desperately needs correcting. Change begins by realizing that science is a rational search for dependable knowledge about the world based upon physical evidence and its logical inferences. Science is inherently limited because of inaccessible evidence, unavailable data and unknown information. The inferred explanations (theories and hypotheses) of science can include natural and intelligent cause. Natural cause is a biased assumption of science rather than a dogmatic requirement. Theories of origin in Christian science education should be taught in the light of (1) the amount of specifically supportive evidence, (2) the amount of discrediting evidence, (3) failed or successful experiments, (4) the ability or inability to plausibly explain important phenomena within the realm of applicable theory, (5) failed or successful predictions and (6) design theory.

A winning strategy is based upon open, intensive and continuous discussion within the Christian community about these issues. Change in Christian schools and colleges should take place through teaching, debate and compelling dialogue rather than fiat decree. With this approach, there is plenty of opportunity for motivated Christians to influence the outcome in a positive manner. In short, if Christians want to change the secular world they must first change their own religious culture. Until the church can change the beliefs of those sympathetic to its cause, it has little possibility of changing those outside the church hostile to its cause. Until the church can clean up origins theology in its own house, there is little hope of influencing ideas about origins in the public arena.

For a winning strategy to be effective, action is needed. In 2 Kings 13:14-18 the story of Elisha's death is recorded. Elisha was one of the great prophets of Israel, and God performed many miracles through him. Elisha fell sick and Jehoash, the king of Israel, came to visit him on his death bed. As king, Jehoash was disobedient to God and led the nation of Israel away from serving Yahweh. In spite of this, Elisha was compassionate and wanted to help the king overcome his adversaries. Before he died, Elisha told Jehoash to take a bow and shoot arrows of victory. He shot three arrows and stopped, which made Elisha angry since each arrow represented a victory in battle that God was to give Israel over its enemies. As an instrument of God's offered deliverance, Jehoash failed to exploit his opportunity. Perhaps personal pride kept him from honoring the request of his chief moral critic. Possibly Jehoash did not respect Elisha because of his lack of education, wealth and political connections. We can suspect Jehoash lacked historical perspective and mature leadership vision. As a king accustomed to privilege and power, he may have been self-satisfied with his personal circumstances and indifferent to the plight of the common people. Regardless, Jehoash had a golden opportunity never offered to any other king, but failed to define himself in history by seizing the moment.

When properly interpreted and translated, the Genesis creation account is an abbreviated, non-technical, historical account that can be accepted as truthful and trustworthy without apology. Although written thousands of years ago, the creation story is logical, accurate, carefully structured and sophisticated when judged by twenty-first century standards. The Genesis account stands significantly above all other religious and philosophical explanations of origins regardless of their age.

The great message of the New Testament is the gospel. We are lost in sin and cannot save ourselves, regardless of our good works. The good news is that God sent His son to die for our sins. If we believe in Jesus Christ, repent of our sins, ask His forgiveness and are baptized in obedience to the New Testament pattern, we can be saved. If we walk in a committed, obedient, love relationship with Jesus Christ, then life can have joy, hope, meaning and purpose. The good news of the New Testament combined with the creation account of the Old Testament make a winning combination.

Appendix: A

The *Exemplar* Translation – Genesis Chapters 1 and 2

God inspired Moses to write a creation story that was easily understood, yet correctly explained the origin of the heavens, Earth and life. God knew that any attempt to explain in ancient Hebrew the complex physics and chemistry of creation would be futile. In contrast, He knew that a story could be easily memorized and handed down through generations. God wanted a straightforward, engaging story humankind could believe during its years of development. On the other hand, He wanted an account that was credible when humanity became more culturally and scientifically advanced. Today we know that simple cultures and children are best taught by stories. In advanced cultures, abstract concepts are best explained, understood and remembered through stories. Good stories have the unique ability to transcend time, culture and language. To this end, God orchestrated the events of the Genesis story to create an abbreviated, non-technical account of our origins that is absolutely true.

The words in italics in the following *exemplar* translation are not in the original Hebrew text and are added for grammatical and explanatory purposes, similar to the King James Version of the Bible.

Chapter 1 – *Undefined time in the past*

1. In the beginning, God created the heavens and the Earth.
2. And the Earth was vacant and empty *of life*. And darkness *was* over the surface of the deep *waters*. And the Spirit of God hovered over the surface of the waters.

Day one of the six-day chronicle

3. And God said, "There is light, and light exists."
4. And God saw that the light *was* good. And God distinguished between the light and between the darkness.
5. And God called the light "day" and called the darkness "night." And evening was, and morning was—day one.

The second day

6. And God said, "There is an expanse in the middle of the water, and it exists to separate between water *above* from water *below*."
7. And God *is the one who* made the expanse and separated between the waters which *were* under the expanse and between the waters which *were* over the expanse. And *it* was so.

8. And God called the expanse "sky." And evening was, and morning was—second day.

The third day

9. And God said, "The waters under the sky wait in one place and the dry ground is seen." And *it* was so.
10. And God called the dry ground "land" and called the collection of waters "seas." And God saw that *it was* good.
11. And God said, "The land is sprouting sprouts: plants seeding seeds and the fruit trees making fruit—to which kind is in the seed—over the land." And *it* was so.
12. "And the land brings forth sprouts: plants seeding seeds *according* to *their* kind, and the trees making fruit—in which seed is to kind." And God saw that *it was* good.
13. And evening was, and morning was—third day.

The fourth day

14. And God said, "There are luminaries in the expanse of the heavens to distinguish between the day and between the night, and *they* are a sign *for* appointed times and days and years."
15. "And luminaries exist in the expanse of the heavens to give light over the Earth." And *it* was so.
16. And God *is the one who* made two great luminaries: the greater luminary to rule the day and the smaller luminary to rule the night and the stars.
17. And God gave them in the expanse of the heavens to give light over the Earth.
18. And to rule in the day and in the night and to distinguish between the light and between the darkness. And God saw that *it was* good.
19. And evening was, and morning was—fourth day.

The fifth day

20. And God said, "The waters swarm *with* swarms of living creatures, and fliers flying over the land and over the surface of the expanse of the sky."
21. And God *is the one who* created the great reptiles and every moving living creature with which the waters swarm, *according* to *their* kinds, and every winged flier *according* to *its* kind. And God saw that *it was* good.
22. And God blessed them saying, "Be fruitful and multiply and fill the waters in the seas and multiply the fliers on the land."
23. And evening was, and morning was—fifth day.

The sixth day

24. And God said, "The land brings forth living creatures *according* to *their* kind: beasts and moving *things* and animals of the land, *according* to *their* kind." And *it* was so.
25. And God *is the one who* made the animals of the Earth *according* to *their* kind:

the beasts *according* to *their* kind and every moving *thing* upon the ground *according* to *its* kind. And God saw that *it was* good.

26. And God said, "*We shall* make man in our image, after our likeness. And *he shall* have dominion on fish of the sea and on fliers in the sky and on beasts and on all the Earth and on all the moving *things* that move over the land."

27. And God created man in His image, in God's image created him, male and female *God* created them.

28. And God blessed them, and God said to them, "Be fruitful and multiply and fill the Earth and subdue it and have dominion on the fish of the sea and on the fliers of the sky and on all living *things* that move over the land."

29. And God said, "Behold, *I* give to you every plant seeding seed which is over the surface of all the land, and every tree in which fruit is seeding seed. As food *they* are for you."

30. "And to every animal of the land and to every flier in the sky and to all *that* moves over the land—in which is the living creature—every green plant is food." And *it* was so.

31. And God saw everything that *He* did, and behold *it was* very good. And evening was, and morning was—the sixth day.

Chapter 2 – *The seventh day*

1. And the heavens and Earth *were* finished, and all their multitude *of miscellaneous things*.

2. And on the seventh day God finished His work that *He* did, and rested on the seventh day from all His work that *He* did.

3. And God blessed the seventh day and sanctified it, because on it *He* rested from all His work—that God did *when* creating.

Introduction to a detailed account of the sixth day

4. The *following is the* account of the heavens and the Earth when created; on the *sixth* day *when* Yahweh God did *things to the* Earth and heavens.

The Garden of Eden

5. And each bush of the field was not yet on the land, and each plant of the field *did* not yet grow, because Yahweh God had not caused rain over the land and there was no man to cultivate the ground *of the garden*.

6. And mists rose up from the land and watered all the surface of the ground *of the garden*.

7. And Yahweh God formed the man *of* dust from the ground and breathed in his nostrils the breath *of* life. And the man existed as a living creature.

8. And Yahweh God planted a garden from the east in Eden, and there *He* put the man whom *He had* formed.

9. And Yahweh God grew from the ground every tree *that was* desirable to the sight

and good for food, and the tree of life in the middle of the garden and the tree of knowledge of good and evil.
10. And a river went out from Eden to water the garden and from there diverged and became four branches.
11. The name of the first is Pishon; it winds through all the land of Havilah, where there is gold.
12. And the gold of that land is good; bdellium and onyx stone *are* there.
13. And the name of the second river is the Gihon. It winds through all the land of Cush.
14. And the name of the third river is the Tigris; it goes east of Asshur. And the fourth river is the Euphrates.

Man in the garden

15. And Yahweh God took the man and put him in the Garden of Eden to cultivate and to keep it.
16. And Yahweh God commanded to the man saying, "From any tree of the garden, you *may* eat-eat."
17. "And from the tree of the knowledge of good and evil, *do* not eat from, because on the day you eat from it *you shall* die-die."
18. And Yahweh God said, "*It is* not good that the man be alone. For him *I shall* make as his counterpart a helper."

Living things of the garden created

19. And Yahweh God formed from the ground each animal of the field and each flier of the sky *within the garden*. And *God* brought *them* to the man to see what he *would* call them and whatever the man called the living creature, that *was its* name.
20. And the man called names to all the beasts and to the fliers of the sky and to all animals of the field *within the garden*. And for Adam *was* not found as his counterpart a helper.

Woman is created

21. And Yahweh God caused a deep sleep to fall over the man and *during* sleep took from one of his ribs and closed up under the flesh.
22. And Yahweh God built—which *He* took *of* the rib from the man—into a woman, and brought her to the man.

Marriage is established

23. And the man said, "This is now bone from my bones and flesh from my flesh; for this *shall* be called woman, because this *was* taken from man."
24. So over *this reason,* a man *will* leave his father and his mother and hold fast on his wife. And *they shall* be as one flesh.
25. And the two were naked, the man and his wife, and were not ashamed.

Appendix: B

The Ten Key Verbs of Genesis Chapter 1

This appendix is an extension of Chapter 5, and the study of seven verses in the Genesis creation account containing ten key verbs identified in Table 5.4. These key verbs are traditionally assigned a jussive verb state and considered jussive/command verbs. According to traditional interpretations, God speaks in these seven verses and commands things to occur. In contrast, the *exemplar* translation views Genesis 1:3, 6, 9, 11, 14, 20 and 24 as descriptive statements.

Hebrew verbs can be divided into the perfect and imperfect state. Verbs can be further subdivided into the imperative, jussive, cohortative, infinitive and participle verb states.[1]

All jussive verbs are a subcategory of imperfect verbs. The number of jussive verbs in the Old Testament identified by the *Old Testament Parsing Guide* is quite small.

The complete Hebrew *verb form* is a combination of its *verb state* (mentioned above) and one of *seven verb stems* (verb stems are identified in Chapter 5). Verb form is determined primarily by spelling and secondarily by context. For verb forms with the same spelling, context is used exclusively to determine the verb state. For example, standard verb tables for strong Hebrew verbs show that imperfect 3ms verbs (third person, masculine, singular) and jussive 3ms verbs have the same spelling except for the hiphil stem.[2, 3] In this instance, context must be used to determine between an imperfect or jussive state.

When the context of a verse is neutral or unclear, classifying a verb as imperfect or jussive is a judgment call. Since spelling is identical for most imperfect 3ms and jussive 3ms verbs, selecting the verb state is a *judgment decision* which in turn can be driven by theology. This is the main issue behind traditional explanations of Genesis 1:3, 6, 9, 11, 14, 20 and 24. The primary driver for treating these verses as commands is traditional young-Earth creation theology, which has been the dominant interpretative position for the past twenty-two hundred years. This explanatory position is now challenged.

As mentioned, two main uninspired works are the basis for current interpretations of the Genesis creation story. The first five books of the Septuagint (the Pentateuch) were translated from Hebrew into Greek around 275 BC by Jewish scribes. Traditional young-Earth creation theology was the dominant belief of the time and embedded into the Old Testament though this translation. The ten key verbs identified in Table 5.4 and their accompanying seven verses were translated as commands in the Septuagint.[4, 5] The Masoretic scribe vocal marks (added starting around 500

AD) support the Septuagint explanation. These two uninspired works control present-day interpretations of the creation account.

Exemplar creation now challenges traditional interpretations of the Genesis story. In doing so, the Septuagint translation and Masoretic vocal marks are subjugated to the original Hebrew text, and the Bible becomes the sole basis of authority.

Genesis 1:3, 6 and 14 – Hebrew verb (יהי) meaning "be" or "exist"

The Hebrew verb (יהי) meaning "be" occurs in Genesis 1:3, 6 and 14, and is one of many verbs based on the root (היה). Hebrew verbs based on the root (היה), identified as Strong's 01961, occur approximately 3,482 times in the Old Testament as shown in Table B.1 of this appendix. The verb (יהי) is considered a shortened version (apocopation) of (יהיה) in speech and writing.[6]

The verb (יהי) occurs three times in Genesis 1:3, 6 and 14 as shown in Table 5.4 of Chapter 5 and is assigned a qal jussive verb form by the *Old Testament Parsing Guide*.[7]

Jussive verbs are traditionally considered command verbs. The *American Heritage College Dictionary* defines jussive as: "A word, mood or form used to express a command."[8] Jussive verbs are used in commands as when a king orders his subject to perform a specific task. In Genesis 1:3, 6 and 14 God speaks and commands things to occur. This is the standard explanation given for jussive verbs.

The verb (יהי) occurs as a stand-alone word 50 times in 49 verses in the Old Testament. The *Old Testament Parsing Guide* classifies all 50 occurrences, listed in Table B.3, as qal jussive. On the surface, the classification of all 50 occurrences of the verb (יהי) as jussive strongly supports explaining Genesis 1:3, 6 and 14 as commands of God. This appears to be an open and shut case.

However, a close examination of the 49 verses that contain (יהי) casts serious doubt on the above conclusion. Subtracting Genesis 1:3, 6 and 14 from the above, leaves 46 verses to examine.

As shown below, a careful analysis identifies only two verses out of 46 that are truly commands. The first verse, Genesis 33:9 can be explained as a mutual agreement between two equal brothers. However, for sake of discussion, consider Esau as the older, stronger brother who is commanding the younger brother Jacob. In the second verse, Ezekiel 45:10, it is quite clear that God is commanding the nation of Israel to be honest in their business dealings. In the verses below, the English word translated from (יהי) is underlined and identified by its Strong's number.

- Genesis 33:9 And Esau said, there is abundance for me brother, what <u>is</u>[01961] yours is for you. (ET) [In this instance, Esau told Jacob to keep his gifts.]
- Ezekiel 45:10 Ye <u>shall have</u>[01961] just balances, and a just ephah, and a just

bath. (KJV) [Yahweh spoke through the prophet Ezekiel and commanded the nation of Israel to have consistent weights and measures.]

In examining the remaining 44 verses containing the verb (יְהִי), two verses are clearly requests based on context.

- Judges 6:39 And Gideon said unto God, Let not thine anger be hot against me, and I will speak but this once: let me prove, I pray thee, but this once with the fleece; let it now be[01961] dry only upon the fleece, and upon all the ground let there be dew. (KJV) [Gideon is requesting that the fleece be dry and the ground be wet as a sign from God.]
- 1 Kings 22:13 And the messenger that was gone to call Micaiah spake unto him, saying, Behold now, the words of the prophets declare good unto the king with one mouth: let thy word, I pray thee, be[01961] like the word of one of them, and speak that which is good. (KJV) [The messenger, an emissary from the King of Israel, spoke to the prophet Micaiah and requested that he agree with the other prophets.]

Of the remaining 42 verses containing the verb (יְהִי), four verses make statements about Yahweh. In each case, the speaker invokes a blessing on God.

- 1 Kings 10:9 Blessed be[01961] the LORD thy God, which delighted in thee, to set thee on the throne of Israel: because the LORD loved Israel for ever, therefore made he thee king, to do judgment and justice. (KJV)
- 2 Chronicles 9:8 Blessed be[01961] the LORD thy God, which delighted in thee to set thee on his throne, to be king for the LORD thy God: because thy God loved Israel, to establish them for ever, therefore made he thee king over them, to do judgment and justice. (KJV)
- Psalm 113:2 Blessed be[01961] the name of the LORD from this time forth and for evermore. (KJV)
- Job 1:21 And said, Naked came I out of my mother's womb, and naked shall I return thither: the LORD gave, and the LORD hath taken away; blessed be[01961] the name of the LORD. (KJV)

An examination of the remaining 38 verses identifies 14 that are simply statements. They are not commands. For example, in Exodus 10:10 when Pharaoh speaks the text is traditionally translated, “And he said unto them, Let the LORD be so with you.” This translation makes it appear as if Pharaoh is commanding a blessing upon Moses and the nation of Israel. A better translation is that Pharaoh is acknowledging that God is with Moses. Pharaoh has just been humiliated by the first seven plagues. Moses has beaten all of Pharaoh’s magicians, and they have conceded that the

Table B.1 – Hebrew Verbs Based on the Root (היה) in the Old Testament
(*hâyâh*, Strong's 01961)

Hebrew Word	First Occurrence	No. Times Occurs	Hebrew Word	First Occurrence	No. Times Occurs
ויהי	Gen. 1:3	808	בהיות	Exod. 5:13	14
יהיה	Gen. 1:29	423	ואהי	Judg. 18:4	13
והיה	Gen. 2:10	408	נהיה	Gen. 38:23	13
היה	Gen. 3:1	274	הייתם	Gen. 22:21	12
היו	Gen. 6:4	173	ואהיה	Gen. 26:3	12
תהיה	Gen. 4:12	165	תהיין	Gen. 41:36	12
והיו	Gen. 1:14	120	היינו	Gen. 42:31	11
ויהיו	Gen. 2:25	117	מהיות	Ruth 1:12	8
היתה	Gen. 1:2	116	שהיה	Ps. 124:1	8
יהיו	Gen. 6:9	115	ותהיינה	Num. 36:11	8
ותהי	Gen. 10:10	101	נהיתה	Exod. 11:6	8
והיתה	Gen. 9:13	93	להית	Exod. 23:1	6
להיות	Gen. 10:8	67	והיינו	Gen. 34:16	5
יהי	Gen. 1:3	50	בהיותם	Gen. 4:8	5
הייתי	Gen. 31:40	49	שיהיה	Eccles. 1:9	5
אהיה	Exod. 3:12	43	בהיותו	Ps. 63:1	4
היית	Gen. 40:13	29	מהית	Exod. 9:28	3
והיית	Gen. 17:4	27	ונהי	Num. 13:33	3
תהי	Gen. 13:8	25	ונהיה	Gen. 47:19	3
תהיינה	Lev. 23:15	20	היותם	Exod. 10:6	3
והייתי	Gen. 4:14	18	בהיתו	1 Kgs. 12:6	3
תהיו	Gen. 34:15	17	ותהיין	Gen. 26:35	2
והייתם	Gen. 3:5	17	ונהיתה	Ezek. 21:12	2
היות	Gen. 2:18	16	היהיה	2 Kgs 7:2	2

Note: There are approximately 3,482 occurrences of words and terms based on the Hebrew root (היה) in the Old Testament identified by reference (1) which represent 73 derivatives.

References:

(1) van der Merwe, Christo, ed. (2004) *The Lexham Hebrew-English Interlinear Bible.*

(2) Brown, Francis and S. R. Driver, Charles A. Briggs, (1907). *A Hebrew and English Lexicon of the Old Testament,* pp224-227.

plagues are due to the finger of God (Exodus 8:19). In Exodus 10:10, Pharaoh is simply acknowledging that God is with Moses. Keep in mind that Pharaoh worshiped the Egyptian gods and it would have been totally out of character for him to command a blessing in the name of Moses' God.

- Genesis 49:17 Dan shall be[01961] a serpent by the way, an adder in the path, that biteth the horse heels, so that his rider shall fall backward. (KJV) [Immediately before he died Jacob told all of his sons, including Dan, what was to happen to them in the future. This verse is a prophecy of coming events rather than a command or wish.]
- Genesis 30:34 And Laban said, Behold, I would it might be[01961] according to thy word. [Jacob had just made a proposition to Laban for his wages, and Laban agrees to the proposal.]
- Exodus 7:9 When Pharaoh shall speak unto you, saying, Shew a miracle for you: then thou shalt say unto Aaron, Take thy rod, and cast it before Pharaoh, and it shall become[01961] a serpent. (KJV) [God told Moses what would happen to his rod when he cast it down before Pharaoh. Moses was commanded to cast down the rod, but the rod itself was not being commanded or wished.]
- Exodus 10:10 "And he said to them, "Yahweh is[01961] with you so that you are bold in trying to gain release for your children. I see that you have some evil purpose in mind." (ET) [The unbelieving Pharaoh said Yahweh was with Moses. The meaning in the first part of this verse in the Hebrew text is quite plain. Pharaoh is simply stating a fact rather than a command, wish or desire.]
- Deuteronomy 32:38 Which did eat the fat of their sacrifices, and drank the wine of their drink offerings? Let them rise up and help you, and be[01961] your protection. (KJV) [This verse is part of a song in Deut. 32:1-44 that Moses sang to the people of Israel.]
- 1 Kings 8:57 The LORD our God be[01961] with us, as he was with our fathers: let him not leave us, nor forsake us. (KJV) [Solomon did not have the authority to command God to be with the nation of Israel. He *asked* God to be with Israel.]
- Job 24:14 The murderer rising with the light killeth the poor and needy, and in the night is[01961] as a thief. (KJV)
- Job 27:7 Let mine enemy be[01961] as the wicked, and he that riseth up against me as the unrighteous. (KJV)
- Psalm 72:16 There shall be[01961] an handful of corn in the earth upon the top of the mountains; the fruit thereof shall shake like Lebanon: and they of the city shall flourish like grass of the earth. (KJV)
- Psalm 72:17 His name shall endure[01961] for ever: his name shall be continued as long as the sun: and men shall be blessed in him: all nations shall call him blessed. (KJV)

• Psalm 104:31 The glory of the LORD <u>shall endure</u>[01961] for ever: the LORD shall rejoice in his works. (KJV)
• Jeremiah 20:14 Cursed be the day wherein I was born: let not the day wherein my mother bare me <u>be</u>[01961] blessed. (KJV)

Table B.2 – Occurrences of the Hebrew Verb (היתה) in the Old Testament
(Strong's 01961, *hâyâh* a stative verb meaning "be" or "exist")

Hebrew Word	KJV Translation	Number of Occurrences	Verse Location
היתה	Was	41	Gen. 1:2, 3:20, 29:17, 36:12; Exod. 16:24, 36:7; Deut. 2:15; Josh. 11:20; Ruth 1:7; 1 Sam. 14:20; 2 Sam. 3:37, 10:9, 14:27; Judg. 2:15; 1 Kgs. 2:15, 12:15, 18:46; 2 Kings 8:18; 1 Chron. 19:10; 2 Chron. 1:11, 10:15, 22:3, 22:7, 22:11, 30:12; Ezra 8:31; Est. 2:20; Ps. 114:2; Isa. 11:16; Jer. 26:24; Ezek. 16:56; 26:17, 31:3, 33:22, 36:17, 37:1, 40:1; Jonah 3:3; Mal. 2:5, 2:6
היתה	There was	15	Gen. 38:21, 38:22; Exod. 8:15; Deut. 2:36, 3:4; Josh. 11:19; 1 Sam. 5:11; 1 Kgs. 14:30, 15:16, 15:32, 15:6, 15:7; 2 Chron. 13:2, 14:14, 15:19
היתה	Had, have or hath	12	Gen. 18:12; Lev. 21:3; Num. 14:24; Josh. 17:6, 17:8; 1 Kgs. 4:11; 2 Chron. 21:6; Est. 8:16; Ps. 119:56; Eccl. 6:3; Isa. 50:11; Ezek. 44:25
היתה	Is	10	Judg. 21:3; 1 Kgs. 11:11; 2 Sam. 17:9; Ps. 118:23; Prov. 31:14; Isa. 64:10; Jer. 12:8, 25:38; Lam. 1:17, 1:8
היתה	Been	9	1 Sam. 4:17, 14:38, 4:7; 2 Sam. 13:32; Ezra 9:2, 9:8; Ps. 42:3; Jer. 32:31; Mal. 1:9
היתה	Become	8	Ps. 118:22; Isa. 1:21; Jer. 50:23, 51:41; Lam. 1:1; Ezek. 36:35; Zeph. 2:15
היתה	Came or come	7	Judg. 21:5; 2 Kgs. 24:3, 24:20; 2 Chron. 15:1, 20:14; Isa. 14:24; Jer. 52:3
היתה	Became	4	Gen. 47:26; Exod. 9:24; Josh. 14:14; 1 Sam. 10:12
היתה	Be	2	Jer. 2:10; Ezek. 21:12
היתה	Pertaineth	2	2 Kgs. 24:7; 1 Sam. 27:6
היתה	Not translated	2	Exodus 16:13; Joel 2:3
היתה	Are	1	Ezek. 36:2
היתה	By reason	1	Ezek. 19:10
היתה	Lay	1	Ezek. 36:34
היתה	Went	1	1 Chron. 7:23

Note: According to reference (1), (היתה) occurs 116 times in 114 verses in the Old Testament.

References:
(1) van der Merwe, Christo, ed. (2004) *The Lexham Hebrew-English Interlinear Bible.*
(2) Holy Bible, King James Version (1981). New York: Cambridge University Press.

• Jeremiah 42:5 Then they said to Jeremiah, The LORD <u>be</u>[01961] a true and faithful witness between us, if we do not even according to all things for the which the LORD thy God shall send thee to us. (KJV) [The people declared, with Yahweh as their witness, they would obey God as He spoke through Jeremiah. This is not a command or wish.]

• Ezekiel 16:15 But thou didst trust in thine own beauty, and playedst the harlot because of thy renown, and pouredst out thy fornications on every one that passed by; his it <u>was</u>[01961]. (KJV)

Of the 46 verses examined, 2 verses are commands, 2 are requests, 4 invoke a blessing on God and 14 make statements. The remaining 24 verses can be considered a mixture of wishes or statements depending on the language of the text. The significant issue is that only 2 verses out of 46 are clearly commands based on context. If we accept the classification of all 50 occurrences of (יְהִי) by the *Old Testament Parsing Guide* as jussive, then we are forced to question the previous dictionary definition of a jussive verb.

Some Hebrew grammar books mention that jussive verbs are a shortened version of imperfect verbs. *Gesenius' Hebrew Grammar* says, "Along with the usual form of the imperfect, there exists also a lengthened form of it (the cohortative), and a shortened form (the jussive)."[9] *Biblical Hebrew* by Harrison agrees and says, "This is in speech a shortened form of the imperfect to express the quick reaction of the mind to a situation, generally in terms of a wish or a command."[10] So if we define a jussive verb as a shortened version of an imperfect verb, then (יְהִי) fits the description perfectly. However, five of the jussive verbs identified in Genesis 1:9, 11, 20 and 24 in Table 5.4 are not shortened versions of the root. So either the definition is wrong or these five verbs are not categorized correctly.

Both Harrison and Pratico in their Hebrew grammar books say a jussive verb indicates the *speaker's command or wish.*[11, 12] There is a huge difference between a command and a wish—they are almost opposites. Any beggar can wish, but only the rich and powerful can command. Since only 2 out of the 46 verses containing (יְהִי) are clearly commands, the remainder must be something else. In truth, an analysis of all 46 verses shows that most are statements, requests, curses, blessings or at best wishes—they are definitely not commands.

The verb (יִהְיֶה) is not a jussive/command verb. As previously mentioned, (יְהִי) is considered an apocopation (shortened version) of (יִהְיֶה) in the Old Testament[13] The verb (יִהְיֶה) occurs 423 times in the Old Testament as shown in Table B.1. Although not all 423 occurrences were examined, the first 100 instances between Genesis 1:29 and Leviticus 25:4 are assigned a qal imperfect verb form by the *Old Testament Parsing Guide*. None are assigned a jussive verb form.

In summary, the Hebrew verb (יְהִי) is used throughout the Old Testament in verses that are clearly not commands. When Genesis 1:3, 6 and 14 are excluded, then

(יְהִי) is used as a command in only 2 out of 46 verses. We can conclude that (1) the verb (יִהְיֶה), of which (יְהִי) is considered a shortened version, is never assigned a jussive state, (2) the verb (יְהִי), excluding Genesis chapter 1, is not used in commands in 44 out of 46 verses in the Old Testament, (3) the context of Genesis 1:3, 6 and 14 do not force (יְהִי) to be a jussive/command verb and (4) the verb (יְהִי) is translated as the English stative verbs "be, have, is and was" 86% of the time in the KJV and (5) dynamic rather than stative verbs are used in commands. Therefore, an imperfect state is assigned to this verb. We can reject dogmatic claims that (יְהִי) is a jussive/command verb in Genesis 1:3, 6 and 14.

Genesis 1:6 – Hebrew term (וִיהִי) meaning "and be" or "and exist"

The Hebrew term (וִיהִי) meaning "and be" occurs in Genesis 1:6 as shown in Table 5.4. The term (וִיהִי) comes from (יְהִי) discussed in the previous section, but

Table B.3 – Occurrences of the Hebrew Verb (יְהִי) in the Old Testament
(Strong's 01961, *hâyâh* a stative verb meaning "be" or "exist")

Hebrew Word	KJV Translation	Number of Occurrences	Verse Location
יְהִי	Be	18	Gen. 30:34, 49:17; Deut. 32:38; Ruth 2:19; 2 Sam. 14:17; 1 Kgs. 1:37, 8:57, 10:9; 2 Kgs. 2:10; 1 Chron. 22:11; 2 Chron 9:8; Ezra 1:3; Job 1:21, 18:12; Ps. 72:16, 113:2, 122:7; Jer. 42:5
יְהִי	Let be	15	Ex. 10:10; Deut. 33:24; Jud. 6:39; 1 Kgs. 22:13; 2 Chron. 19:7; Job 3:4, 3:7, 27:7; Ps. 33:22, 35:6, 109:13, 119:76, 119:80; Prov. 5:18; Jer. 20:14
יְהִי	Let there be	5	Genesis 1:3, 1:6, 1:14; Psalm 109:12 (twice)
יְהִי	Let	2	Psalm 69:25; Jeremiah 50:29
יְהִי	Shall endure	2	Psalm 72:17, 104:31
יְהִי	Have	2	Job 16:18; Ezekiel 45:10
יְהִי	Become	2	Exodus 7:9; Psalm 69:22
יְהִי	Is	2	Job 20:23; 24:14
יְהִי	Keep	1	Genesis 33:9
יְהִי	Was	1	Ezekiel 16:15

Note: The verb (יְהִי) occurs 50 times in 49 verses. In many instances, the Hebrew text must be examined to identify where the Hebrew verb is placed in a verse in order to determine its translation.
Reference:
(1) van der Merwe, Christo, ed. (2004) *The Lexham Hebrew-English Interlinear Bible.*
(2) Holy Bible, King James Version (1981). New York: Cambridge University Press.

with the vav (ו) prefix added. The *Old Testament Parsing Guide* assigns (ויהי) a qal jussive verb form in this verse, which allegedly makes it a command verb.

The term (ויהי) occurs 21 times in Genesis Chapters 1 and 2 as shown in Table B.4. In the 20 occurrences outside of Genesis 1:6, the *Old Testament Parsing Guide* assigns a qal wci form (qal waw consecutive imperfect form) to this verb expression.

The term (ויהי) occurs 808 times in the entire Old Testament as shown in Table B.1. Although not all of these occurrences were examined, a sampling of verses indicates that in the vast majority of instances (ויהי) is assigned a qal wci verb form by the *Old Testament Parsing Guide.* Of the first 50 occurrences in the Old Testament, only 3 instances in Genesis 1:6, 9:26 and 9:27 are assigned a qal jussive. All 47 remaining occurrences are assigned a qal wci verb form.

Vocal marks inserted into the Hebrew text in Genesis 1:6 by the Masoretic scribes make (ויהי) a jussive verb. For example, in Genesis 1:6, 9:26 and 9:27 the vocal marks assigned to (ויהי) make it a jussive verb in contrast to the identical term in Genesis 1:3, 5 and 7. Based on spelling and verse context, there is nothing in the Hebrew text that forces (ויהי) in Genesis 1:6, 9:26 and 9:27 to be a jussive/command verb.

We can conclude that (1) vocal marks inserted by the Masoretes make (ויהי) a jussive verb in Genesis 1:6, (2) vocal marks are not in the original Hebrew text of the Old Testament, (3) the Hebrew term (ויהי) is assigned a non-jussive state in the vast majority of occurrences in the Old Testament and (4) the context of Genesis 1:6 does not force (ויהי) to be a jussive/command verb. Therefore, an imperfect state is assigned to this verb. We can reject claims that (ויהי) is a jussive/command verb in Genesis 1:6.

Genesis 1:9 – Hebrew verb (יקוו) meaning "wait"

The Hebrew verb (יקוו) meaning "wait," identified as Strong's 06960, occurs in Genesis 1:9; Isaiah 51:5 and 60:9 as shown in Table 5.4. A similar word (קוו) occurs in Job 6:19; Psalm 56:6; 119:95; Lamentations 3:25 and Jeremiah 3:17 as shown in Table 5.1. As discussed in Chapter 5, in Genesis 1:9 (יקוו) should be translated "wait" rather than the illegitimate term "let be gathered together."

The verb (יקוו) occurs in Genesis 1:9 and is assigned a niph jusm verb form (rather than niph *jussive* verb form) by the *Old Testament Parsing Guide.* The parsing guide points out that, based on spelling, this verb is not jussive. The "jussive/command" meaning assigned is because of theological influence.

The verb (יקוו) is based on the root (קוה). Of the 49 verbs based on this root in the Old Testament, Job 3:9 is the only one assigned a piel jussive form by the *Old Testament Parsing Guide.* All 48 remaining verbs listed in Table 5.1 are primarily a mixture of the piel imperfect, piel perfect and qal participle verb forms based on their spelling and context.

We can conclude that (1) the spelling of the Hebrew word (יקוו) shows it is not a jussive verb, (2) all verbs based on the root (קוה) listed in Table 5.1 are assigned a non-jussive state except for one, (3) the context of Genesis 1:9 does not force (יקוו) to be a jussive/command verb and (4) a jussive/command interpretation of this verse forces (יקוו) to be mistranslated as "let be gathered together" rather than correctly translated "wait." Therefore, an imperfect state is assigned to this verb. We can reject claims that (יקוו) is a jussive/command verb in Genesis 1:9.

Genesis 1:9 – Hebrew term (ותראה) meaning "and see"

The Hebrew term (ותראה) meaning "and see," identified as Strong's 07200, occurs in Genesis 1:9. The term (ותראה) is the same as (תראה) but with the vav (ו) prefix added. The term (ותראה) occurs in Genesis 1:9 and is assigned a niph jusm verb form (rather than niph *jussive* verb form) by the *Old Testament Parsing Guide*.

Table B.4 – Occurrences of the Term (וַיְהִי) in Genesis Chapters 1 and 2
(Strong's 01961, *hâyâh* and a conjunction meaning "and be" or "and exist")

Verse	Hebrew Term	Number of occurrences	KJV Translation	Assigned Verb Form[2]
Gen. 1:3	וַיְהִי	1	and there was	Qal wci
Gen. 1:5	וַיְהִי	2	and were[1]	Qal wci
Gen. 1:6	וִיהִי	1	let there be	Qal jussive
Gen. 1:7	וַיְהִי	1	and was	Qal wci
Gen. 1:8	וַיְהִי	2	and were[1]	Qal wci
Gen. 1:9	וַיְהִי	1	and was	Qal wci
Gen. 1:11	וַיְהִי	1	and was	Qal wci
Gen. 1:13	וַיְהִי	2	and were[1]	Qal wci
Gen. 1:15	וַיְהִי	1	and was	Qal wci
Gen. 1:19	וַיְהִי	2	and were[1]	Qal wci
Gen. 1:23	וַיְהִי	2	and were[1]	Qal wci
Gen. 1:24	וַיְהִי	1	and was	Qal wci
Gen. 1:30	וַיְהִי	1	and was	Qal wci
Gen. 1:31	וַיְהִי	2	and were[1]	Qal wci
Gen. 2:7	וַיְהִי	1	and became	Qal wci

Notes:
(1) The term (וַיְהִי) is only translated once in these verses in the KJV even though it occurs two times in Genesis 1:5, 8, 13, 19, 23 and 31.
(2) Assigned verb form based on the *Old Testament Parsing Guide* by Beall, Banks and Smith.

Based on spelling, this verb is not jussive. The "jussive/command" meaning assigned is because of theological influence.

The Hebrew verb (תראה) occurs 39 times in the Old Testament as shown in Table B.5. Genesis 1:9 is the only instance where the *Old Testament Parsing Guide* assigns (תראה) a jusm state. All 38 remaining occurrences are non-jussive.

We can conclude that (1) the spelling of (ותראה) shows it is not a jussive verb, (2) Genesis 1:9 is the only instance in the Old Testament where the verb (תראה) is interpreted as a jussive verb, (3) the context of Genesis 1:9 does not force (ותראה) to be a jussive/command verb and (4) all jussive verbs are a subcategory of imperfect verbs. Therefore, an imperfect state is assigned to this verb. We can reject claims that (ותראה) is a jussive/command verb in Genesis 1:9.

Genesis 1:11 – Hebrew verb (תדשא) meaning "sprouting"

The word (תדשא) meaning "sprouting" occurs in Genesis 1:11 and is based on the root (דשא). Hebrew nouns and verbs based on the root (דשא) occur 16 times in the Old Testament. In 14 occurrences, (דשא) is considered a noun and identified as Strong's 01877. In 2 occurrences, (דשא) is considered a verb and identified as Strong's 01876.[14] The two verbs occur in Genesis 1:11 as (תדשא) and in Joel 2:22 as (דשאו) and are respectively assigned a hiphil jussive and qal perfect by the *Old Testament Parsing Guide.*

We can conclude that (1) the spelling of the word (תדשא) is not a shortened form of the root (דשא), (2) Genesis 1:11 is the only instance in the Old Testament where the verb (תדשא) is assigned a jussive state, (3) the context of Genesis 1:11 does not force (תדשא) to be a jussive/command verb and (4) all jussive verbs are a subcategory of imperfect verbs. Therefore, an imperfect state is assigned to this verb identified in Table 5.4. We can reject claims that (תדשא) must be a jussive/command verb in Genesis 1:11.

Genesis 1:20 – Hebrew verb (ישרצו) meaning "swarming"

The word (ישרצו) meaning "swarming" occurs in Genesis 1:20 and is based on the root (שרץ). Hebrew nouns and verbs based on the root (שרץ) occur 29 times in the Old Testament. In 15 occurrences, words based on the root (שרץ), identified as Strong's 08318, are considered nouns. In 14 occurrences, other words based on the root (שרץ), identified as Strong's 08317, are considered verbs.[15]

The 14 verbs based on the root (שרץ) occur in Gen. 1:20, 1:21, 7:21, 8:17, 9:7; Exodus 1:7, 7:28; Lev. 11:29, 11:41, 11:42, 11:43, 11:46; Psalm 105:30 and Ezek. 47:9. The specific verb (ישרצו) occurs in Genesis 1:20 and is assigned a qal jusm verb form (rather than a qal *jussive* verb form) by the *Old Testament Parsing Guide.*

The parsing guide points out that, based on spelling, this verb is not jussive. The "jussive/command" meaning assigned is because of theological influence. All 13 remaining verbs based on the root (שרץ) are assigned a non-jussive state.

We can conclude that (1) the spelling of (ישרצו) shows it is not a jussive verb, (2) Genesis 1:20 is the only instance in the Old Testament where a verb based on the root (שרץ) is assigned a "jusm" state, (3) the context of Genesis 1:20 does not require (ישרצו) to be a jussive/command verb and (4) a command interpretation of this verse has caused (ישרצו) to be mistranslated as "let bring forth abundantly" rather than correctly translated "swarm." Therefore, an imperfect state is assigned to this verb identified in Table 5.4. We can reject claims that (ישרצו) is a jussive/ command verb in Genesis 1:20.

Genesis 1:20 – Hebrew verb (יעופף) meaning "flying"

The word (יעופף) meaning "flying" occurs in Genesis 1:20 and is based on the root (עוף). There are approximately 118 nouns and verbs based on the root (עוף)

Table B.5 – Occurrences of (תראה) in the Old Testament
(Strong's 07200, a verb meaning "see")

Hebrew Term	Basic Translation	Hebrew Word	Assigned Verb Form[1]	Verse Location
ותראה	and (you/she) see	תראה	Niph jusm	Genesis 1:9
ותראה	and (you/she) see	תראה	Qal wci	Jeremiah 3:7
ותראהו	and (you/she) see he	תראה	Qal wci	Exodus 2:6
נתראה	(us) see	תראה	Hith cohm	2 Kings 14:8; 2 Chron. 25:17
תראה	(you/she) see	תראה	Hiph impf	Leviticus 13:57; Isaiah 47:3
תראה	(you/she) see	תראה	Qal impf	Exodus 6:1, 23:5; Lev.20:17; Num. 11:23, 23:13, 23:13; Deut. 3:28, 12:13, 22:1, 22:4, 28:34, 28:67, 32:52; 1 Sam. 1:11; 2 Sam. 3:13; 2 Kings 2:10; Job 10:4, 22:11, 33:28, 38:17, 38:22; Psalm 35:17, 37:34, 91:8; Ecc. 5:7; Isaiah 33:19, 58:7; Ezek. 8:6, 8:13, 8:15, 12:6; Dan. 1:13

Notes:
(1) Assigned verb form based on the *Old Testament Parsing Guide* by Beall, Banks and Smith.
(2) The exact spelling of the Hebrew verb (תראה) from the root (ראה) occurs 39 times in 38 verses in the Old Testament.
(3) Derivatives based on the root (ראה) occur approximately 1,313 times in the Old Testament.
Reference: van der Merwe, Christo, ed. (2004) *The Lexham Hebrew-English Interlinear Bible.*

identified by a Hebrew word search of the Old Testament.[16] Nouns are identified as Strong's 05775. About 30 verbs are identified as Strong's 05774.[17] Verbs based on the root (עוף) have several variations and a number of different spellings. The specific verb (יעופף) occurs in Genesis 1:20 and is assigned a polel jusm verb form (rather than a polel *jussive* verb form) by the *Old Testament Parsing Guide*. The parsing guide points out that, based on spelling, this verb is not jussive. The "jussive/command" meaning assigned is because of theological influence. All 29 remaining verbs based on the root (עוף) are assigned a non-jussive verb state.

We can conclude that (1) the spelling of (יעופף) shows it is not a jussive verb, (2) Genesis 1:20 is the only instance in the Old Testament where a verb based on the root (עוף) is assigned a "jusm" state and (3) verse context does not force (יעופף) to be a jussive/command verb. Therefore, an imperfect state is assigned to this verb identified in Table 5.4. We can reject claims that (יעופף) is a jussive/command verb in Genesis 1:20.

Genesis 1:24 – Hebrew verb (תוצא) meaning "bring forth"

The Hebrew verb (תוצא) meaning "bring forth" occurs in Genesis 1:24 and is based on the root (יצא). Words based on this root occur 1,069 times in the Old Testament with a number of prefixes and suffixes.[18] Of the first 50 occurrences of verbs based on the root (יצא), Genesis 1:24 is the only instance where the *Old Testament Parsing Guide* assigns it a jussive state. In all 49 remaining occurrences, it is non-jussive.

The exact verb (תוצא) only occurs in the Old Testament in Genesis 1:24 and is assigned a hiph jussive verb form by the *Old Testament Parsing Guide*. A similar spelling occurs in Genesis 1:12 and Ruth 2:18 in the form of (ותוצא). Because of the vav (ו) in the prefix, (ותוצא) is assigned a hiph wci verb form in both occurrences.

We can conclude that (1) the spelling of (ותוצא) is not a shortened form of the root (יצא), (2) essentially all verbs based on the root (יצא) are assigned a non-jussive state and (3) the context of Genesis 1:24 does not force (תוצא) to be a jussive/command verb. Therefore, an imperfect state is assigned to this verb identified in Table 5.4. We can reject claims that (תוצא) must be a jussive/command verb in Genesis 1:24.

The Myth of the Jussive/Command Verb

As previously mentioned, jussive verbs are normally explained as command verbs, similar to a king commanding one of his subjects. English dictionaries often define jussive as: "A word, mood or form used to express a command."[19] However, Hebrew grammar books describe the jussive verb quite differently as follows:

• *Gesenius' Hebrew Grammar*: "Along with the usual form of the imperfect, there exists also a lengthened form of it (the cohortative), and a shortened form (the jussive)."[20]

• *Gesenius' Hebrew Grammar*: "The general characteristic of the jussive form of the imperfect is rapidity of pronunciation, combined with a tendency to retract the tone from the final syllable, in order by that means to express the urgency of the command in the very first syllable."[21]

• *Gesenius' Hebrew Grammar*: "The meaning of the jussive is similar to that of the cohortative, except that in the jussive the command or wish is limited almost exclusively to the 2nd or 3rd person."[22]

• *Biblical Hebrew* by Harrison: "This is in speech a shortened form of the imperfect to express the quick reaction of the mind to a situation, generally in terms of a wish or a command. It coincides with the imperfect in form in all parts of the regular verb except the hiphil (or causative) but is used only in the second and third persons."[23]

• *Biblical Hebrew* by Harrison: A jussive verb indicates the speaker's command or wish.[24]

• *Basics of Biblical Hebrew Grammar* by Pratico and Van Pelt: "The next three conjugations (Imperative, Cohortative and Jussive) are volitional conjugations, meaning they are used to express some type of command, wish or desire."[25]

• *Basics of Biblical Hebrew Grammar* by Pratico and Van Pelt: "The jussive conjugation is also used to express either some type of mild command or strong wish. Strictly speaking, it occurs only in the third person, singular and plural."[26]

• *Learn Biblical Hebrew* by Dobson: Jussive is "A form of the imperative with a prefixed (י) (he, they) or (ת) (you)."[27]

• *Learn Biblical Hebrew* by Dobson: In most instances jussive forms are the same as imperfective forms. So it is only from the context that we can tell whether a word is imperfective or jussive.[28]

• *Learn Biblical Hebrew* by Dobson: What someone wishes, longs or prays for may be expressed using the imperfective or jussive state.[29]

From these four Hebrew grammar books, there appears to be little consensus as to the meaning of a jussive verb. About all we can conclude is that: (1) A jussive verb is a shortened form of an imperfect verb. The shortened form applies to both writing and speaking. (2) A jussive verb can be used to express a command or wish. (3) Context must be used to determine if a verb is jussive or imperfect.

It is important to notice what the above grammar books *do not say.* None of them say that a jussive verb *must be* a command verb. None of them say that jussive verbs *are only used* within the context of a command. We should keep in mind that comments in the above books are dominated by the theological interpretation given to Genesis Chapter 1. Quotations from the creation account, such as Genesis 1:3 (let there be light), are often given to illustrate jussive verbs.

As we examine the usage of jussive verbs in the Old Testament, a much different and clearer picture emerges. Part of this information has already been presented in the above discussion of the ten key verbs of Table 5.4. Additional verses shed more light on this subject.

An analysis of verses beginning with the expression "And God said" shows the fallacy of the "jussive/command" verb. The Hebrew expression, "And God said" occurs at the start of the seven key verses being examined. The same expression also occurs 16 other different times where God is directly speaking. In 8 verses God issues a command, while the remaining 8 are statements. The command verses are Genesis 6:13, 17:9, 17:15, 17:19, 21:12, 35:1; Exodus 3:14 and Numbers 22:12. If jussive verbs are command verbs, then one would expect all eight of these verses to contain jussive verbs. However, none contain a single jussive verb based on word spelling according to the *Old Testament Parsing Guide*. Context and language define these verses as commands—not jussive verbs.

• Genesis 6:13-14 And God said unto Noah, The end of all flesh is come before me; for the earth is filled with violence through them; and, behold, I will destroy them with the earth. Make thee an ark of gopher wood; rooms shalt thou make in the ark, and shalt pitch it within and without with pitch. (KJV) [God commands Noah to make an ark.]
• Genesis 17:9 And God said unto Abraham, Thou shalt keep my covenant therefore, thou, and thy seed after thee in their generations. (KJV) [God commands Abraham to keep the covenant.]
• Genesis 17:15 And God said unto Abraham, As for Sarai thy wife, thou shalt not call her name Sarai, but Sarah shall her name be. (KJV) [God commands a name change for Sarai.]
• Genesis 17:19 And God said, Sarah thy wife shall bear thee a son indeed; and thou shalt call his name Isaac: and I will establish my covenant with him for an everlasting covenant, and with his seed after him. (KJV) [God commands what the child's name is to be.]
• Genesis 21:12 And God said unto Abraham, Let it not be grievous in thy sight because of the lad, and because of thy bondwoman; in all that Sarah hath said unto thee, hearken unto her voice; for in Isaac shall thy seed be called. (KJV) [God commands Abraham to listen to Sarah.]
• Genesis 35:1 And God said unto Jacob, Arise, go up to Bethel, and dwell there: and make there an altar unto God, that appeared unto thee when thou fleddest from the face of Esau thy brother. (KJV) [God commands Jacob to go to Bethel.]
• Exodus 3:14 And God said unto Moses, I AM THAT I AM: and he said, Thus shalt thou say unto the children of Israel, I AM hath sent me unto you. (KJV) [God tells Moses what he should say.]
• Numbers 22:12 And God said unto Balaam, Thou shalt not go with them; thou

> shalt not curse the people: for they are blessed. (KJV) [God tells Balaam the course of action to take.]

If jussive verbs are truly command verbs, then one would expect them to be found in important commands in the Bible. For example, Exodus 20:3-17 and Deuteronomy 5:7-21 record when God gave the Ten Commandments to the nation of Israel. God directly commanded the Hebrew people regarding things they were to do and not do. However, none of the verbs in these thirty verses are identified as jussive by the *Old Testament Parsing Guide*.

In addition to the Ten Commandments, the Bible is full of other verses where God gives commands to the nation of Israel. For example, Leviticus 18:8-22 lists fifteen commands to abstain from specific forms of nakedness and sexual behavior. Leviticus 19:9-19 lists ten miscellaneous commands about business and personal relationships. None of these twenty-five verses contain jussive verbs. Of the many different commands given by God throughout Leviticus and Deuteronomy, essentially none of them contain jussive verbs.

In summary, in Genesis Chapter 1 there are ten key verbs (representing eight different words) in seven verses traditionally treated as jussive/command verbs. Four of these words (ישרצו, ותראה, יקוו and יעופף) are not jussive verbs based on spelling. The remaining four words (תדשא, ויהי, יהי and תוצא) are identified as jussive verbs by the *Old Testament Parsing Guide*, but are seldom used in commands in the Old Testament. Outside of Genesis Chapter 1, more than 97% of the time these four verbs are used in other than a command context in the Old Testament.

The "jussive/command" verb is mainly due to theological influence and is not well supported by scripture. The link between jussive verbs and commands based on context is exceptionally weak. Therefore, (1) the belief that jussive verbs are only command verbs is patently false, (2) the belief that any verse containing a jussive verb must be a command is false and (3) treating (תדשא, ויהי, יהי and תוצא) in Genesis Chapter 1 as command verbs is not well supported elsewhere in the Old Testament. The jussive/command verb is primarily myth.

Conclusion

Sound hermeneutics and exegesis of scripture lead to the conclusion that the ten key verbs identified in Table 5.4 should be assigned an imperfect verb state rather than jussive (with no requirement for a command). When we lay aside the Septuagint translation and Masoretic vocal marks, nothing is inherent within Genesis 1:3, 6, 9, 11, 14, 20 and 24 that require them to be commands. At best, treating these verses as commands was a bad judgment call. At worst, scripture was deliberately manipulated (although with good intentions) and countless generations have been misled. In this regard, the Septuagint translators and Masoretic scribes have been eminently successful in promoting their theology through the guise of scripture.

We should be aware of the current state of biblical scholarship in regard to this discussion of jussive verbs. All biblical Hebrew parsing guides show the ten key verbs in these seven verses as jussive (or jusm). All Hebrew grammar books which discuss these ten verbs show them as jussive. All commentaries on the Genesis creation story treat these seven verses as commands. All English Bibles treat these ten verbs as commands through use of the word "let." All biblical Hebrew scholars, theologians and their predecessors for countless generations have been taught these are jussive/command verbs.

Biblical scholarship in support of traditional explanations of the Genesis creation story appears overwhelming. One would expect no less from a paradigm that has dominated biblical origins thinking for more than twenty-two hundred years. Theological influence has built a significant barrier to correctly understanding these ten key verbs.

Once a paradigm has been established, it is difficult to change. Even when error has been clearly exposed, it is not easy to go against the flow of opinion. Rock solid evidence is needed to counteract doctrinal inertia built up over hundreds of years. This evidence is now provided by (1) an appeal to the original God inspired text of the Old Testament and (2) powerful support from general revelation.

When all of the material is carefully examined, we can confidently accept the verbs of Table 5.4 as imperfect—with no requirement for a command. These ten imperfect verbs are used in seven descriptive statements within the six-day Genesis chronicle. In Genesis 1:3-25, when God spoke during the first five days and part of the sixth He was viewing a mature creation that had existed for a significant length of time. Yahweh created the Earth and all therein (except for man, the garden and animals within the garden) in the undefined past prior to Genesis 1:3.

Appendix: C

Comparing Translations – Genesis Chapters 1 and 2

***Exemplar* Translation**
Chapter 1

In the beginning
1. In the beginning, God created the heavens and the Earth.
2. And the Earth was vacant and empty *of life*. And darkness *was* over the surface of the deep *waters*. And the Spirit of God hovered over the surface of the waters.

Day one
3. And God said, "There is light, and light exists."
4. And God saw that the light *was* good. And God distinguished between the light and between the darkness.
5. And God called the light "day" and called the darkness "night." And evening was, and morning was—day one.

Second day
6. And God said, "There is an expanse in the middle of the water, and it exists to separate between water *above* from water *below*."
7. And God *is the one who* made the expanse and separated between the waters which *were* under the expanse and between the waters which *were* over the expanse. And *it* was so.
8. And God called the expanse "sky." And evening was, and morning was—second day.

English Standard Version
Chapter 1

In the beginning
1. In the beginning, God created the heavens and the earth.
2. The earth was without form and void, and darkness was over the face of the deep. And the Spirit of God was hovering over the face of the waters.

Day one
3. And God said, "Let there be light," and there was light.
4. And God saw that the light was good. And God separated the light from the darkness.
5. God called the light Day, and the darkness he called Night. And there was evening and there was morning, the first day.

Second day
6. And God said, "Let there be an expanse in the midst of the waters, and let it separate the waters from the waters."
7. And God made the expanse and separated the waters that were under the expanse from the waters that were above the expanse.
And it was so.
8. And God called the expanse Heaven. And there was evening and there was morning, the second day.

Third day
9. And God said, "The waters under the sky wait in one place and the dry ground is seen."
And *it* was so.
10. And God called the dry ground "land" and called the collection of waters "seas." And God saw that *it was* good.
11. And God said, "The land is sprouting sprouts: plants seeding seeds and the fruit trees making fruit—to which kind is in the seed—over the land."
And *it* was so.
12. "And the land brings forth sprouts: plants seeding seeds *according* to *their* kind, and the trees making fruit—in which seed is to kind."
And God saw that *it was* good.
13. And evening was, and morning was—third day.

Fourth day
14. And God said, "There are luminaries in the expanse of the heavens to distinguish between the day and between the night, and *they* are a sign *for* appointed times and days and years."
15. "And luminaries exist in the expanse of the heavens to give light over the Earth." And *it* was so.
16. And God *is the one who* made two great luminaries: the greater luminary to rule the day and the smaller luminary to rule the night and the stars.
17. And God gave them in the expanse of the heavens to give light over the Earth.
18. And to rule in the day and in the night and to distinguish between the light and between the darkness.
And God saw that *it was* good.
19. And evening was, and morning was—fourth day.

Third day
9. And God said, "Let the waters under the heavens be gathered together into one place, and let the dry land appear." And it was so.
10. God called the dry land Earth, and the waters that were gathered together he called Seas. And God saw that it was good.
11. And God said, "Let the earth sprout vegetation, plants yielding seed, and fruit trees bearing fruit in which is their seed, each according to its kind, on the earth." And it was so.
12. The earth brought forth vegetation, plants yielding seed according to their own kinds, and trees bearing fruit in which is their seed, each according to its kind. And God saw that it was good.
13. And there was evening and there was morning, the third day.

Fourth day
14. And God said, "Let there be lights in the expanse of the heavens to separate the day from the night. And let them be for signs and for seasons, and for days and years,
15. and let them be lights in the expanse of the heavens to give light upon the earth." And it was so.
16. And God made the two great lights—the greater light to rule the day and the lesser light to rule the night—and the stars.
17. And God set them in the expanse of the heavens to give light on the earth,
18. to rule over the day and over the night, and to separate the light from the darkness.
And God saw that it was good.
19. And there was evening and there was morning, the fourth day.

Fifth day
20. And God said, "The waters swarm *with* swarms of living creatures, and fliers flying over the land and over the surface of the expanse of the sky."
21. And God *is the one who* created the great reptiles and every moving living creature with which the waters swarm, *according* to *their* kinds, and every winged flier *according* to *its* kind. And God saw that *it was* good.
22. And God blessed them saying, "Be fruitful and multiply and fill the waters in the seas and multiply the fliers on the land."
23. And evening was, and morning was—fifth day.

Sixth day
24. And God said, "The land brings forth living creatures *according* to *their* kind: beasts and moving *things* and animals of the land, *according* to *their* kind." And *it* was so.
25. And God *is the one who* made the animals of the Earth *according* to *their* kind: the beasts *according* to *their* kind and every moving *thing* upon the ground *according* to *its* kind.
And God saw that *it was* good.
26. And God said, "*We shall* make man in our image, after our likeness. And *he shall* have dominion on fish of the sea and on fliers in the sky and on beasts and on all the Earth and on all the moving *things* that move over the land."
27. And God created man in His image, in God's image created him, male and female *God* created them.
28. And God blessed them, and God said to them, "Be fruitful and multiply and fill the Earth and subdue it and have dominion on the fish of the sea and on the fliers of the sky and on

Fifth day
20. And God said, "Let the waters swarm with swarms of living creatures, and let birds fly above the earth across the expanse of the heavens."
21. So God created the great sea creatures and every living creature that moves, with which the waters swarm, according to their kinds, and every winged bird according to its kind. And God saw that it was good.
22. And God blessed them, saying, "Be fruitful and multiply and fill the waters in the seas, and let birds multiply on the earth."
23. And there was evening and there was morning, the fifth day.

Sixth day
24. And God said, "Let the earth bring forth living creatures according to their kinds—livestock and creeping things and beasts of the earth according to their kinds." And it was so.
25. And God made the beasts of the earth according to their kinds and the livestock according to their kinds, and everything that creeps on the ground according to its kind.
And God saw that it was good.
26. Then God said, "Let us make man in our image, after our likeness. And let them have dominion over the fish of the sea and over the birds of the heavens and over the livestock and over all the earth and over every creeping thing that creeps on the earth."
27. So God created man in his own image, in the image of God he created him; male and female he created them.
28. And God blessed them. And God said to them, "Be fruitful and multiply and fill the earth and subdue it and have dominion over the fish of the sea and over the birds of the heavens and over

all living *things* that move over the land."
29. And God said, "Behold, *I* give to you every plant seeding seed which is over the surface of all the land, and every tree in which fruit is seeding seed. As food *they* are for you."
30. "And to every animal of the land and to every flier in the sky and to all *that* moves over the land—in which is the living creature—every green plant is food."
And *it* was so.
31. And God saw everything that *He* did, and behold *it was* very good. And evening was, and morning was—the sixth day.

Chapter 2

The Seventh day
1. And the heavens and Earth *were* finished, and all their multitude *of miscellaneous things*.
2. And on the seventh day God finished His work that *He* did, and rested on the seventh day from all His work that *He* did.
3. And God blessed the seventh day and sanctified it, because on it *He* rested from all His work—that God did *when* creating.

Introduction to a detailed account of the sixth day
4. The *following is the* account of the heavens and the Earth when created; on the *sixth* day *when* Yahweh God did *things to the* Earth and heavens.

Garden of Eden
5. And each bush of the field was not yet on the land, and each plant of the field *did* not yet grow, because Yahweh God had not caused rain over the land

every living thing that moves on the earth."
29. And God said, "Behold, I have given you every plant yielding seed that is on the face of all the earth, and every tree with seed in its fruit.
You shall have them for food.
30. And to every beast of the earth and to every bird of the heavens and to everything that creeps on the earth, everything that has the breath of life, I have given every green plant for food."
And it was so.
31. And God saw everything that he had made, and behold, it was very good. And there was evening and there was morning, the sixth day.

Chapter 2

The Seventh day
1. Thus the heavens and the earth were finished, and all the host of them.
2. And on the seventh day God finished his work that he had done, and he rested on the seventh day from all his work that he had done.
3. So God blessed the seventh day and made it holy, because on it God rested from all his work that he had done in creation.

Introduction to a detailed account of the sixth day
4. These are the generations of the heavens and the earth when they were created, in the day that the LORD God made the earth and the heavens.

Garden of Eden
5. When no bush of the field was yet in the land and no small plant of the field had yet sprung up—for the LORD God had not caused it to rain on the

and there was no man to cultivate the ground *of the garden.*
6. And mists rose up from the land and watered all the surface of the ground *of the garden.*
7. And Yahweh God formed the man *of* dust from the ground and breathed in his nostrils the breath *of* life. And the man existed as a living creature.
8. And Yahweh God planted a garden from the east in Eden, and there *He* put the man whom *He had* formed.
9. And Yahweh God grew from the ground every tree *that was* desirable to the sight and good for food, and the tree of life in the middle of the garden and the tree of knowledge of good and evil.
10. And a river went out from Eden to water the garden and from there diverged and became four branches.
11. The name of the first is Pishon; it winds through all the land of Havilah, where there is gold.
12. And the gold of that land is good; bdellium and onyx stone *are* there.
13. And the name of the second river is the Gihon. It winds through all the land of Cush.
14. And the name of the third river is the Tigris; it goes east of Asshur. And the fourth river is the Euphrates.

Man in the garden
15. And Yahweh God took the man and put him in the Garden of Eden to cultivate and to keep it.
16. And Yahweh God commanded to the man saying, "From any tree of the garden, you *may* eat-eat."
17. "And from the tree of the knowledge of good and evil, *do* not eat from, because on the day you eat from it *you shall* die-die."

land, and there was no man to work the ground,
6. and a mist was going up from the land and was watering the whole face of the ground—
7. then the LORD God formed the man of dust from the ground and breathed into his nostrils the breath of life, and the man became a living creature.
8. And the LORD God planted a garden in Eden, in the east, and there he put the man whom he had formed.
9. And out of the ground the LORD God made to spring up every tree that is pleasant to the sight and good for food. The tree of life was in the midst of the garden, and the tree of the knowledge of good and evil.
10. A river flowed out of Eden to water the garden, and there it divided and became four rivers.
11. The name of the first is the Pishon. It is the one that flowed around the whole land of Havilah, where there is gold.
12. And the gold of that land is good; bdellium and onyx stone are there.
13. The name of the second river is the Gihon. It is the one that flowed around the whole land of Cush.
14. And the name of the third river is the Tigris, which flows east of Assyria. And the fourth river is the Euphrates.

Man in the garden
15. The LORD God took the man and put him in the garden of Eden to work it and keep it.
16. And the LORD God commanded the man, saying, "You may surely eat of every tree of the garden,
17. but of the tree of the knowledge of good and evil you shall not eat, for in the day that you eat of it you shall surely die."

18. And Yahweh God said, "*It is* not good that the man be alone. For him *I shall* make as his counterpart a helper."

Living things of the garden created
19. And Yahweh God formed from the ground each animal of the field and each flier of the sky *within the garden.* And *God* brought *them* to the man to see what he *would* call them and whatever the man called the living creature, that *was its* name.
20. And the man called names to all the beasts and to the fliers of the sky and to all animals of the field *within the garden.* And for Adam *was* not found as his counterpart a helper.

Woman is created
21. And Yahweh God caused a deep sleep to fall over the man and *during* sleep took from one of his ribs and closed up under the flesh.
22. And Yahweh God built—which *He* took *of* the rib from the man—into a woman, and brought her to the man.

Marriage established
23. And the man said, "This is now bone from my bones and flesh from my flesh; for this *shall* be called woman, because this *was* taken from man."
24. So over *this reason,* a man *will* leave his father and his mother and hold fast on his wife. And *they shall* be as one flesh.
25. And the two were naked, the man and *his* wife, and *were* not ashamed.

18. Then the LORD God said, "It is not good that the man should be alone; I will make him a helper fit for him."

Living things of the garden created
19. Now out of the ground the LORD God had formed every beast of the field and every bird of the heavens and brought them to the man to see what he would call them. And whatever the man called every living creature, that was its name.
20. The man gave names to all livestock and to the birds of the heavens and to every beast of the field. But for Adam there was not found a helper fit for him.

Woman is created
21. So the LORD God caused a deep sleep to fall upon the man, and while he slept took one of his ribs and closed up its place with flesh.
22. And the rib that the LORD God had taken from the man he made into a woman and brought her to the man.

Marriage established
23. Then the man said, "This at last is bone of my bones and flesh of my flesh; she shall be called Woman, because she was taken out of Man."
24. Therefore a man shall leave his father and his mother and hold fast to his wife, and they shall become one flesh.
25. And the man and his wife were both naked and were not ashamed.

Appendix: D

The Septuagint Translation – Genesis Chapters 1 and 2

The following English translation of the Septuagint is from *The Septuagint with Aprocrypha: Greek and English* by Sir Lancelot C. L. Brenton originally published in 1851 by Samuel Bagster & Sons Ltd., London, England. The Brenton translation can be compared with *A New English Translation of the Septuagint* by Albert Pietersma and Benjamin J. Wright published in 2007 or to the Orthodox Study Bible published in 2008 (all three English translations of the Septuagint are available on the Internet).

The first five books of the Septuagint, the Pentateuch, were translated from Hebrew to Greek around 275 BC by Hebrew scribes. This translation established the interpretative paradigm for the Genesis creation story which has continued unchallenged until the present. The Septuagint translation has numerous errors as the following illustrates. Significant words that are mistranslated or poorly translated are underlined in the text below. Significant words added to the original text are highlighted in **bold**. Significant words that were left out and not translated in any manner have been added to the text and are enclosed in [brackets].

Chapter 1

1. In the beginning God made the heaven and the earth.

2. But the earth was unsightly and unfurnished, and darkness was over [the surface] the deep, and the Spirit of God moved over [the surface] the water.

3. And God said, Let there be light, and there was light.

4. And God saw the light that it was good, and God divided between the light and [between] the darkness.

5. And God called the light Day, and the darkness he called Night, and there was evening and there was morning, the first day.

6. And God said, Let there be a firmament in the midst of the water, and let it be a division between water and water, **and it was so**.

7. And God made the firmament, and God divided between the water which was under the firmament and the water which was above the firmament. [And it was so.]

8. And God called the firmament Heaven, **and God saw that it was good**, and there was evening and there was morning, the second day.

9. And God said, Let the water which is under the heaven be collected into one place, and let the dry land appear, and it was so. **And the water which was under the heaven was collected into its places, and the dry land appeared.**

10. And God called the dry land Earth, and the gatherings of the waters he called

Seas, and God saw that it was good.
11. And God said, Let the earth bring forth the herb of grass bearing seed **according to its kind and according to its likeness**, and the fruit-tree bearing fruit whose seed is in it, according to its kind on the earth, and it was so.
12. And the earth brought forth the herb of grass bearing seed according to its kind **and according to its likeness**, and the fruit tree bearing fruit whose seed is in it, according to its kind **on the earth**, and God saw that it was good.
13. And there was evening and there was morning, the third day.
14. And God said, Let there be lights in the firmament of the heaven to **give light upon the earth,** to divide between day and [between] night, and let them be for signs and for seasons and for days and for years.
15. And let **them** [luminaries] be for light in the firmament of the heaven, **so as** to shine upon the earth, and it was so.
16. And God made the two great lights, the greater light for regulating the day and the lesser light for regulating the night, the stars also.
17. And God placed them in the firmament of the heaven, **so as** to shine upon the earth,
18. and to regulate day and night, and to divide between the light and [between] the darkness. And God saw that it was good.
19. And there was evening and there was morning, the fourth day.
20. And God said, Let the waters bring forth reptiles having life, and **winged** creatures flying above the earth in the firmament of heaven, and it was so.
21. And God made great whales, and every [moving] living reptile, which the waters brought forth according to their kinds, and every **creature** that flies with wings according to its kind, and God saw that they were good.
22. And God blessed them saying, Increase and multiply and fill the waters in the seas, and let the creatures that fly be multiplied on the earth.
23. And there was evening and there was morning, the fifth day.
24. And God said, Let the earth bring forth the living creature according to its kind, quadrupeds and reptiles and wild beasts of the earth according to their kind, and it was so.
25. And God made the wild beasts of the earth according to their kind, and cattle according to their kind, and all the reptiles of the earth according to their kind, and God saw that they were good.
26. And God said, Let us make man according to our image and likeness, and let them have dominion over the fish of the sea, and over the flying creatures of heaven, and over the cattle and all the earth, and over all the reptiles that creep on the earth.
27. And God made man, according to the image of God he made him, male and female he made them.
28. And God blessed them, [and God] saying, Increase and multiply, and fill the earth and subdue it, and have dominion over the fish of the seas and flying **creatures**

of heaven, **and all the cattle and all the earth**, and all the reptiles that creep on the earth.
29. And God said, Behold I have given to you every seed-bearing herb sowing seed which is upon [the surface of] all the earth, and every tree which has in itself the fruit of seed that is sown, to you it shall be for food.
30. And to all the wild beasts of the earth, and to all the flying creatures of heaven, and to every reptile creeping on the earth, which has in itself the breath of life, even every green plant for food; and it was so.
31. And God saw all the things that he had made, and, behold, they were very good. And there was evening and there was morning, the sixth day.

Chapter 2
1. And the heavens and the earth were finished, and the whole world of them.
2. And God finished on the sixth day his works which he made, and he ceased on the seventh day from all his works which he made.
3. And God blessed the seventh day and sanctified it, because in it he ceased from all his [creating] works which God **began to** do.
4. This is the book of the generation of heaven and earth, when they were made, in the day in which the Lord God made the heaven and the earth,
5. and every herb of the field before it was on the earth, and all the grass of the field before it sprang up, for [Yahweh] God had not rained on the earth, and there was not a man to cultivate [the ground] it.
6. But there rose a fountain out of the earth, and watered the whole face of the earth.
7. And [Yahweh] God formed the man of dust of the earth, and breathed upon his face the breath of life, and the man became a living soul.
8. And [Yahweh] God planted a garden eastward in Edem, and placed there the man whom he had formed.
9. And [Yahweh] God made to spring up also out of the earth every tree beautiful to the eye and good for food, and the tree of life in the midst of the garden, and the tree of **learning** the knowledge of good and evil.
10. And a river proceeds out of Edem to water the garden, thence it divides itself into four heads.
11. The name of the one, Phisom, this it is which encircles the whole land of Evilat, where there is gold.
12. And the gold of that land is good, there also is carbuncle and emerald.
13. And the name of the second river is Geon, this it is which encircles the whole land of Ethiopia.
14. And the third river is Tigris, this is that which flows forth over against the Assyrians. And the fourth river is Euphrates.
15. And the Lord God took the man **whom he had formed**, and placed him in the garden of Delight, to cultivate and keep it.

16. And the Lord God gave a charge to Adam, saying, Of every tree which is in the garden thou mayest freely eat,
17. but of the tree of the knowledge of good and evil—of it ye shall not eat, but in whatsoever day ye eat of it, ye shall surely die.
18. And the Lord God said, It is not good that the man should be alone, let us make for him a help suitable to him.
19. And [Yahweh] God formed **yet farther** out of the earth all the wild beasts of the field, and all the birds of the sky, and he brought them to Adam, to see what he would call them, and whatever Adam called any living creature, that was the name of it.
20. And Adam gave names to all the cattle and to all the birds of the sky, and to all the wild beasts of the field, but for Adam there was not found a help like to himself.
21. And [Yahweh] God brought a [deep] trance upon Adam, and he slept, and he took [from] one of his ribs, and filled up the flesh instead thereof.
22. And [Yahweh] God formed the rib which he took from Adam into a woman, and brought her to Adam.
23. And Adam said, This now is bone of my bones, and flesh of my flesh; she shall be called woman, because she was taken out of her husband.
24. Therefore shall a man leave his father and his mother and shall cleave to his wife, and they **two** shall be one flesh.
3:1. (2:25) And the two were naked **both** Adam and his wife, and were not ashamed.

References

Chapter 1 – *Exemplar* Creation

1. Castagno, Joseph M., ed. (2006). *The New Book of Popular Science*, vol. 2. Scholastic Library Publishing Inc. p14.
2. Hancock, Paul L. and Brian J. Skinner, eds. (2000). *The Oxford Companion to the Earth.* New York: Oxford University Press. p230.
3. Emsley, John (1998). *The Elements,* 3rd ed. Oxford: Clarendon Press. p289.
4. Dasch, E. Julius, ed. (1996). *Encyclopedia of Earth Sciences*, vol.1. New York: Simon & Schuster Macmillan. p134.
5. Hancock. *The Oxford Companion to the Earth.* p278.
6. Darling, David (2004). *The Universal Book of Astronomy.* Hoboken, NJ: John Wiley & Sons Inc. p340.
7. Angelo, Joseph A. Jr. (2006). *Encyclopedia of Space and Astronomy.* New York: Facts on File. p403.
8. National Aeronautics and Space Administration (1999, Apr.). NASA's Goddard Space Flight Center: The First Forty Years. *NASA Facts*, FS-1999 (01)-003GSFC. p4.
9. Faulkner, Danny (2004). *Universe by Design.* Green Forest, AR: Master Books. p126.
10. Johnston, Sarah Iles, ed. (2004). *Religions of the Ancient World: A Guide.* Cambridge, MA: Harvard University Press. p395.
11. Bergeron, Lou (1997, Aug. 30). Deep Waters. *New Scientist*, 155(2097):22–26.
12. Swenson, Herbert (2008). Why is the Ocean Salty? U.S. Geological Survey Publication. Retrieved on March 12, 2008 from http://www.palomar.edu/oceanography/salty_ocean.htm
13. Schneider, S. H. (1996). Water resources by P. H. Gleick. *Encyclopedia of Climate and Weather*, vol. 2. New York: Oxford University Press. pp817-823.
14. Dalrymple, G. Brent (1991). *Age of the Earth.* Stanford, CA: Stanford University Press. p207, 223.
15. Dasch. *Encyclopedia of Earth Sciences.* pp496-498.
16. Doniger, Wendy, ed. (1999). *Merriam-Webster's Encyclopedia of World Religions.* Springfield, MA: Merriam-Webster Inc. p910.
17. Ibid. p1095.
18. Ibid. p637.
19. MacArthur, John (2001). *Battle for the Beginning: Creation, Evolution and the Bible.* Nashville, TN: Thomas Nelson, Inc. p165.
20. Eliyah.com (2006). Strong's Concordance - King James Version. Retrieved on May 20, 2006 from http://www.eliyah.com/lexicon.html

Chapter 2 – Understanding Science

1. Sarfati, Jonathan (2004). *Refuting Compromise.* Green Forest, AR: Master Books. p17.
2. Komonchak, Joseph A. and Mary Collins, Dermot A. Lane, eds. (1987). *The New Dictionary of Theology.* Wilmington, DE: Michael Glazier Inc. p247.
3. Borchert, Donald M., ed. (1996). *The Encyclopedia of Philosophy, Supplement.* New York: Simon and Schuster-Macmillan. p372.
4. Dembski, William A. (1996, Apr. 1). What Every Theologian Should Know About Creation, Evolution and Design. *The Princeton Theological Review.* p10. Retrieved on May 20, 2008 from http://www.discovery.org/scripts/viewDB/index.php?

command=view&id=122
5. Gove, Phillip B., ed. (2002). Science. *Webster's Third New International Dictionary of the English Language Unabridged.* Springfield, MA: Merriam-Webster. p2032.
6. Ibid. p2032.
7. Ibid. p2032.
8. Parker, Sybil P., ed. (1994). Science. *McGraw-Hill Dictionary of Scientific and Technical Terms*, 5th ed. New York: McGraw-Hill Book Company. p1765.
9. Morris, Christopher, ed. (1992). Science. *Academic Press Dictionary of Science and Technology.* San Diego, CA: Academic Press. p1926.
10. Dalrymple, Brent G. (1981, Dec. 3). Deposition of Dr. G. Brent Dalrymple of the U. S. Geological Survey, Menlo Park, CA. in the case of *McLean v. Arkansas Board of Education*, U. S. District Court. p38.
11. Behe, Michael J. (2007). *The Edge of Evolution: The Search for the Limits of Darwinism.* New York: Free Press. p235.
12. Parker. Empirical science. *McGraw-Hill Dictionary of Scientific and Technical Terms.* p674.
13. Strahler, Arthur N. (1992). *Understanding Science: An Introduction to Concepts and Issues.* Buffalo, NY: Prometheus Books. p76.
14. Cromer, Alan (1993). *Uncommon Sense: The Heretical Nature of Science.* New York, NY: Oxford University Press. p6.
15. Ibid. p6.
16. Tolson, Jay (2006, Oct. 23). Is There Room for the Soul? *U.S. News & World Report*, vol. 141, no. 15.
17. National Academy of Sciences, Steering Committee on Science and Creationism (1999). *Science and Creationism: A View from the National Academy of Sciences*, 2nd ed. Washington, DC: National Academy Press. p2.
18. Purves, William K. and David Sadava, Gordon H. Orians, H. Craig Heller (2001). *Life-The Science of Biology*, 6th ed. Gordonsville, VA: W. H. Freeman and Company. p13.
19. *American Heritage College Dictionary,* 4th ed. (2004). Boston, MA: Houghton Mifflin Company. p498.
20. Castell, Alburey and Donald M. Borchert, Arthur Zucker (1994). *An Introduction to Modern Philosophy: Examining the Human Condition,* 6th ed. New York: Macmillian College Publishing Company. p1.
21. Hazen, Robert M. (2005). *Genesis: The Scientific Quest for Life's Origin.* Washington, DC: Joseph Henry Press. pp33-37.
22. National Academy of Sciences, Working Group on Teaching Evolution (1998). *Teaching About Evolution and the Nature of Science.* Washington, DC: National Academy Press. p5.
23. Humsavar (2007, May 15). Index of Human Polymorphisms and Disease Mutations. Uniprot/Swiss Prot Protein Knowledgebase, Release 52.5. Retrieved on May 16, 2007 from http://www.expasy.org/cgi-bin/lists?humsavar.txt.
24. National Academy of Sciences and Institute of Medicine, Committee on Revising Science and Creationism (2008). *Science, Evolution and Creationism.* Washington, DC: The National Academies Press. p10.
25. Mayr, Ernst (1997). *This is Biology: The Science of the Living World.* Cambridge, MA: Harvard University Press. p25.
26. National Academy of Sciences. *Teaching About Evolution and the Nature of Science*. p5.
27. Trefil, James (2003). *The Nature of Science: An A-Z Guide to the Laws & Principles Governing Our Universe*. Boston, MA: Houghton Mifflin Company. pxiii.
28. de Vries, Paul (1986). Naturalism in the Natural Sciences. *Christian Scholar's Review*, 15, 388-396.
29. Asimov, Isaac (1982). *Asimov's Biographical Encyclopedia of Science and Technology*, 2nd ed. Garden City, NY: Doubleday & Company Inc. p59 and 99.
30. *Webster's New World College Dictionary,* 4th ed. (2004). Naturalism, philosophical.

Cleveland, OH: Wiley Publishing Co. p960.
31. Borchert, Donald M., ed. (1996). *The Encyclopedia of Philosophy, Supplement.* New York: Simon and Schuster-Macmillan. p372.
32. Honderich, Ted, ed. (2005). *The Oxford Companion to Philosophy*, 2nd ed. New York: Oxford University Press. p640.
33. Mautner, Thomas, ed. (1996). *A Dictionary of Philosophy.* Cambridge, MA: Blackwell Publishers, Ltd. p285.
34. Gould, Stephen Jay (1994). *Hen's Teeth and Horse's Toes*, reissued. New York: W. W. Norton & Company. pp253-255.
35. Dembski, William A. (1999). *Intelligent Design: The Bridge Between Science & Theology.* Downers Grove, IL: InterVarsity Press. p76.
36. Ibid. p107.

Chapter 3 – Biblical Creation Theology

1. MacArthur, John (2001). *Battle for the Beginning: Creation, Evolution and the Bible.* Nashville, TN: W Publishing Group. p27.
2. Würthwein, Ernst (1988). *The Text of the Old Testament*, 5th ed. (translated by Erroll F. Rhodes and published in English in 1995) Grand Rapids, MI: William B. Eerdmans Publishing Co. pp107-122.
3. Borchert, Donald M., ed. (1996). *The Encyclopedia of Philosophy, Supplement.* New York: Simon and Schuster-Macmillan. p372.
4. Asimov, Isaac (1982). *Asimov's Biographical Encyclopedia of Science and Technology*, 2nd ed. Garden City, NY: Doubleday & Company Inc. p4.
5. Ibid. p10.
6. Wilkins, John S. (2004, April 26). Spontaneous Generation and the Origin of Life. The TalkOrigins Archive. Retrieved on May 15, 2007 from http://www.talkorigins.org/faqs/abioprob/spontaneous-generation.html
7. Asimov. *Asimov's Biographical Encyclopedia of Science and Technology.* p36.
8. Ibid. p37.
9. Templeton, Charles (1999). *Farewell to God: My Reason for Rejecting the Christian Faith.* Toronto: McClelland & Steward Inc. p30.
10. Ibid. p31.
11. Ussher, James (1658). *The Annals of the World.* (Ussher's timeline chart by Paul Hansen, published in *Creation*, Sept. 2005). Retrieved on March 3, 2006 from http://www.answersingenesis.org/creation/v27/i4/TimelineOfTheBible.pdf
12. Josephus, Flavius (1987). *The Works of Josephus: Complete and Unabridged*, 13th printing. (translated by William Whiston) Peabody, MA: Hendrickson Publishers. p29.
13. Philo (1993). *The Works of Philo: Complete and Unabridged*, 4th printing. (translated by C. D. Yonge) Peabody, MA: Hendrickson Publishers. p6.
14. Ibid. p7.
15. Ibid. p9.
16. Ibid. p10.
17. Freedman, David Noel (1992). *The Anchor Bible Dictionary*, vol. 6. New York: Doubleday. pp819-823. A history of Bible translations only—no discussion of theological influence.
18. Ussher. *The Annals of the World.*
19. Freedman. *The Anchor Bible Dictionary*, vol. 4. p599.
20. *Britannica, The New Encyclopedia*, vol. 7, 15th ed. (2005). Masoretic Text. Chicago: Encyclopedia Britannica Inc. p914.
21. Würthwein. *The Text of the Old Testament.* p21.
22. Fahlbusch, Erwin and Jan Lochman, John Mbiti, Jaroslav Pelikan, Lukas Vischer, eds. (2005). *The Encyclopedia of Christianity*, vol. 4. Grand Rapids, MI: William B. Eerd-

mans Publishing Co. p913.

23. Freedman. *The Anchor Bible Dictionary*, vol. 5. p1094.
24. Pietersma, Albert and Benjamin J. Wright (2007). *A New English Translation of the Septuagint.* New York, NY: Oxford University Press. Retrieved on May 20, 2008 from http://ccat.sas.upenn.edu/nets/edition/
25. Morris, Henry (1976). *The Genesis Record.* Grand Rapids, MI: Baker Book House. p45.
26. Batten, Don, ed. and Ken Ham, Jonathan Sarfati, Carl Wieland (1990). *The Revised & Expanded Answers Book*, 24th printing. Green Forest, AR: Master Books. p58.
27. Custance, Arthur C. (1970). *Without Form and Void.* Brookville, Canada: Pub. by Arthur C. Custance. pxi.
28. Ross, Hugh (2004). *A Matter of Days*. Colorado Springs, CO: NavPress. pp142-144.
29. Ibid. p140.
30. Collins, C. John (2006). *Genesis 1-4: A Linguistic, Literary, and Theological Commentary*. Phillipsburg, NJ: P & R Publishing Co. p123.
31. Ibid. p124.
32. Dembski, William A. (1996, Apr. 1). What Every Theologian Should Know About Creation, Evolution and Design. *The Princeton Theological Review*. Retrieved on May 20, 2008 from http://www.discovery.org/scripts/viewDB/index.php?command=view&id=122. p4.
33. Collins, Francis S. (2006). *The Language of God.* New York, NY: Free Press. p200.
34. Ibid. p193.

Chapter 4 – Evidence of Age

1. Stobel, Nick (2005). Astronomy Notes. Retrieved on October 10, 2005 from http://www.astronomynotes.com/starprops/s1.html
2. European Space Agency (2008). The Hipparcos Space Astronomy Mission. Retrieved on May 27, 2008 from http://www.rssd.esa.int/index.php?project=HIPPARCOS
3. Angelo, Joseph A. Jr. (2006). *Encyclopedia of Space and Astronomy*. New York: Facts on File. p522.
4. Ribas, Ignasi and Carme Jordi, Francesc Vilardell, Edward L. Fitzpartick, Ron W. Hilditch, Edward F. Guinan (2005, Nov.). First Determination of the Distance and Fundamental Properties of an Eclipsing Binary in The Andromeda Galaxy. *Astrophysical Journal Letters*, 635 (2005) L37-L40.
5. Central Bureau for Astronomical Telegrams (2007). List of Recent Supernovae. Retrieved on May 7, 2007 from http://www.cfa.harvard.edu/iau/lists/RecentSupernovae.html
6. Batten, Don, ed. and Ken Ham, Jonathan Sarfati, Carl Wieland (1990). *The Revised & Expanded Answers Book*, 24th printing. Green Forest, AR: Master Books. p97.
7. Humphreys, D. Russell (2004). *Starlight and Time: Solving the Puzzle of Distant Starlight in a Young Universe*, 9th printing. Green Forest, AR: Master Books. p26.
8. Ibid. pp32-38.
9. Batten. *The Revised & Expanded Answers Book.* pp98-101.
10. Ibid. p97.
11. Morris, Henry (2003, June). The Uncertain Speed of Light, *Back to Genesis,* no. 174.
12. Dasch, E. Julius, ed. (1996). *Encyclopedia of Earth Sciences*, vol. 1. New York: Simon & Schuster Macmillan. p494.
13. Dalrymple, G. Brent (1991). *Age of the Earth.* Stanford, CA: Stanford University Press. p207, 223.
14. McDowell, Jonathan (2007, Mar. 30). A merge of NASA-RP-1097 (Anderson and Whitaker 1982) and USGS lunar crater data. Retrieved on April 26, 2007 from http://host.planet4589.org/astro/lunar/
15. Hines, Jason (2007, Feb. 16). Earth Impact Database, Planetary and Space Science Centre

at the University of New Brunswick, Canada. Retrieved on April 23, 2007 from http://www.unb.ca/passc/ImpactDatabase/

16. Ussher, James (1658). *The Annals of the World.* (Ussher's timeline chart by Paul Hansen, published in *Creation*, Sept. 2005). Retrieved on March 3, 2006 from http://www.answersingenesis.org/creation/v27/i4/TimelineOfTheBible.pdf
17. Lorius, C. and L. Merlivat, J. Jouzel, M. Pourchet (1979). A 30,000 Year Isotope Climatic Record from Antarctic Ice. *Nature*, 280:644-648.
18. Petit, J. R. and J. Jouzel, D. Raynaud, N. I. Barkov, J-M. Barnola, I. Basile, et al (1999, June 3). Climate and Atmospheric History of the past 420,000 years from the Vostok ice core, Antarctica. *Nature*, vols. 399, 6735.
19. Stauffer, Bernhard and Jacqueline Flückiger, Eric Wolff, Piers Barnes (2004). The EPICA Deep Ice Cores: First Results and Perspectives, *Annals of Glaciology*, 39 2004.
20. Parrenin, F. and J. M. Barnola, J. Beer, T. Blunler, E. Castellano, J. Chappellaz, et al (2007, Mar. 12). The EDC3 Chronology for the EPICA Dome C Ice Core. *Climate of the Past*, 3, 575-606.
21. Cross, M., compiler (2003, Jan.). Greenland summit ice cores. Boulder, CO: National Snow and Ice Data Center in association with the World Data Center for Paleoclimatology at NOAA-NGDC and the Institute of Arctic and Alpine Research. CD-ROM.
22. Ibid.
23. Andersen, Katrine (2004). North Greenland Ice Core Project. Retrieved on January 1, 2007 from http://www.gfy.ku.dk/~www-glac/ngrip/index_eng.htm
24. Svensson, A. and S. W. Nielsen, S. Kipfstuhl, S. J. Johnsen, J. P. Steffensen, M. Bigler, U. Ruth, R. Röthlisberger (2005, Jan. 21). Visual Stratigraphy of North Greenland Ice Core Project (NorthGRIP) Ice Core During the Last Glacial Period. *Journal of Geophysical Research*, vol. 110, D02108.
25. White, James (2004, Aug 13). Greenland Ice Core Project Yields Probable Ancient Plant Remains. Retrieved on April 1, 2007 from http://www.eurekalert.org/pub_releases/2004-08/uoca gic081304.php
26. Oard, Michael J. (2005). *The Frozen Record.* El Cajon, CA: Institute for Creation Research. p73.
27. Oard, Michael J. (2007, Apr.-Jun.). Setting the Stage for an Ice Age, *Answers*, vol. 2, no. 2.
28. Oard. *The Frozen Record.* pp45-49.
29. *Antarctica* (1985). Surrey Hills, NSW: Reader's Digest Services Pty. Ltd. p175.
30. Ibid. p28.
31. Walton, D. W. H. (1987). *Antarctic Science*. New York: Cambridge University Press. p41, 186.
32. McKelvey, B. C. and N. C. N. Stephenson (1990). A Geological Reconnaissance of the Radok Lake Area, Amery Oasis, Prince Charles Mountains. *Geological Magazine*, vol. 134 and *Antarctic Science*, 1990.
33. Hammer, William R. and William J. Hickerson (1994, May 6). A Crested Theropod Dinosaur from Antarctica. *Science,* vol. 264, issue 5160, 828–830.
34. Lubick, Noami (2004, Mar. 8). Dinosaurs in the Cold, Dark Antarctic, *Geotimes*.
35. Dalrymple. *Age of the Earth.* p14.
36. Wiens, Roger C. (2005). Radiometric Dating - A Christian Perspective. American Scientific Affiliation. Retrieved on September 17, 2005 from http://www.asa3.org/ASA/resources/Wiens.html
37. Dasch. *Encyclopedia of Earth Sciences*, vol. 2, p767.
38. Dalrymple. *Age of the Earth.* p230.
39. Ibid. p239.
40. Ibid. p287.
41. Ibid. p305.
42. Ibid. p357.
43. Dean, John A. (1999). *Lange's Handbook of Chemistry*, 15th ed. New York: McGraw-

Hill Inc. p7.124.
44. Taylor, R. E. (2007). Radiocarbon Dating. *McGraw-Hill Encyclopedia of Science & Technology,* vol. 15, 10th ed. New York: McGraw-Hill. p146.
45. Ibid. p151.
46. Suess, H. E. and T. W. Linick (1990). The ^{14}C Record in Bristlecone Pine Wood of the past 8000 Years Based on the Dendrochronology of the Late C. W. Ferguson. *Philosophical Transactions of the Royal Society of London. Series A, Mathematical and Physical Sciences*, vol. 330, issue 1615, pp. 403-412.
47. Reimer, P. J. and M. G. L. Baillie, E. Bard, A. Bayliss, J. W. Beck, C. J. H. Bertrand, P. G. Backwell, et al (2004). IntCal04 Terrestrial Radiocarbon Age Calibrations, 0-26ka BP. *Radiocarbon*, 46, 1029-1058.
48. Ibid. p1029-1058.
49. Fairbanks, Richard G. and Richard A. Mortlock, Tzu-Chien Chiu, Li Cao, Aley Kapan, Thomas P. Guilderson, et al (2005, Apr.). Radiocarbon Calibration Curve Spanning 0 to 50,000 Years BP Based on Paired ^{230}Th /^{234}U/^{238}U/ and ^{14}C Dates on Pristine Corals. *Quaternary Science Reviews* 24:1781-1796.
50. Kitagawa, H. and J. van der Plicht (1998, Feb. 20). Atmospheric Radiocarbon Calibration to 45,000 yr B.P.: Late Glacial Fluctuations and Cosmogenic Isotope Production. *Science,* vol. 279, 1187-1190.
51. Taylor. *McGraw-Hill Encyclopedia of Science & Technology*. p151.
52. Maslin, M. A. and Thomas, E. (2003, Apr.). Balancing the Deglacial Global Carbon Budget: The Hydrate Factor. *Quaternary Science Reviews*, 22:p1733.
53. Arens, Nan Crystal and A. Hope Jahren, Ronald Amundson (2000). Can C3 Plants Faithfully Record the Carbon Isotopic Composition of Atmospheric Carbon Dioxide? *Paleobiology* 26(1), p147.
54. Turney, C. S. M. and D. Wheeler, Allan R. Chivas (2006). Carbon Isotope Fractionation in Wood During Carbonization. *Geochimica et Cosmochimica Acta*, 70, p962.
55. Lowe, J. J. and M. J. C. Walker (1997). Chapter 5, Dating Methods. *Reconstructing Quaternary Environments*. Prentice Hall.
56. Higham, Tom (2002, May 16). Sample isotopic fractionation. Radiocarbon Laboratory, University of Waikato, New Zealand. Retrieved on February 24, 2007 from http://www.c14dating.com/frac.html
57. Taylor. *McGraw-Hill Encyclopedia of Science & Technology*. p151.
58. *Britannica, The New Encyclopedia*, vol. 5, 15th ed. (2005). Giza, Pyramids of. Chicago: Encyclopedia Britannica Inc. p288.

Chapter 5 – The Words of Genesis

1. Brown, Francis and S. R. Driver, Charles A. Briggs (1907). *A Hebrew and English Lexicon of the Old Testament,* 1978 printing. Oxford: Oxford Clarendon Press. p135.
2. Strong, James (1890). *Strong's Exhaustive Concordance of the Bible*, 31st printing. New York: Abingdon Press.
3. Brown. *A Hebrew and English Lexicon of the Old Testament*. p793.
4. Eliyah.com (2006). Strong's Concordance - King James Version. Retrieved on May 20, 2006 from http://www.eliyah.com/lexicon.html
5. Gesenius, Wilhelm (1846). *Hebrew and Chaldee Lexicon to the Old Testament Scriptures*, 1st ed. (translated by Samuel P. Tregelles) London: Samuel Bagster & Sons Ltd. Retrieved on May 20, 2008 from http://www.eliyah.com/lexicon.html
6. Brown. *A Hebrew and English Lexicon of the Old Testament*. p96.
7. Eliyah. Strong's Concordance - King James Version.
8. Brown. *A Hebrew and English Lexicon of the Old Testament*. p1062.
9. Eliyah. Strong's Concordance - King James Version.

10. Young, Robert (1898). Holy Bible, Young's Literal Translation. Retrieved on May 20, 2008 from http://www.blueletterBible.org/
11. Brown. *A Hebrew and English Lexicon of the Old Testament*. p1062.
12. Eliyah. Strong's Concordance - King James Version.
13. Darby, John N. (1890). Holy Bible, The Darby Translation. Retrieved on May 20, 2008 from http://www.blueletterBible.org/
14. Freedman, David Noel (1992). *The Anchor Bible Dictionary*, vol. 2. New York: Doubleday. p125.
15. Eliyah. Strong's Concordance - King James Version.
16. Crosswalk.com (2006). Word search for "sky" in the NIV using the online study Bible. Retrieved on May 20, 2008 from http://bible.crosswalk.com/
17. Eliyah. Strong's Concordance - King James Version.
18. Snoke, David (2006). *A Biblical Case for an Old Earth*. Grand Rapids, MI: Baker Books. p199-202.
19. Brown. *A Hebrew and English Lexicon of the Old Testament*. p874.
20. Gesenius, Wilhelm (1846). *Hebrew and Chaldee Lexicon to the Old Testament Scriptures*, 1st ed. (translated by Samuel P. Tregelles) London: Samuel Bagster & Sons Ltd. Retrieved on May 20, 2008 from http://www.eliyah.com/lexicon.html
21. Eliyah. Strong's Concordance - King James Version.
22. Ibid.
23. Ibid.
24. Ibid.
25. Ibid.
26. Ibid.
27. Pietersma, Albert and Benjamin J. Wright (2007). *A New English Translation of the Septuagint*. New York, NY: Oxford University Press. Retrieved on May 20, 2008 from http://ccat.sas.upenn.edu/nets/edition/
28. Webster, Noah (1833). Holy Bible, Noah Webster's Translation. Retrieved on May 20, 2008 from http://www.blueletterBible.org/
29. Brown. *A Hebrew and English Lexicon of the Old Testament*. p876.
30. Eliyah. Strong's Concordance - King James Version.
31. Gesenius. *Hebrew and Chaldee Lexicon to the Old Testament Scriptures*.
32. Brown. *A Hebrew and English Lexicon of the Old Testament*. p226.
33. Chalker, Sylvia and Edmund Weiner (1994). *The Oxford Dictionary of English Grammar*. Oxford: Clarendon Press. p128.
34. Ibid. p372.
35. *American Heritage College Dictionary*, 4th ed. (2004). Boston, MA: Houghton Mifflin Company. p122.
36. Ibid. p75.
37. Ibid. p489.
38. van der Merwe, Christo, ed. (2004). *The Lexham Hebrew-English Interlinear Bible*. (translated by the University of Stellenbosch, Stellenbosch, South Africa) Bellingham, WA: Logos Research Systems Inc. Based on a search of the Hebrew text for the root word (הָיָה).
39. Strong. *Strong's Exhaustive Concordance of the Bible*.
40. van der Merwe. *The Lexham Hebrew-English Interlinear Bible*. Based on a search of the Hebrew text for the root word (הָיָה).
41. Scofield, Cyrus I. (1917). Scofield notes for Genesis 1:2. Holy Bible, King James Version, Scofield Reference. Oxford University Press.
42. Batten, Don, ed. and Ken Ham, Jonathan Sarfati, Carl Wieland (1990). *The Revised & Expanded Answers Book*, 24th printing. Green Forest, AR: Master Books. p58.
43. Custance, Arthur C. (1970). *Without Form and Void*. Brookville, Canada: Pub. by Arthur C. Custance. pxi.

44. Harrison, R. K. (1955). *Biblical Hebrew*. Berkshire, England: Cox & Wyman Ltd. p90.
45. Dobson, John H. (2005). *Learn Biblical Hebrew*, 2nd ed. Grand Rapids, MI: Baker Publishing Group. pp157, 182, 194.
46. Pratico, Gary D. and Miles V. Van Pelt (2001). *Basics of Biblical Hebrew Grammar*. Grand Rapids, MI: Zondervan. p137.
47. Ibid. p137.
48. Dobson. *Learn Biblical Hebrew*. pp354-355.
49. Ibid. pp339, 354-355.
50. Harrison, R. K. (1955). *Biblical Hebrew*. Berkshire, England: Cox & Wyman Ltd. p160, 161.
51. Beall, Todd S. and William A. Banks, Colin Smith (2000). *Old Testament Parsing Guide*, Revised and Updated Edition. Nashville, TN: Broadman & Holman Publishers. p1.
52. van der Merwe. *The Lexham Hebrew-English Interlinear Bible*. Count is based on a search of the Hebrew text for the exact expression "and God said" in Genesis 1:3.
53. Beall, *Old Testament Parsing Guide*. pviii.

Chapter 6 – The Genesis Creation Story

1. Pratico, Gary D. and Miles V. Van Pelt (2001). *Basics of Biblical Hebrew Grammar*. Grand Rapids, MI: Zondervan. p129.
2. Ibid. p129-130.
3. van der Merwe, Christo, ed. (2004). *The Lexham Hebrew-English Interlinear Bible*. (translated by the University of Stellenbosch, Stellenbosch, South Africa) Bellingham, WA: Logos Research Systems Inc.
4. Brenton, Lancelot C. L., trans. (1851). *The Septuagint with Apocrypha: Greek and English*. London: Samuel Bagster & Sons. Retrieved on May 20, 2008 from http://www.ccel.org/bible/brenton/
5. Waltke, Bruce K. (2001). *Genesis: A Commentary*. Grand Rapids, MI: Zondervan. p58.
6. Morris, Henry M. (1976). *The Genesis Record*, 10th printing. Grand Rapids, MI: Baker Book House. p40.
7. Ibid. p41.
8. Ibid. p41.
9. Sailhamer, John H. (1996). *Genesis Unbound: A Provocative New Look at the Creation Account*. Sisters, OR: Multnomah Books. p38-40.
10. Eliyah.com (2006). Strong's Concordance - King James Version. Retrieved on May 20, 2006 from http://www.eliyah.com/lexicon.html
11. Ibid.
12. Brown, Francis and S. R. Driver, Charles A. Briggs (1907). *A Hebrew and English Lexicon of the Old Testament*, 1978 printing. Oxford: Oxford Clarendon Press. p912.
13. Eliyah. Strong's Concordance - King James Version.
14. Brenton. *The Septuagint with Apocrypha: Greek and English*.
15. Tyndale, William (1530). *William Tyndale's Five Books of Moses Called the Pentateuch*. (by Jacob Isidor Mombert, 1884, Anson D. F. Randolph & Co. New York).
16. Coverdale, Myles (1535). Myles Coverdale Bible. Retrieved on May 20, 2008 from http://www.studylight.org/
17. Waltke. *Genesis: A Commentary*. p59.
18. Ibid. p59.
19. Sailhamer. *Genesis Unbound: A Provocative New Look at the Creation Account*. p23.
20. Ibid. p23.
21. Ibid. p23.
22. Morris. *The Genesis Record*. p50.
23. Custance, Arthur C.(1970). *Without Form and Void*. Brookville, Canada: Pub. by Arthur

C. Custance. p130.
24. Brenton. *The Septuagint with Apocrypha: Greek and English.*
25. Beall, Todd S. and William A. Banks, Colin Smith (2000). *Old Testament Parsing Guide: Revised and Updated Edition.* Nashville, TN: Broadman & Holman Publishers. p1.
26. Waltke. *Genesis: A Commentary.* p76.
27. van der Merwe. *The Lexham Hebrew-English Interlinear Bible.* Based on a word search of the Hebrew text.
28. Eliyah. Strong's Concordance - King James Version.
29. Ibid.
30. Gesenius, Wilhelm (1846). *Hebrew and Chaldee Lexicon to the Old Testament Scriptures*, 1st ed. (translated by Samuel P. Tregelles) London: Samuel Bagster & Sons. Retrieved on May 20, 2008 from http://www.eliyah.com/lexicon.html
31. Strong, James (1890). *Strong's Exhaustive Concordance of the Bible*, 31st printing. New York: Abingdon Press. p19.
32. Morris. *The Genesis Record.* p55.
33. Ibid. p41.
34. Pratico, *Basics of Biblical Hebrew Grammar.* p111.
35. Ibid. p111.
36. Eliyah. Strong's Concordance - King James Version.
37. Ibid.
38. Brenton. *The Septuagint with Apocrypha: Greek and English.*
39. Beall. *Old Testament Parsing Guide: Revised and Updated Edition.* p1.
40. Eliyah. Strong's Concordance - King James Version.
41. van der Merwe. *The Lexham Hebrew-English Interlinear Bible.* Based on a word search of the Hebrew text.
42. Pietersma, Albert and Benjamin J. Wright (2007). *A New English Translation of the Septuagint.* New York, NY: Oxford University Press. Retrieved on May 20, 2008 from http://ccat.sas.upenn.edu/nets/edition/
43. Beall. *Old Testament Parsing Guide: Revised and Updated Edition.* p1.
44. Eliyah. Strong's Concordance - King James Version.
45. Beall. *Old Testament Parsing Guide: Revised and Updated Edition.* p1.
46. Eliyah. Strong's Concordance - King James Version.
47. van der Merwe. *The Lexham Hebrew-English Interlinear Bible.* Based on a word search of the Hebrew text.
48. Eliyah. Strong's Concordance - King James Version.
49. Ibid.
50. Ibid.
51. Ibid.
52. Beall. *Old Testament Parsing Guide: Revised and Updated Edition.* p1.
53. Ibid. p1.
54. Pietersma. *A New English Translation of the Septuagint.*
55. Eliyah. Strong's Concordance - King James Version.
56. Ibid.
57. Pietersma. *A New English Translation of the Septuagint.*
58. Beall. *Old Testament Parsing Guide: Revised and Updated Edition.* p1.
59. Eliyah. Strong's Concordance - King James Version.
60. Ibid.
61. Ibid.
62. Ibid.
63. Ibid.
64. Ibid.
65. Ibid.

66. Ibid.
67. Ibid.
68. Ibid.
69. Pietersma. *A New English Translation of the Septuagint.*
70. Beall. *Old Testament Parsing Guide: Revised and Updated Edition.* p1.
71. van der Merwe. *The Lexham Hebrew-English Interlinear Bible.* Based on a combined search of (חית) and "animal."
72. van der Merwe. *The Lexham Hebrew-English Interlinear Bible.* Based on a word search of the Hebrew text.
73. Beall. *Old Testament Parsing Guide: Revised and Updated Edition.* Occurrences where *ramas* is translated as the verbs "creep" or "move."
74. van der Merwe. *The Lexham Hebrew-English Interlinear Bible.* Based on a word search of the Hebrew text.
75. Eliyah. Strong's Concordance - King James Version.
76. Beall. *Old Testament Parsing Guide: Revised and Updated Edition.* p1.
77. Brenton. *The Septuagint with Apocrypha: Greek and English.*
78. Eliyah. Strong's Concordance - King James Version.
79. Ham, Ken, ed. (2006). *The New Answers Book.* Green Forest, AR: Master Books. p48.
80. Sarfati, Jonathan (2004). *Refuting Compromise.* Green Forest, AR: Master Books. p195.
81. Morris. *The Genesis Record.* pp118-126.
82. Ham. *The New Answers Book.* pp204-205.
83. Morris. *The Genesis Record.* p118.
84. Sarfati. *Refuting Compromise.* p205, 211.
85. Ibid. p195.
86. Collins, C. John (2006). *Genesis 1-4: A Linguistic, Literary, and Theological Commentary.* Phillipsburg, NJ: P & R Publishing Co. p183.
87. Sarfati. *Refuting Compromise.* p212.
88. Snoke, David (2004, June). Why were Dangerous Animals Created? *Perspectives on Science and Christian Faith,* vol. 56, no. 2, p125.
89. Eliyah. Strong's Concordance - King James Version.
90. Brenton. *The Septuagint with Apocrypha: Greek and English.*
91. van der Merwe. *The Lexham Hebrew-English Interlinear Bible.* Based on a word search of the Hebrew text.
92. Sarfati. *Refuting Compromise.* p71-72.
93. Eliyah. Strong's Concordance - King James Version.
94. Ibid.
95. Ibid.
96. Collins. *Genesis 1-4: A Linguistic, Literary, and Theological Commentary.* p137.
97. Ibid. p126.
98. Eliyah. Strong's Concordance - King James Version.
99. Ibid.
100. Ibid.
101. Pietersma. *A New English Translation of the Septuagint.*
102. Eliyah. Strong's Concordance - King James Version.
103. Ibid.
104. van der Merwe. *The Lexham Hebrew-English Interlinear Bible.* Based on a word search of the Hebrew text.
105. Young, Robert (1898). Holy Bible, Young's Literal Translation, 3rd ed. Retrieved on May 20, 2008 from http://www.blueletterBible.org/
106. Brown. *A Hebrew and English Lexicon of the Old Testament.* p617.
107. Henry, Matthew (1706). *Matthew Henry's Commentary on the Whole Bible*, vol. 1. Fleming H. Revell Company. p20.
108. van der Merwe. *The Lexham Hebrew-English Interlinear Bible.* Based on a word search of the Hebrew text.
109. Eliyah. Strong's Concordance - King James Version.

Chapter 7 – Design Theory

1. Muller, Jim (2007). Private conversation. Schnucks Supermarket, St. Louis, Missouri.
2. Wal-Mart (2007). Wal-Mart Retail Divisions. Retrieved on September 7, 2007 from http://www.walmartfacts.com/articles/2502.aspx
3. *Wikipedia, The Free Encyclopedia* (2008). Catalase. Retrieved on May 8, 2008 from http://en.wikipedia.org/wiki/Catalase
4. Behe, Michael J. (1996). *Darwin's Black Box: The Biochemical Challenge to Evolution.* New York: Simon & Schuster. pp144-149.
5. Zirin, Harold (2007). Sun. *McGraw-Hill Encyclopedia of Science & Technology*, vol. 17, 10th ed. New York: McGraw-Hill. p672.
6. Gonzalez, Guillermo and Jay W. Richards (2004). *The Privileged Planet.* Washington, DC: Regnery Publishing, Inc. pp68-70.

Chapter 8 – The Myth of Chemical Evolution

1. National Academy of Sciences, Steering Committee on Science and Creationism (1999). *Science and Creationism: A View from the National Academy of Sciences*, 2nd ed. Washington, DC: National Academy Press. p6.
2. Gillispie, Charles C. (1971). *Dictionary of Scientific Biography*, vol. 10. New York: Charles Scribner's Sons. pp351-388.
3. Hazen, Robert M. (2005). *Genesis: The Scientific Quest for Life's Origin.* Washington, DC: Joseph Henry Press. pxiii.
4. Ibid. p11.
5. Ibid. p12.
6. Ibid. p15, 17, 22.
7. Ibid. p87.
8. Ibid. p91.
9. Ibid. p93.
10. Ibid. p155.
11. Ibid. p98.
12. Ibid. p109.
13. Ibid. pp146-148.
14. Ibid. pp35-37.
15. Ibid. p148.
16. Ibid. pp157-158.
17. Behe, Michael J. (1996). *Darwin's Black Box: The Biochemical Challenge to Evolution.* New York: Simon & Schuster. p172.
18. Purves, William K. and David Sadava, Gordon H. Orians, H. Craig Heller (2001). *Life-The Science of Biology*, 6th ed. Gordonsville, VA: W. H. Freeman and Company.
19. Denton, Michael (1985). *Evolution: A Theory in Crisis.* Bethesda, MD: Adler & Adler. p250.
20. Hazen. *Genesis: The Scientific Quest for Life's Origin.* p131.
21. Ibid. p9.
22. Ibid. p93.
23. Ibid. p93.
24. Ibid. p110.
25. Denton. *Evolution: A Theory in Crisis.* p261.
26. Gish, Duane T. (2007, Jan.). A Few Reasons an Evolutionary Origin of Life is Impossible. *Institute for Creation Research, Impact No. 403.* p1.
27. Behe. *Darwin's Black Box: The Biochemical Challenge to Evolution.* p169.
28. Ibid. p171.
29. Hazen. *Genesis: The Scientific Quest for Life's Origin.* p167.

30. Rosenthal, Anne M. (2003, Feb. 12). Murchison's Amino Acids: Tainted Evidence? *Astrobiology Magazine.*
31. Fasman, Gerald D. (1989). *Practical Handbook of Biochemistry and Molecular Biology.* Boca Raton, FL: CRC Press.
32. Hazen. *Genesis: The Scientific Quest for Life's Origin.* p131.
33. Ibid. p127.
34. *Wikipedia, The Free Encyclopedia* (2008). Carbonic anhydrase. Retrieved on May 8, 2008 from http://en.wikipedia.org/wiki/Carbonic_anhydrase
35. *Wikipedia, The Free Encyclopedia* (2008). Catalase. Retrieved on May 8, 2008 from http://en.wikipedia.org/wiki/Catalase
36. Schomburg, D. (2008). Brenda, The Comprehensive Enzyme Information System, Retrieved on May 20, 2008 from http://www.brenda-enzymes.info/
37. Uniprot (2007, May 15). Statistics. Uniprot/Swiss-Prot Protein Knowledgebase, Release 52.5. Retrieved on May 16, 2007 from http://www.expasy.org/sprot/relnotes/relstat.html
38. Bowie, J. U. and J. F. Reidhaar-Olson, W. A. Lim, R. T. Sauer (1990). Deciphering the Message in Protein Sequences: Tolerance to Amino Acid Substitution. *Science* 247, 1306-1310.
39. Behe, Michael J. (1999, June 7). Experimental Support for Regarding Functional Classes of Proteins to be Highly Isolated from Each Other. *The Weekly Standard.*
40. TIGR (2007). Comprehensive Microbial Resource. The Institute of Genomic Research (TIGR). Retrieved on May 16, 2007 from http://cmr.tigr.org/tigr-scripts/CMR/CmrHomePage.cgi
41. Ibid.
42. Ibid.
43. Dawkins, Richard (1996). *The Blind Watchmaker: Why the Evidence of Evolution Reveals a Universe Without Design.* New York: W. W. Norton & Company. p46.
44. Smith, John Maynard (1986). *The Problems of Biology.* Oxford: Oxford University Press. p5.

Chapter 9 – The Illusion of Biological Evolution

1. Darwin, Charles (2003). *The Origin of Species and The Voyage of The Beagle* (reprint of *On the Origin of the Species by Means of Natural Selection*, 1st ed. 1859). New York: Alfred A. Knopf. p635.
2. National Academy of Sciences and Institute of Medicine, Committee on Revising Science and Creationism (2008). *Science, Evolution and Creationism.* Washington, DC: The National Academies Press. p17.
3. Smith, John Maynard and Eörs Szathmáry (1999). *The Origins of Life: From the Birth of Life to the Origin of Language.* Oxford: Oxford University Press. p122.
4. Carroll, Sean B. (2006). *The Making of the Fittest: DNA and the Ultimate Forensic Record of Evolution.* New York: W. W. Norton & Company. p79.
5. National Academy of Sciences and Institute of Medicine. *Science, Evolution and Creationism.* p30.
6. Darwin. *The Origin of Species and The Voyage of The Beagle.* p689.
7. Humsavar (2007). Index of Human Polymorphisms and Disease Mutations. Uniprot/Swiss Prot Protein Knowledgebase, Release 52.5. Retrieved on May 16, 2007 from http://www.expasy.org/cgi-bin/lists?humsavar.txt
8. OMIM (2008). Online Mendelian Inheritance in Man, John Hopkins University. Retrieved on May 20, 2008 from http://www.ncbi.nlm.nih.gov/Omim/mimstats.html
9. Cooper, D. N. and E. V. Ball, P. D. Stenson, A. D. Phillips, K. Howells, M. E Mort (2008). The Human Gene Mutation Database. Institute of Medical Genetics in Cardiff. Retrieved on May 20, 2008 from http://www.hgmd.cf.ac.uk/ac/hahaha.php
10. Carroll. *The Making of the Fittest.* p53.

11. Ibid. p54.
12. Ibid. p56.
13. Ibid. p63.
14. Ibid. p128C.
15. *Wikipedia, The Free Encyclopedia* (2008). Evolution. Retrieved on May 8, 2008 from http://en.wikipedia.org/wiki/Evolution
16. Hendry, A. P. and M. T. Kinnison (2001). An Introduction to Microevolution: Rate, Pattern, Process. *Genetica*. 112-113, 1-8.
17. Carroll. *The Making of the Fittest*. p209.
18. Ibid. p208.
19. Spetner, Lee M. (1997). *Not By Chance: Shattering the Modern Theory of Evolution*. Brooklyn, NY: The Judaica Press. pp181-192.
20. Ibid. pp187-190.
21. Ibid. pp192-194.
22. Behe, Michael J. (1996). *Darwin's Black Box: The Biochemical Challenge to Evolution*. New York: Simon & Schuster. pp81-87.
23. Ibid. pp144-149.
24. Ibid. p18-21.
25. National Academy of Sciences, Steering Committee on Science and Creationism (1999). *Science and Creationism: A View from the National Academy of Sciences*, 2nd ed. Washington, DC: National Academy Press. p22.
26. Zimmer, Carl (2006, Nov.). A Fin is a Limb is a Wing, How Evolution Fashioned its Masterworks. *National Geographic*, vol. 210, no. 5.
27. Carroll. *The Making of the Fittest*. p198.
28. Purves, William K. and David Sadava, Gordon H. Orians, H. Craig Heller (2001). *Life-The Science of Biology*, 6th ed. Gordonsville, VA: W. H. Freeman and Company. p443.
29. Behe, Michael J. (2007). *The Edge of Evolution: The Search for the Limits of Darwinism*. New York: Free Press. p67.
30. Ibid. p73.
31. TIGR (2007). Comprehensive Microbial Resource. The Institute of Genomic Research (TIGR). Retrieved on May 16, 2007 from http://cmr.tigr.org/tigr-scripts/CMR/CmrHomePage.cgi
32. Purves. *Life-The Science of Biology*. p410.
33. UniProt (2007). Statistics. UniProt/Swiss-Prot Protein Knowledgebase, Release 52.5. Retrieved on May 16, 2007 from http://www.expasy.org/sprot/relnotes/relstat.html
34. Sanford, John C. (2005). *Genetic Entropy & The Mystery of the Genome*. Lima, NY: Ivan Press. pp47-56.
35. Ibid. p49.
36. Dawkins, Richard (1996). *The Blind Watchmaker: Why the Evidence of Evolution Reveals a Universe Without Design*. New York: W. W. Norton & Company. p48.
37. Carroll. *The Making of the Fittest*. p60.
38. Behe. *The Edge of Evolution: The Search for the Limits of Darwinism*. p57.
39. Humsavar. Uniprot/Swiss Prot Protein Knowledgebase, Release 52.5.
40. Sanford. *Genetic Entropy & The Mystery of the Genome*. p45.
41. Humsavar. Uniprot/Swiss Prot Protein Knowledgebase, Release 52.5.
42. Humsavar. Uniprot/Swiss Prot Protein Knowledgebase, Release 52.5.
43. Carroll. *The Making of the Fittest*. pp60-61.
44. Ensembl (2008). Ensembl Mouse Genome, Release 49. Retrieved on May 16, 2008 from http://www.ensembl.org/Mus_musculus/index.html
45. Carroll. *The Making of the Fittest*. p60.
46. Behe. *The Edge of Evolution: The Search for the Limits of Darwinism*. p110.
47. Ensembl. Release 49.
48. Genomics (2007). Human Genome Project Information. Retrieved on May 25, 2007 from http://www.ornl.gov/sci/techresources/Human_Genome/project/info.shtml

49. HapMap (2007). International HapMap Project. Retrieved on May 25, 2007 from http://www.hapmap.org/whatishapmap.html
50. *Wikipedia, The Free Encyclopedia* (2007). Genetic disorder. Retrieved on May 14, 2007 from http://en.wikipedia.org/wiki/Genetic_disorder
51. Emory University School of Medicine (2008). The Sickle Cell Information Center. Retrieved on May 20, 2008 from http://www.scinfo.org/faqtrait.htm
52. LabCorp (2008). Cystic Fibrosis, Laboratory Corporation of America. Retrieved on May 20, 2008 from http://www.labcorp.com/genetics/genetic_disorders/cystic_fibrosis.html
53. Behe. *The Edge of Evolution: The Search for the Limits of Darwinism*. pp141-142.
54. Sanford. *Genetic Entropy & The Mystery of the Genome*. p144.
55. Wood, R. D. and M. Mitchell, J. Sgouros, T. Lindahl (2001, Feb. 16). Human DNA Repair Genes. *Science*, 291:1284.
56. Sanford. *Genetic Entropy & The Mystery of the Genome*. p131.
57. Gregory, T. Ryan (2007). Animal Genome Size Database. Retrieved on May 16, 2007 from http://www.genomesize.com
58. Ibid.
59. Gerstein, Mark and Deyou Zheng (2006, Apr.). The Real Life of Pseudogenes. *Scientific American*.
60. Ibid.
61. Ibid.
62. Carroll. *The Making of the Fittest*. p79.
63. Gerstein. The Real Life of Pseudogenes. *Scientific American*.
64. National Academy of Sciences and Institute of Medicine. *Science, Evolution and Creationism*. p30.

Chapter 10 – The Origin of Man

1. Chimpanzee Sequencing and Analysis Consortium (2005, Sept. 1). Initial Sequence of the Chimpanzee Genome and Comparison with the Human Genome. *Nature*, vol. 437. p73.
2. Chimpanzee. *Nature*. p75.
3. Chimpanzee. *Nature*. p76.
4. Glazko, Galina and Vamsi Veeramachaneni, Masatoshi Nei, Wojciech Makałowski (2005, Feb. 14). Eighty Percent of Proteins are Different Between Humans and Chimpanzees. *Gene*, vol. 346:215-219.
5. Chimpanzee. *Nature*. p78.
6. Chimpanzee. *Nature*. p81.
7. Begley, Sharon (2007, Mar. 19). Beyond Stones and Bones. *Newsweek*. vol. CXLIX, no. 12.
8. Genomics (2008). Human Genome Project Information. Retrieved on May 20, 2008 from http://www.ornl.gov/sci/techresources/Human_Genome/project/info.shtml
9. NCBI (2008). National Center for Biotechnology Information, Build 36.3. Retrieved on May 20, 2008 from http://www.ncbi.nlm.nih.gov/mapview/maps.cgi?TAXID=9606&MAPS=ideogr,ugHs,genes&CHR=MT
10. Johanson, Donald and Blake Edgar (1996). *From Lucy to Language*. Italy: Nevraumont Publishing Company. p41.
11. Vardiman, Larry (May 2008). Mitochondrial DNA and a Recent Human Origin. *ICR Acts & Facts*. p6.
12. *Wikipedia The Free Encyclopedia* (2008). Great Ape. Retrieved on May 8, 2008 from http://en.wikipedia.org/wiki/Hominidae
13. Foley, Jim (2002). Creationist Arguments: Brain Sizes. The TalkOrigins Archive. Retrieved on December 15, 2006 from http://www.talkorigins.org/faqs/homs/a_brains.html
14. Burenhult, Goran (1993). *The First Humans: Human Origins and History to 10,000 BC*.

New York: HarperCollins.

15. Hrdlicka, Ales (1939). Normal Micro and Macrocephaly in America. *American Journal of Physical Anthropology*, 25:1-91.
16. McHenry, T. M. (1994, Jul. 19). Tempo and Mode in Human Evolution, Table I. *Proceedings of the National Academy of Sciences, USA*, 91(15): 6780–6786.
17. Gish, Duane T. (1995). *Evolution: The Fossils Still Say No*. El Cajon, CA: Institute for Creation Research. p277.
18. Cáceres, Mario and Joel Lachuer, Matthew A. Zapala, John C. Redmond, et al (2003, Oct. 28). Elevated Gene Expression Levels Distinguish Human from Non-Human Primate Brains, *Proceedings of the National Academy of Sciences*, 100(22):13030–13035.
19. Khaitovich, Philipp and Bjoern Muetzell, Xinwei She, Michael Lachmann, Ines Hellmann, et al (2004, Jun. 8). Regional Patterns of Gene Expression in Human and Chimpanzee Brains, Table 2. *Genome Research,* 14:1462-1473.
20. Tattersall, Ian (1993). *The Human Odyssey: Four Million Years of Human Evolution*. New York: Prentice Hall. p120.
21. Foley, Jim (2005). Prominent Hominid Fossils. The TalkOrigins Archive. Retrieved on November 15, 2005 from http://www.talkorigins.org/faqs/homs/specimen.html
22. Ibid.
23. Tattersall. *The Human Odyssey: Four Million Years of Human Evolution*. p103.
24. Ibid. p74.
25. Johanson. *From Lucy to Language*. p124.
26. *Wikipedia The Free Encyclopedia* (2008). Lucy. Retrieved on May 8, 2008 from http://en.wikipedia.org/wiki/Lucy_(Australopithecus)
27. Tattersall. *The Human Odyssey: Four Million Years of Human Evolution*. p96.
28. Leakey, Richard E. (1994). *The Origin of Humankind*. New York: Basic Books. p71.
29. Johanson. *From Lucy to Language*. p91.
30. Tattersall. *The Human Odyssey: Four Million Years of Human Evolution*. p99.
31. Ibid. p102.
32. Gish. *Evolution: The Fossils Still Say No*. p281.
33. Walker, Alan and Pat Shipman (1996). *The Wisdom of the Bones: In Search of Human Origins*. New York: Alfred A. Knopf. p34, 38.
34. Gish. *Evolution: The Fossils Still Say No*. p287.
35. Ibid. p295.
36. Walker. *The Wisdom of the Bones: In Search of Human Origins*. p65.
37. Ibid. p55.
38. Ibid. pp58-59.
39. Ibid. pp59-61.
40. Ibid. p68.
41. Gish. *Evolution: The Fossils Still Say No*. p287.
42. Walker. *The Wisdom of the Bones: In Search of Human Origins*. p77.
43. Ibid. p212.
44. Ibid. p194.
45. Ibid. p180-15.
46. Ibid. p218.
47. Tattersall. *The Human Odyssey: Four Million Years of Human Evolution*. p125.
48. *Wikipedia, The Free Encyclopedia* (2009). List of Neanderthal Sites. Retrieved on Jan. 5, 2009 from http://en.wikipedia.org/wiki/List_of_neanderthal_sites
49. Tattersall. *The Human Odyssey: Four Million Years of Human Evolution*. p124.
50. *Wikipedia, The Free Encyclopedia*. Neanderthal.
51. Sawyer, Gary J. and Blaine Maley (2005, Mar.). Neanderthal Reconstructed, *The Anatomical Record, Part B: The New Anatomist*, vol. 283B, issue 1.
52. *Wikipedia, The Free Encyclopedia*. Neanderthal.
53. Tattersall. *The Human Odyssey: Four Million Years of Human Evolution*. p118.
54. Wild, Eva M. and Maria Teschler-Nicola, Walter Kutschera, Peter Steier, Erik Trinkaus,

Wolfgang Wanek (2005, May 19). Direct Dating of Early Upper Palaeolithic Human Remains from Mladec. *Nature*, vol. 435:332-335.

55. Johanson. *From Lucy to Language*. p163.
56. Aston, Mick and Tim Taylor (1998). *The Atlas of Archaeology,* 1st American ed. New York: DK Publishing. pp136-157.
57. Schiller, Ronald (1973, Aug.). New Findings on the Origin of Man. *Reader's Digest*, vol. 103, no. 616.
58. Butzer, K. W. and P. B. Beaumont, J. C. Vogel (1978). Lithostratigraphy of Border Cave, KwaZulu, South Africa: A Middle Stone Age Sequence Beginning c. 195,000 BP. *Journal of Archaeological Science*, pp317-341.
59. Moore, A. M. T. and G. C. Hillman, A. J. Legge (2000). *Village on the Euphrates.* New York: Oxford University Press.
60. Weninger, Bernhard and Olaf Jöris, Uwe Danzeglocke (2005, Sept.). CalPal, Cologne Radiocarbon Calibration & Paleoclimate Research Package. Retrieved on May 20, 2008 from http://www.calpal-online.de/
61. *Britannica, The New Encyclopedia*, vol. 5, 15th ed. (2005). Giza, Pyramids of. Chicago: Encyclopedia Britannica Inc. p288.
62. Bonani G, and H. Haas, Z. Hawass, M. Lehner, S. Nakhla, J. Nolan, R. Wenke, W. Wölfli (2001). Radiocarbon Dates of Old and Middle Kingdom Monuments in Egypt. *Radiocarbon* 43, no. 3, 1297-1320(24).
63. Johanson. *From Lucy to Language*. p27.
64. Kuzmin, Y. V. and G. S. Burr, A. J. T. Jull, L. D. Sulerzhitsky (2004, Aug.). Proceedings of the Ninth International Conference on Accelerator Mass Spectrometry. *Nuclear Instruments and Methods in Physics Research Section B: Beam Interactions with Materials and Atoms*, vol. 223-224. pp731-734.
65. Wild. Direct Dating of Early Upper Palaeolithic Human Remains from Mladec. *Nature*.
66. Snelling, Andrew (2000, Mar.). Geological Conflict: Young Radiocarbon Date for Ancient Fossil Wood Challenges Fossil Dating. *Creation ex nihilo,* 22(2):44–47.
67. Taylor, R. E. and Martin J. Aitken, eds. (1997). *Chronometric Dating in Archaeology.* New York: Plenum Press. p66.
68. Taylor. *Chronometric Dating in Archaeology*. p82.
69. Ibid. p91.
70. Morton, Glenn R. (2002). History of Human Technology. Retrieved on December 27, 2006 from http://home.entouch.net/dmd/chron.htm
71. Taylor. *Chronometric Dating in Archaeology*. p191.
72. Ibid. p183, 197.
73. Ibid. p221.
74. Skinner, Anne (2007). Electron Spin Resonance Dating at Williams College. Retrieved on February 22, 2007 from http://www.williams.edu/Chemistry/askinner/Website/WCESR/WilliamsCollegeESR.htm
75. Lugan, Matt (2007). Electron Spin Resonance Dating. Retrieved on February 22, 2007 from http://www.mnsu.edu/emuseum/archaeology/dating/dat_esr.html
76. Skinner. Electron Spin Resonance Dating at Williams College.
77. Begley. Beyond Stones and Bones. *Newsweek*.
78. Aston. *The Atlas of Archaeology*. pp136-157.

Chapter 11 – The Bible and the Origin of Man

1. Ussher, James (1658). *The Annals of the World.* (Ussher's timeline chart by Paul Hansen, published in *Creation*, Sept. 2005). Retrieved on March 3, 2006 from http://www.answersingenesis.org/creation/v27/i4/TimelineOfTheBible.pdf
2. Ibid.
3. Sarfati, Jonathan (2004). *Refuting Compromise*. Green Forest, AR: Master Books. p290.

4. Ussher. *The Annals of the World.*
5. Brown, Francis and S. R. Driver, Charles A. Briggs (1907). *A Hebrew and English Lexicon of the Old Testament,* 1978 printing. Oxford: Oxford Clarendon Press. p408.
6. Morris, Henry (1976). *The Genesis Record.* Grand Rapids, MI: Baker Book House. p45.
7. Ham, Ken, ed. (2006). *The New Answers Book.* Green Forest, AR: Master Books. p26.
8. Collins, C. John (2006). *Genesis 1-4: A Linguistic, Literary, and Theological Commentary.* Phillipsburg, NJ: P & R Publishing Co. p202.
9. Moore, A. M. T. and G. C. Hillman, A. J. Legge (2000). *Village on the Euphrates.* New York: Oxford University Press.
10. Hitchcock, Don (2007). Musical Instruments. Retrieved on September 15, 2007 from http://donsmaps.com/musicalinstruments.html
11. Foley, Jim (2007). Fossil Hominids: Mitochondrial DNA. Retrieved on Sept. 12, 2007 from http://www.talkorigins.org/faqs/homs/mtDNA.html
12. Hall, Stephen S. (2008, Oct.). Last of the Neanderthals. *National Geographic*, vol. 214, no. 4. p51.
13. Ibid. pp39-40.
14. Ibid. p46.
15. Ibid. p51. (Map, listing number of Neanderthal sites)
16. Ibid. p58.
17. Smith, Fred H. and Erik Trinkaus, Paul B. Pettitt, Ivor Karavanic, Maja Paunovic (1999, Oct. 26). Direct Radiocarbon Dates for Vindija G1 and Velika Pećina Late Pleistocene Hominid Remains, *Proceedings of the National Academy of Sciences,* vol. 96, no. 22, 12281-12286.
18. Johanson, Donald and Maitland Edey (1981). *Lucy: The Beginnings of Humankind.* Simon & Schuster. p363.
19. Ibid. p363.

Chapter 12 – The Fantasy of Cosmological Evolution

1. National Aeronautics and Space Administration (1999, Apr.). NASA's Goddard Space Flight Center: The First Forty Years. *NASA Facts*, FS-1999 (01) -003GSFC. p4.
2. Darling, David (2004). *The Universal Book of Astronomy.* Hoboken, NJ: John Wiley & Sons Inc. p340.
3. Angelo Jr., Joseph A. (2006). *Encyclopedia of Space and Astronomy.* New York: Facts on File. p403.
4. Pickering, Edward C. (1912, Mar. 3). Periods of 25 Variable Stars in the Small Magellanic Cloud. *Harvard College Observatory*, Circular 173.
5. Angelo. *Encyclopedia of Space and Astronomy.* p83.
6. Ibid. p83.
7. Darling. *The Universal Book of Astronomy.* p63.
8. Angelo. *Encyclopedia of Space and Astronomy.* p84.
9. Ibid. p84.
10. Tyson, Neil De Grasse and Charles Liu, Robert Irion (2000). *One Universe: At Home in the Cosmos.* Washington, DC: Joseph Henry Press. p81.
11. Angelo. *Encyclopedia of Space and Astronomy.* p84.
12. Darling. *The Universal Book of Astronomy.* p63.
13. Murdin, Paul, ed. (2001). *Encyclopedia of Astronomy and Astrophysics.* London: Nature Publishing Group. p174.
14. Tyson. *One Universe: At Home in the Cosmos.* p82.
15. Angelo. *Encyclopedia of Space and Astronomy.* p442.
16. Arp, Halton C. (1998). *Seeing Red: Redshifts, Cosmology and Academic Science.* Montreal: Apeiron. p251.
17. Ibid. p7.

18. Schroeder, Gerald L. (2001). *The Hidden Face of God.* New York: The Free Press. p55.
19. Ibid. p55.
20. Huchra, John P. (2008). The Hubble Constant. Retrieved on June 2, 2008 from http://cfa-www.harvard.edu/~huchra/hubble/
21. Faulkner, Danny (2004). *Universe by Design.* Green Forest, AR: Master Books. p93.
22. Darling. *The Universal Book of Astronomy.* p173.
23. Angelo. *Encyclopedia of Space and Astronomy.* p49.
24. *Britannica, The New Encyclopedia*, vol. 1, 15th ed. (2005). Astronomy. Chicago: Encyclopedia Britannica Inc. p656.
25. Faulkner. *Universe by Design.* p12.
26. Angelo. *Encyclopedia of Space and Astronomy.* p152.
27. Berlinski, David (2008). *The Devil's Delusion.* New York: Crown Publishing Group. p212-213.
28. Arp. *Seeing Red: Redshifts, Cosmology and Academic Science.* p270.

Chapter 13 – A Comparative Analysis

1. Morris, Henry (1976). *The Genesis Record.* Grand Rapids, MI: Baker Book House. p45.
2. Hazen, Robert M. (2005). *Genesis: The Scientific Quest for Life's Origin.* Washington, DC: Joseph Henry Press. p48.
3. National Academy of Sciences and Institute of Medicine, Committee on Revising Science and Creationism (2008). *Science, Evolution and Creationism.* Washington, DC: The National Academies Press. p21.
4. Hazen. *Genesis: The Scientific Quest for Life's Origin.* p59.
5. Wilson, Don E. and DeeAnn. M. Reeder, eds. (2005). *Mammal Species of the World: A Taxonomic and Geographic Reference,* 3rd ed. Johns Hopkins University. Press. Retrieved database on May 17, 2008 from http://nmnhgoph.si.edu/msw/ Based on a sorting and deletion of duplicate species.
6. Peterson, Alan P., ed. (2008). Zoonomen Zoological Nomenclature Resource, Birds of the World. Retrieved on May 17, 2008 from http://www.zoonomen.net.
7. AmphibiaWeb (2008). Current number of amphibian species. Retrieved on May 17, 2008 from http://amphibiaweb.org/
8. Uetrz, Peter, ed. (2008). The TIGR Reptile Database. Retrieved on May 17, 2008 from http://www.reptile-database.org/
9. Froese, R. and D. Pauly, eds. (2008). FishBase World Wide Web. Retrieved on May 17, 2008 from http://www.fishbase.org/search.php
10. Ramel, Gordon (2008). Wonderful World of Insects. Retrieved on May 17, 2008 from http://www.earthlife.net/insects/six.html
11. Tangley, Laura (1997, Aug. 18). How Many Species are There? *U. S. News and World Report*, vol. 123, no. 7.
12. Jeffreys, David (2008). Plant Species Numbers, Botanic Gardens Conservation International. Retrieved on May 17, 2008 from http://www.bgci.org/conservation/1521/
13. National Aeronautics and Space Administration (1999, Apr.). NASA's Goddard Space Flight Center: The First Forty Years. *NASA Facts*, FS-1999 (01) -003GSFC. p4.
14. Darling, David (2004). *The Universal Book of Astronomy.* Hoboken, NJ: John Wiley & Sons Inc. p199.
15. Murdin, Paul, ed. (2001). *Encyclopedia of Astronomy and Astrophysics.* London: Nature Publishing Group. p895.
16. Darling. *The Universal Book of Astronomy.* p466.
17. Murdin. *Encyclopedia of Astronomy and Astrophysics.* p3017.
18. Darling. *The Universal Book of Astronomy.* p473.
19. Rigden, John S. (1996). *Macmillan Encyclopedia of Physics*, vol. 2. New York: Simon & Schuster Macmillan. p652.

20. Linde, David R., ed. (2004). *CRC Handbook of Chemistry and Physics*, 85th ed. New York: CRC Press. p6-142.
21. Murdin. *Encyclopedia of Astronomy and Astrophysics*. p174.
22. Linde. *CRC Handbook of Chemistry and Physics*. p6-142.
23. Angelo Jr., Joseph A. (2006). *Encyclopedia of Space and Astronomy*. New York: Facts on File. p86.
24. Murdin. *Encyclopedia of Astronomy and Astrophysics*. p180.
25. Ibid. p182.
26. Darling. *The Universal Book of Astronomy*. p340.
27. Angelo. *Encyclopedia of Space and Astronomy*. p403.
28. National Aeronautics and Space Administration. *NASA Facts*. p4.
29. Driver, Simon (2003, July 23). Speech at General Assembly of the International Astronomical Union meeting in Sydney, Australia. Estimate of the number of stars in the universe by astronomers at the Australian National University. Retrieved on September 19, 2006 from http://www.cnn.com/2003/TECH/space/07/22/stars.survey/index.html
30. Murdin. *Encyclopedia of Astronomy and Astrophysics*.p2762.
31. Ibid. p2762.
32. Postgate, John (2007). Nitrogen fixation. *McGraw-Hill Encyclopedia of Science & Technology,* vol. 11, 10th ed. New York: McGraw-Hill. p790.
33. Ibid. p789.
34. Ibid. p790.
35. Batten, Don, ed. and Ken Ham, Jonathan Sarfati, Carl Wieland (1990). *The Revised & Expanded Answers Book*, 24th printing. Green Forest, AR: Master Books. pp95-102.
36. Sarfati, Jonathan (2004). *Refuting Compromise*. Green Forest, AR: Master Books. pp189-193.
37. Ham, Ken, ed. (2006). *The New Answers Book*. Green Forest, AR: Master Books. pp245-254.
38. McDowell, Jonathan (2007, Mar. 30). A merge of NASA-RP-1097 (Anderson and Whitaker 1982) and USGS lunar crater data. Retrieved on April 26, 2007 from http://host.planet4589.org/astro/lunar/
39. Ham. *The New Answers Book*. p119.
40. Ibid. p121.
41. Ibid. p122.
42. Snelling, Andrew (2000, Mar.). Geological Conflict: Young Radiocarbon Date for Ancient Fossil Wood Challenges Fossil Dating. *Creation ex nihilo,* 22(2):44–47.
43. Ham. *The New Answers Book*. p85.
44. Weiland, Carl (2004, Mar.). Radiometric Dating Breakthroughs. *Creation*, 26(2):42-44.
45. Arnett, William A. (2008). The Nine Planets, A Multimedia Tour of the Solar System. Retrieved on May 17, 2008 from http://www.nineplanets.org/
46. Weissman, Paul R. and Lucy-Ann McFadden, Torrence V. Johnson, eds. (1999). *Encyclopedia of the Solar System*. New York: Academic Press. pp430-434.
47. Douthitt, Bill (2006, Dec.). Beautiful Stranger: Saturn's Mysteries Come to Light. *National Geographic*, vol. 210, no. 6.
48. Arnett. The Nine Planets, A Multimedia Tour of the Solar System.
49. Weissman. *Encyclopedia of the Solar System*. p471.
50. Morton, Glen (2005). The Geologic Column and its Implications for the Flood. Talk of the Origins Archive. Retrieved on November 4, 2005 from http://www.talkorigins.org/faqs/geocolumn/#column.
51. IHS (2008). IHS Basin Monitors. Retrieved on March 13, 2008 from http://energy.ihs.com/Products/Basinmonitors/index.htm?gclid=CO_Rj-3Q0JICFQP4lgodFnlWGw
52. Bond, C. Gerard and Michelle A. Kominz (2007). Basins. *McGraw-Hill Encyclopedia of Science & Technology*, vol. 2, 10th ed. New York: McGraw-Hill. p634.
53. Hazen. *Genesis: The Scientific Quest for Life's Origin*. p92.

54. Ibid. p110.
55. Denton, Michael (1985). *Evolution: A Theory in Crisis*. Bethesda, MD: Adler & Adler. p261.
56. Gish, Duane T. (2007, Jan.). A Few Reasons an Evolutionary Origin of Life is Impossible. *Institute for Creation Research, Impact No. 403*. p1.
57. Emsley, John (1998). *The Elements,* 3rd ed. Oxford: Clarendon Press. p289.
58. Anderson, J. A. R. (1964). The Structure and Development of the Peat Swamps of Sarawak and Brunei. *Journal of Tropical Geography*, 18:7-16.
59. Damberger, Heinz H. (2007). Coal. *McGraw-Hill Encyclopedia of Science & Technology,* vol. 4, 10th ed. New York: McGraw-Hill. p309.
60. Ibid. p309.
61. Nevins, Stuart E. (1976, Nov.). The Origin of Coal. *Institute for Creation Research, Impact No. 41.*
62. Damberger. *McGraw-Hill Encyclopedia of Science & Technology.* p309.
63. Johnson, Kirk (2007, Jun. 14). Forests Frozen in Time. *Nature*, vol. 447, p786-787.
64. Nevins. The Origin of Coal. *Impact No. 41.*
65. Ussher, James (1658). *The Annals of the World.* (Ussher's timeline chart by Paul Hansen, published in *Creation*, Sept. 2005). Retrieved on March 3, 2006 from http://www.answersingenesis.org/creation/v27/i4/TimelineOfTheBible.pdf.
66. Damberger. *McGraw-Hill Encyclopedia of Science & Technology.* p317.
67. Gordon Jr., Raymond G., ed. (2005). *Ethnologue: Languages of the World*, 15th ed. Dallas, TX: SIL International. p15.
68. Ibid. pp17-23.
69. Premack, David (1985). Gavagai! or the Future History of the Animal Language Controversy. Cognition, 19, p281-282.
70. Dawkins, Richard (1986). *The Blind Watchmaker: Why the Evidence of Evolution Reveals a Universe Without Design.* New York: W.W. Norton & Company. pix.
71. Ibid, p1.
72. Ham, Ken (2008, Oct.-Dec.). Part 1: The Created World, Mature for Her Age. *Answers*, vol. 1, no. 4. p32.
73. Ibid, p32.
74. Webb, George E. (1994). *The Evolution Controversy in America.* Lexington, KY: The University Press of Kentucky.

Chapter 14 – Paradigm Shift

1. Miller, H. S. (1940). *General Biblical Introduction: From God to Us*. Houghton, NY: The Word-Bearer Press. p223.
2. *Newsweek* Poll (2007). *Newsweek* Poll March 31, 2007: Conducted by Princeton Survey Research Associates International. Retrieved on May 5, 2007 from http://www.msnbc.msn.com/id/17875540/site/newsweek/
3. Newport, Frank (2007, Jun. 11). Majority of Republicans Doubt Theory of Evolution. Gallup, Politics and Government. Retrieved on May 20, 2008 from http://www.gallup.com/poll/27847/Majority-Republicans-Doubt-Theory-Evolution.aspx
4. CBS News Polls (2005, Oct. 23). Poll: Majority Reject Evolution. CBS News. Retrieved on May 20, 2008 from http://www.cbsnews.com/stories/2005/10/22/opinion/polls/main965223.shtml
5. Mitchell, Susan (1996). *The Official Guide to American Attitudes*. Ithaca, NY: New Strategist Publications Inc.
6. Niemi, Richard and John Mueller, Tom Smith (1989). *Trends in Public Opinion: A Compendium of Survey Data*. New York: Greenwood Press.

Appendix B – The Ten Key Verbs of Genesis Chapter One

1. Pratico, Gary D. and Miles V. Van Pelt (2001). *Basics of Biblical Hebrew Grammar.* Grand Rapids, MI: Zondervan. p137.
2. Dobson, John H. (2005). *Learn Biblical Hebrew*, 2nd ed. Grand Rapids, MI: Baker Publishing Group. pp339, 354-355.
3. Harrison, R. K. (1955). *Biblical Hebrew*. Berkshire, England: Cox & Wyman Ltd. pp158-161.
4. Brenton, Lancelot C. L., trans. (1851). *The Septuagint with Apocrypha: Greek and English*. London: Samuel Bagster & Sons. Retrieved on May 20, 2008 from http://www.ccel.org/bible/brenton/
5. Pietersma, Albert and Benjamin J. Wright (2007). *A New English Translation of the Septuagint.* New York, NY: Oxford University Press. Retrieved on May 20, 2008 from http://ccat.sas.upenn.edu/nets/edition/
6. Dobson. *Learn Biblical Hebrew*. p334.
7. Beall, Todd S. and William A. Banks, Colin Smith (2000). *Old Testament Parsing Guide*, Revised and Updated Edition. Nashville, TN: Broadman & Holman Publishers. p1.
8. *American Heritage College Dictionary,* 4th ed. (2004). Boston, MA: Houghton Mifflin Company. p753.
9. Kautzsch, E. (1909). *Gesenius' Hebrew Grammar*, 2nd ed. (translated by A. E. Cowley from the 28th German edition) Oxford: Clarendon Press. p129.
10. Harrison. *Biblical Hebrew*. p81.
11. Ibid. p81.
12. Pratico. *Basics of Biblical Hebrew Grammar*. p130.
13. Dobson. *Learn Biblical Hebrew*. p334.
14. Eliyah.com (2006). Strong's Concordance - King James Version. Retrieved on May 20, 2006 from http://www.eliyah.com/lexicon.html
15. Ibid.
16. van der Merwe, Christo, ed. (2004). *The Lexham Hebrew-English Interlinear Bible.* (translated by the University of Stellenbosch, Stellenbosch, South Africa) Bellingham, WA: Logos Research Systems Inc.
17. Eliyah. Strong's Concordance - King James Version.
18. van der Merwe. *The Lexham Hebrew-English Interlinear Bible*. Based on a word search of the Hebrew text.
19. *American Heritage College Dictionary.* p753.
20. Kautzsch. *Gesenius' Hebrew Grammar*. p129.
21. Ibid. p131.
22. Ibid. p131.
23. Harrison. *Biblical Hebrew*. p81.
24. Ibid. p81.
25. Pratico. *Basics of Biblical Hebrew Grammar*. p130.
26. Ibid. p130.
27. Dobson. *Learn Biblical Hebrew*. p338.
28. Ibid. p123.
29. Ibid. p213.

Index

Printed in the United States
145870LV00003B/1/P

9 780615 267289